Die Grundlehren der mathematischen Wissenschaften

in Einzeldarstellungen
mit besonderer Berücksichtigung
der Anwendungsgebiete

Band 137

Handbook for
Automatic Computation

Edited by
F. L. Bauer · A. S. Householder · F. W. J. Olver
H. Rutishauser · K. Samelson · E. Stiefel

Volume I · Part b

A. A. Grau · U. Hill · H. Langmaack

Translation of ALGOL 60

Chief editor
K. Samelson

Springer-Verlag Berlin Heidelberg GmbH 1967

Prof. A. A. Grau

Mathematics and Engineering Sciences Departments, Northwestern University, Evanston, Ill.

Dipl.-Math. U. Hill

Mathematisches Institut der Technischen Hochschule München

Priv.-Doz. Dr. H. Langmaack

Computer Sciences Department, Purdue University, Lafayette, Indiana/USA
und Mathematisches Institut der Technischen Hochschule München

Geschäftsführende Herausgeber:

Prof. Dr. B. Eckmann

Eidgenössische Technische Hochschule Zürich

Prof. Dr. B. L. van der Waerden

Mathematisches Institut der Universität Zürich

Additional material to this book can be downloaded from http://extras.springer.com.

ISBN 978-3-642-86939-6   ISBN 978-3-642-86937-2 (eBook)

DOI 10.1007/978-3-642-86937-2

Titel-Nr. 5120

Preface

Problem oriented programming languages as they have developed over the last ten years essentially serve two purposes which somewhat crudely can be described by the terms man-man communication and man-machine communication, respectively. As a carrier of information between humans, the problem oriented programming language is designed to express the essence of an algorithm in a way which is unambiguous and concise as well as independent of (and therefore meaningful without any reference to) the changing details of computing machinery. As a carrier of information from man to computer, the language permits the human programmer to express his computational needs in a compact way adapted to the general characteristics of computers, but freed from the burdening details of specific computer facilities. This presupposes the existence of algorithms, or programs, which permit the computer itself to transform efficiently programs written in the problem oriented language into machine programs. Thus the entire computing community profits from the work of the individual programmer.

The primary purpose of the Handbook is to present a set of algorithms of broad utility from the domain of numerical mathematics written in the problem oriented language ALGOL 60. Therefore, volumes I a and I b are in a sense supplementary as they serve to introduce this language. Volume I a gives a description of the language proper and of its use for writing correct programs. Thus, volume I a primarily covers the aspect of man-man communication by means of ALGOL 60. By contrast, the present volume I b is devoted exclusively to the aspect of man-machine communications. It presents a complete model of a translator program for ALGOL 60, an outline of the general principles applied, and a description of the operation of the program.

The underlying translator philosophy is an outgrowth of experience gained within the ALCOR group (an international group of computing centers and computer manufacturers collaborating in the development and application of ALGOL translators) which, over the years, has constructed ALGOL translators for so diversified a set of computers as ERMETH-ETH Zürich, ORACLE-Oak Ridge National Laboratory, PERM-TH Munich, Zuse Z 22/23, Siemens 2002, Telefunken TR4, Control Data 1604/3400/3600, IBM 7040/7090.

A large number of ALGOL translators has been developed in the past in different places with different starting points and different basic philosophies. No attempt has been made to give an overall survey, as this would have exceeded the scope of this volume. Considering the status of the art of translation of programming languages the book addresses the uninitiated who want to obtain access to methods of compiler writing; the expert may yet find some interesting material.

Volume I a restricts the discussion essentially to the IFIP SUBSET ALGOL 60. Volume I b, on the other hand, is far less restricted because it addresses a different set of readers, and because the subset excludes some features of ALGOL 60 which have proven to be of particular interest to programmers.

The model translator program itself is written in a language which is a slight extension of ALGOL 60, and therefore should be intelligible to anybody acquainted with this language. The object code produced by the translator is an abstract machine code modelled as close as seemed possible to the instruction codes for, in a certain sense, "average" present-day computers.

The authors wish in particular to express their thanks to Professor K. SAMELSON for his interest in the work and his constant readiness to discuss questions which arose during the writing of the manuscript. Thanks are also due to many colleagues in Oak Ridge and Munich who gave us valuable suggestions. Last but not least we are very much indebted to the secretarial staffs of our institutions; without their capable help the preparation of the manuscript would have been impossible.

Evanston, Munich, Lafayette, May 23, 1967

A. A. GRAU
U. HILL
H. LANGMAACK

Contents

1. Introduction

In volume I a an introduction to the algorithmic language ALGOL 60 was given appropriate for users of the handbook algorithms. It was pointed out that ALGOL fulfils a twofold function. An algorithmic language like ALGOL furnishes a convenient and readable means of publishing and exchanging computing algorithms; in addition, it may also be used directly for programming a modern high-speed stored-program computer, that is, for designing, testing, and using numerical algorithms. However, a modern computer can not interpret ALGOL programs directly. It is therefore necessary to transform the ALGOL statements into a form suitable for the computer. This may be done by hand or by the machine itself. For the latter case, an *interpretive system* for the particular machine in question is needed. In the commonly used type of system (load and run system) the source language (that is, ALGOL) program is handled in two stages: (1) the *source language* program is translated into a corresponding *target (machine) language* program; (2) the generated machine language program is executed. The latter program may be kept for subsequent re-execution without further reference to the translator or to the source language program.

The aim of this volume is the description of the structure of such systems and to show how ALGOL programs can be handled thereby. As a consequence, this book is not primarily addressed to ALGOL programmers, but to those who must construct a system or who otherwise are interested in knowing how such a system works.

Various methods of dealing with the problem of translating ALGOL programs into machine code have been devised. In recent years progress has been made towards developing from these models a systematic theory of translation of formal algorithmic languages. In view of the aims of this book, however, we shall not treat the general theory and its adaptation to our special problem, but restrict the discussion to one single method of translating ALGOL allowing some variants, on which a number of existing implementations on different computers are based[1]. A general outline of this translation method is presented in chapter 2. The reader interested in the logical foundations is referred to linguistic research publications.

The aim in translating ALGOL is to produce by means of computers machine programs for individual computers. However, the machine

---

[1] E.g., the compilers built by the ALC OR group.

languages of existing computers differ. In order to avoid being limited to the instruction code of a specific computer type we introduce as target language of the translation a symbolic instruction code resembling present day symbolic assembly languages. The translation itself must ultimately be formulated as a program written in a suitable language. The language adopted here for this purpose is ALGOL augmented by some facilities for string handling. The elements of the target language and the additions to ALGOL are described in chapter 3.

In chapter 4, the correspondence between constituents of ALGOL programs and elements of the target language is described. For the simple imperative constituents such as assignment statements, go to statements, etc., the equivalent sequences of target language instructions are listed. In particular, procedure calls are discussed in detail. Declarations, apart from their imperative content, carry the information required for the handling of types, for which several different methods are treated, and for the storage allocation mechanism which is tied to the structure represented by blocks and procedures.

In chapter 4 optimization possibilities are not considered. Chapter 5 treats a specific case of optimization, which is important in problems of linear algebra, and which concerns the subscripted variables in for statements. In most cases occurring in practice these subscript expressions may be evaluated recursively.

At this point the description of translation methods has been completed. In chapter 6 follows the description of the action of target language instructions which usually have no simple machine equivalent. These are the interpretation of the array declaration, the instructions connected with procedure calls and formal parameters, and the go to statement which also involves the storage allocation mechanism. These instructions are essentially macros, and can be considered to be subroutines programmed independently of the given source language program for the specific machine. In other words, the execution of the translated program requires auxiliary organizational programs which constitute the run time system, which in turn is a part of the full interpretive system.

The last two chapters contain a complete ALGOL translator described in an extension of ALGOL. In chapter 7, necessary explanations with regard to the program, notations, and lists used are made. The program follows in chapter 8. This model translator is written in accordance with chapters 4, 5, and 6. With the exceptions listed below, it handles all features of ALGOL 60 which are clear and unambiguous in the Revised Report [75], so that the description of the translator may in effect constitute a formal interpretation of the language, insofar as it is applicable to numerical algorithms.

One of the aims of ALGOL was that it should serve as a universal and convenient programming tool, not only for professional programmers, but also for engineers and scientists who wish to use modern machines in their work, but who do not wish to become deeply involved in machine hardware considerations. Practical considerations become important here. An implementor with these in view is strongly tempted to put limitations on the language in the interests of target efficiency, e.g., by eliminating language features which require extensive repeated run time interpretation. At this point he is faced by a basic dilemma: If the effort is made to provide all features of the full language, the generated target programs may in some cases have execution times that are open to criticism on a production basis, even if the translator is complicated by efforts to treat many special cases. On the other hand, the features that cause this may be seldom used. Restrictions placed on the language in the interests of efficiency leave him open to the accusation that he has not implemented the full language.

The question of whether the availability of certain features (e.g. dynamic own arrays and recursive procedures) offsets the increased complexity of the translator or the increase in target execution time, or both, must be answered in terms of the primary uses to which the translator will be put. Restrictions to the language (as opposed to changes) made in accordance with local needs are analogous to the restrictions placed on natural languages by individuals and groups. For a particular computer, some of the generality of ALGOL is lost, but this only in connection with features that it will not use much anyway. It is desirable, however, in all cases to have facilities for relatively complete checks on syntax of programs which are handled, so that automatic indications can be given that certain features used in programs originating elsewhere are not within the scope of the given translator.

The Handbook policy has been to make certain restrictions in the language (IFIP-subset of ALGOL) for publication purposes which do not sacrifice any of the real power of the language [*100*]. A translator which restricts the language to no greater degree will, therefore, be able to handle without change all algorithms published in the series.

The restrictions for our model translator are:

(1) **own** variables, **own** arrays, and value listed formal labels are not handled.

(2) Jumps to undefined switch components lead to undefined actions of the generated program.

(3) If specifications of formal parameters are given, then declarations or specifications of associated actual parameters must be compatible with these[1].

---

[1] In sections 4.3 and 4.10 the full discussion follows.

(4) Function procedures are assumed to be free of side effects[1].

Present computers are for the most part sequential in operation, that is, the machine language is such that programs are executed by performing one machine instruction after another in the sequence in which they are written, though this sequence may be interrupted by a jump instruction. The discussion of translation is limited to such machines. Computer design today is tending away from the simple sequential. For machines with non-sequential features, the translation process and the formal translator may require modification, unless the non-sequential features are handled by suitable monitor systems and assembly programs. It is useless in this volume to attempt to anticipate either the possible innovations or the principles which will be required to adapt the translation process satisfactorily to them.

2. Principles of ALGOL translation

In recent years many papers have been published which deal with the handling of algorithmic languages and especially with the translation of ALGOL. Among these, the papers of H. RUTISHAUSER [88] and C. BÖHM [12] appear to have been the first. The following sections and the formal translator are based on the publications of K. SAMELSON and F. L. BAUER [94] and of A. A. GRAU [40]. The principles expounded in these shall now be summarized.

2.1. Basic linguistic definitions

In the sense used here, the term *language* may be defined as a set of symbols, called the *alphabet* of the language, a set of *syntactic rules* which determine how valid sentences in the language are constructed from the alphabetic symbols, and the set of *sentences* so constructed[2]. An ALGOL program is a sequence of symbols each of which is a basic symbol, that is, an element of the alphabet of ALGOL. The task of an ALGOL translator is that of producing from the ALGOL program a program in a specific target language. The corresponding target program is a sequence of alphabetic symbols of the target language. Thus, a translator is a mechanism which performs operations on sequences of symbols by which elements of the source language ALGOL are transformed into elements of the target language. This may be regarded as a mapping of the set of sentences in ALGOL into the set of sentences in the target language.

---

[1] This point must be regarded when handling arithmetic operations and using the commutative law. It is also of importance for the optimization in chapter 5 when testing for loops suitable for recursive address calculation.

[2] The syntactic rules are sometimes called *production rules*.

The translation problem as a whole consists, on the one hand, in *analyzing* the *source language* sentences and, on the other hand, in *generating target language* sentences. The analysis of a sentence in ALGOL (a program) is the process of finding out which syntactic rules have been used for producing the given sequence. Generating a corresponding sentence is simplest if the production rules of the target language are in a certain sense compatible with the ALGOL productions. This is the case for the machine languages or symbolic machine languages which are of a simple sequential structure such that to each production rule of ALGOL found by the analysis corresponds a certain sequence of production rules used for generating the target program. Thus, program generation is not a serious problem for us since we restrict our discussions to a target language having machine language structure (see elementary language, section 3.2). The primary problem in the translation of an algorithmic language like ALGOL then is that of analysis. In the following, the analysis according to the formally described rules of ALGOL is also called the *decomposition* of the syntactical structure.

Before starting with decomposition methods, we discuss in more detail the structure of the production rules of our source language ALGOL in relation to machine languages.

2.2. The Backus normal form

The structure of many constituents of an ALGOL program, such as variable, number, arithmetic expression, and statement, are most conveniently defined *recursively*. The Report [75] does this by means of the syntactic skeletons given in *Backus normal form*. In this, each production rule is of the form

$$S ::= T_1 T_2 \ldots T_m.$$

The left part variable S is called a *syntactical variable*. The T_i's ($i = 1, \ldots, m$) may be elements of the alphabet of ALGOL (which may occur only in right parts) or syntactical variables; m is called the *length* of the given production rule. Two production rules

$$S ::= T_1 T_2 \ldots T_m \quad \text{and} \quad S ::= U_1 U_2 \ldots U_n$$

may be combined in the shorter form

$$S ::= T_1 T_2 \ldots T_m | U_1 U_2 \ldots U_n.$$

This notation permits recursion, since in the production rules

$$S ::= T_1 T_2 \ldots T_m; \quad R ::= U_1 U_2 \ldots U_n$$

a given T_i or U_j may be identical with S or R, respectively, or T_i may be R and U_j may be S.

The structure "simple arithmetic expression" may be used to illustrate this. Its definition in Backus normal form is ([75], 3.3.1):

⟨simple arithmetic expression⟩ ::= ⟨term⟩ | ⟨adding operator⟩⟨term⟩ | ⟨simple arithmetic expression⟩⟨adding operator⟩⟨term⟩.

Of the three parts of this definition, the first two are in terms of the concepts "term" and "adding operator" that are themselves separately defined. The third uses the defined concept itself. In less formal language this is equivalent to saying that a simple arithmetic expression may consist of a substructure, that is itself a valid arithmetic expression, followed by an adding operator and a term. It is precisely this feature that renders the definition of "simple arithmetic expression" a recursive one. The use of recursive definitions in ALGOL is an extension of a similar use of this important device in mathematics and mathematical logic.

ALGOL differs from all extensively used earlier algorithmic languages in that "statement" also has this property. In this case ([75], 4) the syntactic description is more elaborate than that above.

Briefly, assignment statements, procedure statements, and dummy statements, each of which is separately defined as a non-recursive entity, are examples of the structure "statement". Other statements, having statements as constituents, may be recursively formed by (1) compounding, (2) forming blocks, (3) forming conditional statements, and (4) forming for statements.

Thus, we can say that a basic characteristic of ALGOL is that the instructions for computation are expressed within the frame-work of an elaborate nested *bracket structure*. This structure is derived partly from that of the *algebraic language* used in mathematics, and partly from that in *natural language*. The latter is used to express the dynamic requirements of an algorithmic computation. The use of parentheses and the observance of the usual precedence rules in arithmetic and Boolean expressions are examples of bracket structure of the former type. The for statement and the conditional statement

for $v := E1$ **step** $E2$ **until** $E3$ **do** S

and

if B **then** $S1$ **else** $S2$

are examples of the latter type.

On the other hand, a machine language generally does not have a bracket structure as involved as this. Its production rules contain recursiveness only in a trivial way.

2.3. The analyzing process

The presence of recursively defined syntactic structures in ALGOL precludes the use of *sequential methods* of translation that may serve satisfactorily, for instance, in the translation of symbolic machine languages into machine languages. These methods consist in taking the characters of a program, one by one, in a definite order, usually in the order in which they are written, and immediately using them to produce a bracket-free decomposition or target program.

A good illustration of the problem is the handling of expressions not containing conditions. The results may be applied also to other syntactical elements and especially to statements in a simple way.

In arithmetic expressions, subexpressions are grouped together by means of parentheses, or of the well known rules of *precedence* of arithmetic operators $\big($e.g. $a \times b + c \times d$ means $(a \times b) + (c \times d)\big)$.

These precedence rules must be observed by the translator. The rules for constructing arithmetic and Boolean expressions:

$$\vdots$$

$\langle \text{term} \rangle ::= \langle \text{factor} \rangle \mid \langle \text{term} \rangle \langle \text{multiplying operator} \rangle \langle \text{factor} \rangle$
$\langle \text{simple arithmetic expression} \rangle ::= \langle \text{term} \rangle \mid \langle \text{adding operator} \rangle \langle \text{term} \rangle \mid$
  $\langle \text{simple arithmetic expression} \rangle \langle \text{adding operator} \rangle \langle \text{term} \rangle$

$$\vdots$$

prescribe the definite order in which the operations are to be executed. The course of *decomposition* and *generation* of target language orders must correspond to it exactly.

We formulate the working method of the analysis in the following informal way. A sequence $T_1 T_2 \ldots T_n$ of alphabetic symbols or syntactical variables is called a *constituent* of an ALGOL sentence if this sequence is obtained, in the course of producing the sentence by means of the given production rules, as a replacement, immediate or indirect, of a syntactical variable.

For example, $\langle \text{term} \rangle + (c \times d)$ is a constituent of the expression $(a \times b) + (c \times d)$ whereas $\langle \text{term} \rangle +$ is not a constituent[1]. A constituent consisting of at least two symbols[2] and which does not contain any other constituent is called a *primitive constituent*.

For example: $a \times b$ is a primitive constituent of the above expression. Each primitive constituent is associated with a certain syntactical variable.

The task of analysis is finding primitive constituents in a given ALGOL program which may then be used to generate target language equivalents and to replace them in the ALGOL program by the corresponding syntactical variable. This variable is given unambiguously by the sequence of productions from which the primitive constituent may be obtained.

---

[1] In order to avoid ambiguities we denote the unary $+$ and $-$ operators by $+_u$ and $-_u$.

[2] For the sake of simplicity we can assume an identifier or a number to be an element of the alphabet of ALGOL. Thus, each is considered to be a single symbol.

Since the source language string is shortened by this, the process of repeated application of the search for primitive constituents will be finished after a finite number of steps.

In most cases, generation of certain target language instructions is connected with each primitive constituent. The explanation in the following chapters shows that this correspondence is simple for expressions and statements. By contrast, declarations lead to the general problems of allocation of storage to data, the implementation of type handling, and the problem of handling procedures and procedure calls.

There are several practical ways of scanning a program for primitive constituents. All methods must necessarily be derived from the syntactic skeletons defining ALGOL. This may, of course, be fairly well hidden, as in a *multiple scan method*, or completely explicit, as in the so-called *syntax-directed type* of compiler [121, 116], where the analysis makes explicit use of the production rules written in Backus normal form. It is not necessary in the case of ALGOL to fall back on these. It is sufficient to extract certain rules, which we shall call precedence rules, from the production rules and to base the translator on them.

2.4. The method of the "Klammergebirge"

A useful illustration of a mechanism constructed according to the analyzing principle outlined in the last section is the use of the "*Klammergebirge*" given by RUTISHAUSER [90]. Here, when one primitive constituent is found, it is removed and processed, the program simplified, and the process repeated.

For the sake of simplicity we only consider expressions consisting of variables $a, b, c, \ldots$ and delimiters $(,), +, -, \times, /, \uparrow$. If each expression is included in round brackets, then its structure may be represented by a continuous line drawn as follows:

> ╱ when (occurs
> ╲ when) occurs
> ╲╱ when an operator occurs.

The elements ╱, ╲, and ╲╱ must be joined together from left to right. In addition each term and each factor must be included in round brackets. These implicitly existing brackets are denoted by (' and ') and handled in the same way as (and).

As example we use the expression described in [11]:

$$a + \quad b \times a / \quad (\quad c \quad \times d \quad + e \,) \quad + \quad g \times h \times j$$

The peaks correspond to variables and the valleys to operators. A chain of summits of equal height ∧∧∧ corresponds to a sequence of operators of equal rank, while the differences of levels represent the precedence of certain operations before others.

The rules for taking down the Klammergebirge are:

replace	$V1$ $V2$ ∧∨∧ ω	by	$V3 = V1\,\omega\,V2$ ∧	and produce the appropriate machine instructions
replace	V ∧	by	V ∧	pushing together the rest of the Klammergebirge.

The original set of rules given by RUTISHAUSER required: search for the absolutely highest peak which corresponds to the innermost primitive constituent; apply the substitution rules and then search for the new highest peak and so on.

For conditions and statements the situation is similar. It is also possible to have any degree of bracketing, and this means that at first the operations corresponding to the primitive statement constituents must be generated.

2.5. Recursive sequential methods and push down lists

Methods like the one just described require in general many passes through the material, so that they tend to be relatively slow. However, the levelling of the Klammergebirge can be executed in a certain sense *"sequentially"* and thus combined with a single reading of the sequence of alphabetic symbols. This works in the following way: one symbol after the other is read. At the same time the corresponding line is drawn and before reading the next symbol, it is determined whether parts of the Klammergebirge already produced, can be taken down. After this is done the next character is read and so on.

Most methods of analyzing ALGOL programs used today are based on this principle. We speak of *recursive sequential methods*, since they retain many of the advantages of the simple sequential scan. Multiple scans and passes are avoided here by the systematic storage of scanned material in a single pass in such a way that it is conveniently available when it is needed and can be processed. Basic to these methods is the concept of a *push down list* (or simlpy push down) introduced by K. SAMELSON and F. L. BAUER [94] and NEWELL and SHAW [80] which is a list of variable length operating in a last-in first-out fashion.

The push down is used in such a way that, whenever a symbol occurs which cannot be processed immediately, a corresponding new entry is added to the list. If the last entries of the push down are re-cognized to denote a primitive constituent then they are removed from the list and processed; an entry corresponding to the syntactic variable associated with the primitive constituent processed may be put into the push down list.

The decomposition process of a sequence χ of ALGOL symbols with the aid of a push down list σ may be looked upon to be an *infinite automaton*, where the *states* are the entries of σ and the inputs the χ_k's. Input and state of the automaton determine a transition to a new state and an output.

2.5.1. The decoding matrix

The analyzing system can be set up in such a way that the transition of the push down to a new state depends only on the uppermost state symbol σ_d in σ, which we shall call now the uppermost state of σ, and the incoming symbol χ_k. More precisely, a state transition takes place only when the incoming symbol χ_k is a *delimiter*, whereas identifiers, numbers, truth values, and strings, which together constitute the class of *entities*, are pushed down immediately, and are not considered to be state symbols. Thus, the pairs (σ_d, χ_k) determine uniquely whether a primitive constituent is isolated or not[1]. Analyzing reduces to comparing the incoming symbol with the uppermost state in σ, delivering primitive constituents to the generator and pushing down or eliminating elements.

We call "decoding" the selection of the parts of the translator which must be executed depending on σ_d and χ_k. If each incoming identifier and constant leads to the same action of the translator, it is only necessary to regard pairs of states and incoming delimiters.

One of the simplest methods of formulating the decoding is by the use of a *decoding matrix*. A table of the form

σ_d \ χ_k	*	*	* ...
*	*	*	* ...
*	*	*	* ...
*	*	*	* ...
:	:	:	: .

may then be incorporated in the translator. The row entrances are all possible push down states, and the column entrances are all possible incoming characters χ_k. Each matrix field leads to that part of the program to which control is transferred during translation upon encountering the corresponding pair (σ_d, χ_k).

---

[1] One identifier may be above σ_d in the push down. This means that the language belongs to the "bounded context" class (1, 1) in the sense of Irons [124].

It is, of course, possible to identify states with the symbols of the source language. For the sake of simplicity one even may use the same code (as is done within the model translator). This means that a symbol χ_k can be pushed down immediately without further coding.

While this method is indeed simple to execute, it is possibly wasteful with respect to storage space (e.g. for our model translator with 67 possible rows and 60 possible columns the matrix would require 4020 elements).

In considering the full matrix it is found that certain rows or columns lead to the same or nearly the same actions of the translator program. In some cases, the same subprogram of the translator can be executed for different states (see the binary operators e.g.) and a special decoding is necessary only in a later stage. In some cases, the action of the translator does not depend on the special incoming χ_k (e.g. if χ_k is a delimiter closing an expression). Such states of the push down σ or such χ_k's may be summed up into new states or classes which can be used as new entrances of a new matrix. The table can be reduced considerably in this way. Certain decodings are still necessary within certain program parts. The model of a translator based on such a reduced matrix has been given by A. A. GRAU [40].

2.5.2. Precedence rules and orders

A further reduction of the size of the matrix is possible by generalizing the *precedence rules* for operators to obtain an ordering of the set of states, and of the set of characters. The order of each state may be given by a natural number. This can be chosen in such a way that the sign of the difference of the *state order* of σ_d and of the *character order* of χ_k determines the main course of the action of the analyzer [95, 49, 116].

While it is possible to identify character order and state order for the operators this is not expedient for the opening bracket symbols. A purely closing bracket [such as) | ; | **end**] has no corresponding state and because of this no state order need be defined for closing brackets. But certain delimiters (as **then** | **step**) may appear as closing brackets, when incoming, and behave as opening brackets afterwards; these must have the same state order as the opening brackets.

The order numbers derived from the precedence rules of operators lead to the following complete table, see p. 12.

The somewhat artificial distinction between expression and statement brackets made here gives certain advantages for syntax checking.

For the sublanguage of arithmetic expressions the comparison of state order and character order leads to a straightforward triple branch of the process of analysis.

char- acter order	symbol	state order	state
1	⟨expression opening bracket⟩	1	·/.
2	⟨statement opening bracket⟩	2	·/.
3	↑	3	↑
4	⟨multiplying operator⟩	4	⟨multiplying operator⟩
5	⟨adding operator⟩	5	⟨adding operator⟩
6	⟨relational operator⟩	6	⟨relational operator⟩
7	⌐	7	⌐
8	∧	8	∧
9	∨	9	∨
10	⊃	10	⊃
11	≡	11	≡
12	⟨expression closing bracket⟩	12	⟨expression opening bracket⟩
13	⟨statement closing bracket⟩	13	⟨statement opening bracket⟩

If the order of the incoming character χ_k is less than the order of the uppermost state σ_d, a new state corresponding to χ_k is entered into the push down.

If the orders are equal, the state σ_d with the neighbouring entities is extracted from the push down, handed over to the program generator for further processing and is replaced in the push down by an entity (which may be generated by the system). Then again, a new state corresponding to χ_k is entered into the push down.

If the character order is greater than the state order, the state σ_d is treated the same way as in the last mentioned case. After this however, the process of comparison is repeated with the same character χ_k and the now uppermost state σ_d.

One may notice that the "sequential" working with the Klammergebirge corresponds immediately to the work with the push down. Parentheses and operators are put into the push down in sequence. Instead of relative differences of heights we have the orders related to the delimiters pushed down, and the comparison of orders corresponds to testing points of the Klammergebirge.

For the rest of the language the process of analysis by order comparison works in a similar way. However, a number of special cases depending on state σ_d and/or character χ_k must be distinguished.

From the above considerations, it is clear that only one push down list is required for the intermediate storage requirements of the translator. For convenience this may be split into several different lists. Thus, a state push down may be used to hold the state symbols, and other information associated with syntactic structure, while an auxiliary push down may be used for entities. The combined push down may be obtained from them by a suitable interleaving.

For the model translator in chapter 8 we shall use a single push down. States and entities are stored with marks **'delimiter'** and **'entity'** respectively. For the decoding one always has to check the uppermost state, that means the element on top, or next to the top, of the push down.

2.6. Example for the use of two push down lists and of precedence rules

As an illustration of the use of the push down lists and the precedence rules we demonstrate the decomposition of an assignment statement with an arbitrary expression into simple bracket-free ALGOL statements.

An arbitrary expression may be decomposed into simple expressions in a variety of ways. These, naturally, must generate machine language programs which have to be necessarily equivalent[1]. The methods used differ from each other primarily in the manner in which the reduction of bracket structure takes place, and in the manner in which symbols are represented and compared. The decomposition method given here follows the rules given in the ALGOL 60 Report, but allowing the use of the commutative law. We discuss the following statement (see also [94]):

$$a := (b + c \times d)/(a - d) + b;$$

The individual symbols of this statement are regarded as components $\chi_1, \ldots, \chi_G$ of a vector χ and the target language instructions generated as the components π_p of a vector π. To effect the decomposition we shall use a state push down σ and an auxiliary push down α. Each of these initially contain nothing but a symbol $\varnothing$.

Translation then proceeds basically as described in 2.5.2: the symbols of the statement to be translated are read in sequence one by one. A variable, or number is immediately stored in α. A delimiter is compared with the uppermost entry in σ. On the basis of the precedence rules in 2.5.2 a definite action of the translator is thereby determined. Either the incoming symbol is added to σ as a new uppermost element or the existing uppermost entry in σ is removed. In the former case, a new symbol is read in. In the latter case, subject to certain conditions, a target language instruction may be generated. At the same time the uppermost entries in α are removed.

The bracket-free sequential nature of the target language demands certain new variables to be generated in order to store intermediate results. They are denoted by $v1, v2, \ldots$. The result of a binary operation

---

[1] We mean here "mathematically equivalent". But, as for a computer the associative law is not valid in general, the mathematically equivalent expressions (1) $a + b + c$ and (2) $b + c + a$ evaluated from left to right will not always lead to the same result and therefore are not equivalent from the point of view of a computer.

must be assigned to a new variable vi. As soon as this vi is used as an operand it is free for further intermediate results.

In the following table are shown the characters read, the contents of the push down lists, and the results of the operations performed in sequence through the incoming symbols. The symbols 'a', '$:=$' with character order 1, '(' with character order 1, 'b', '$+$' with character order 5, 'c', '$\times$' with character order 4, 'd' are simply entered into α or σ as the case may be.

The next symbol ')' with character order 12 encounters '$\times$' with state order 4 in σ. As the character order is not less than the state order the operations connected with '$\times$' are performed. This means that

(1) the target language instruction '$v1 := c \times d;$' is generated and stored in π,

ALGOL program			push down			target language program	
g	χ_g	char- acter order (χ_g)	σ	state order (σ_s)	α	p	π_p
			$\varnothing$	$\cdot/.$	$\varnothing$		
1	a				$\varnothing\ a$		
2	$:=$	1	$\varnothing\ :=$	12			
3	(	1	$\varnothing\ := ($	12			
4	b				$\varnothing\ a\ b$		
5	$+$	5	$\varnothing\ := (+$	5			
6	c				$\varnothing\ a\ b\ c$		
7	$\times$	4	$\varnothing\ := (+\times$	4			
8	d				$\varnothing\ a\ b\ c\ d$		
9	)	12	$\varnothing\ := (+$	5	$\varnothing\ a\ b\ v1$	1	$v1 := c \times d;$
			$\varnothing\ := ($	12	$\varnothing\ a\ v1$	2	$v1 := b + v1;$
			$\varnothing\ :=$	12			
10	/	4	$\varnothing\ := /$	4			
11	(	1	$\varnothing\ := / ($	12			
12	a				$\varnothing\ a\ v1\ a$		
13	$-$	5	$\varnothing\ := / (-$	5			
14	d				$\varnothing\ a\ v1\ a\ d$		
15	)	12	$\varnothing\ := / ($	12	$\varnothing\ a\ v1\ v2$	3	$v2 := a - d;$
			$\varnothing\ := /$	4			
16	$+$	5	$\varnothing\ :=$	12	$\varnothing\ a\ v1$	4	$v1 := v1/v2;$
			$\varnothing\ := +$	5			
17	b				$\varnothing\ a\ v1\ b$		
18	;	13	$\varnothing\ :=$	12	$\varnothing\ a\ v1$	5	$v1 := v1 + b;$
			$\varnothing$	$\cdot/.$	$\varnothing$	6	$a := v1;$[1]

[1] The instructions 5 and 6 could, of course, be replaced by $a := v1 + b;$ but we do not want to discuss any optimization here.

(2) 'c' and 'd' are deleted from α, and the generated variable $v1$ is entered;

(3) '$\times$' is deleted from σ. ')' then encounters '$+$' with state order 5 as the uppermost state symbol. Another target language instruction is generated here, and so on. The translation is finished when the vector χ has been exhausted and σ and α reach their original state $\varnothing$.

2.7. The concept of recursive translation

There are other ways of constructing the analyzing and generating part of a translator. The translator may e.g. work in analogy to the productions of ALGOL. This means that the translator contains *subprograms* corresponding to each syntactical variable, or production.

Each subprogram is called into action to process the corresponding type of structure. Thus, the statement subprogram is called to process statements; expressions are processed by the expression subprogram. The subprograms behave like closed subroutines, which in some cases they are; this means that before entry to any of them, provision in some manner is made for a return to a place specified in advance and depending on the program after the subprogram has been executed.

The recursive nature of the definitions of some ALGOL structure requires that the corresponding subprograms be recursive, that is, within each of these it must be possible to call the subprogram itself on a new level without the loss of information on the original level. Thus, the subprogram which handles simple arithmetic expressions must be able to handle a sub-expression before it has finished with the original expression, since a constituent of a simple arithmetic expression may be another arbitrary simple arithmetic expression. The recursive call of a subprogram may in fact be indirect, that is, other subprograms may be called between the two nested calls of the given routine. It follows that recursiveness is in general a mutual property of a whole class of such subprograms.

The essential problem with *recursive subprograms* or subroutines is the storage of intermediate results, including the storage of the return address. Such results may be considered as local variables in the subprograms; they come into existence during execution and are no longer meaningful upon exit. It is quite clear from the discussion above that such information cannot be placed into fixed locations. Instead, it is preferable to store information of this kind relative to an arbitrary location which is known only at execution time. The problem of intermediate storage for the routines is nicely solved by a push down list, that is, by setting aside a block of storage for the use of these routines, together with a counter which keeps track of the beginning of free storage in this block. After making allowance for its own needs, each routine of the set advances the counter for use by subsequently called routines; on exit, it restores the counter to the value it had on entry, since all space used by the routine and its subroutines will then have been vacated.

Such a push down list adequately meets the storage requirements of a set of mutually recursive subroutines. Consequently it also meets the needs for intermediate storage of a translator. The order of information in the list at any time corresponds to the nesting of subroutines in effect at that time.

From this point on, however, we shall not be interested further in regarding the translator explicitly as a collection of recursive sub-routines. A slightly different, though theoretically equivalent, formulation proves somewhat more useful for reason of speed and efficiency, for our concrete task.

2.8. Organization of the translator

A translator can be designed to carry out all its functions in a single *pass* through the source program. By a pass is meant a detailed sequential handling and examination of a source program or its later equivalent. This type of translation will require extensive storage of material while other information needed to complete it is located. The difficulty here is primarily related to declarations and specifications; thus, quantities may appear which are declared only later. A case in point is the label in a forward jump.

The need for elaborate provision for intermediate storage is reduced by making more than one pass. The use of several passes is generally necessary in the case of machines with small high-speed memory for that reason alone. In all cases, however, there is the advantage that a suitable separation of functions disentangles the translation process. These functions include the systematic accumulation of information, editing of the program, and the production of the target program. The language used in the passes may change from one pass to the next. It is useful to list the following tasks of a translator explicitly; each may actually be performed by a pass of its own:

1. The conversion of hardware ALGOL into a suitable internal source language.

2. The decomposition of the ALGOL bracket structure.

3. Optimization of loops with regard to subscript evaluation.

4. The handling of variable types.

5. Allocation of data storage by means of type, array, and procedure declarations.

6. The conversion of the bracket-free target language into the language of a specific machine.

Our model translator consists of three passes: Pass one contains the conversion of hardware ALGOL into internal source language listed as task 1. above. Pass two handles the optimization of loops given as task 3. Pass three combines tasks 2., 4., and 5., decomposition of bracket structure, type handling, and storage allocation. The conversion into machine language listed as task 6. is essentially the task of a standard assembler, and is not included into the model translator for this reason.

3. Languages involved in the translation process

After having represented in the preceding chapter the problem of translating ALGOL into machine language and after the more general discussion of possibilities of solution we shall now start with the description of a specific ALGOL translator and of the target language program to be produced. This requires that terms like source language and target language at the outset must be specified by defining exactly in each case the language used. This is the task of the present chapter.

Conversion of programs written in ALGOL into programs in machine language, together with a description of the process, involve the following languages:

1. The *source language*, i.e. ALGOL.

2. The *target language*, i.e., in the ultimate sense, either an actual or symbolic machine language.

3. The language in which the translation program itself may be written and described.

Between 1. and 2. intermediate language forms are inserted to facilitate the conversion.

3.1. Source language

For our purposes the source language is, of course, ALGOL. However, most present-day computers do not have an input character set which includes all of the ALGOL reference symbols. In practice, therefore, it is necessary to design and use a hardware representation of ALGOL as source language. The problem in this is that of overcoming the shortage of machine characters; devices for this include the increased use of context in processing, heavy reliance on declarations, and the use of "escape" symbols to permit the extension of the machine character set.

By depending on context during processing, it is often possible to represent two or more ALGOL reference symbols by one machine character. For example, the delimiters **step** and **until**, for which there are seldom suitable machine characters, occur only in for clauses, and even more specifically, in for list elements. The reference ALGOL colon (:)

cannot occur in such elements[1]; it serves in different contexts two main purposes: it designates labels and it separates subscript bounds in array declarations. If the colon is a machine character it is possible to use it not only in its reference contexts but also for either or both **step** and **until**. Context alone then must be used by the processor to determine which of the four functions is intended in a particular case.

Since a hardware character set is always strictly bounded and often severely restricted, it is important to have a systematic method for representing an arbitrary number of new symbols. This can satisfactorily be done by reserving a pair of hardware characters for use as "escape" parentheses. A symbol for which there is no reasonable representation may be represented by a pair of these parentheses enclosing a string of other hardware characters. Thus, if ⟨ and ⟩ represent the two escape symbols, the delimiters **if**, **go to**, and **begin** may be represented quite naturally by ⟨if⟩, ⟨go to⟩, and ⟨begin⟩, the binary Boolean operations ∧ and ∨, for which machine characters usually do not exist, by ⟨and⟩ and ⟨or⟩, and the relations <, ≦, and ≠ by ⟨less⟩, ⟨not greater⟩, and ⟨not equal⟩; these are mnemonically acceptable since the resulting hardware language can be read without much explanation. The use of escape symbols preserves the one-to-one correspondence between reference symbols and their hardware representations; it therefore avoids difficulties inherent in other devices. In addition, it provides an open-ended system, since new symbols can be added at any time, and the selection of a representation for a new hardware character set becomes practically automatic.

A single hardware character may well in practice serve as both the left and the right escape parenthesis. Then the function of the hardware character so selected becomes that of changing modes in the hardware language. This can be between, on the one hand, the normal mode, in which a reference character is represented uniquely by a hardware character, and on the other, the extended symbol mode, in which a reference character is uniquely represented by the string of letters up to the point where the mode is changed again.

A special input hardware representation of ALGOL is made necessary by the restricted nature of a given input character set; no such restriction, however, applies to the internal representation of ALGOL. Each reference symbol may, if desired, be represented by a uniquely chosen set of bits or decimal digits in standard format. Practically, it is not desirable in this way to reconstruct ALGOL programs internally; some preliminary processing may well be done on the input phase without

---

[1] We assume that each parameter delimiter) ⟨letter string⟩ :(has already been replaced by a comma.

loss of time and with an increase in overall efficiency. For these reasons, a modified internal language may be used, which we shall call *process language*. This differs from both the hardware representation of ALGOL and reference ALGOL in the following respects:

1. A delimiter which in different contexts has different syntactic functions may in these contexts be represented by different internal delimiters. Thus one representation may be used for **begin** when it is the beginning of a compound statement and another when it is the beginning of a block.

2. Constants (i.e., numbers, strings, and truth values) may be removed from the program and replaced by generated identifiers. The constants themselves with the identifiers that represent them are placed in a separate list which must be available for further use. The standard function names may be added.

This list constitutes the initial part of what we shall call the *identifier list*. During processing identifiers current to the ALGOL program are added to it: with each identifier it contains in coded form characteristics derived from the corresponding declaration, implicit label declaration, or specification including the equivalent of the associated program or data storage address; it is therefore composed of abstracts of declarations and specifications. The process language program must retain those components of declarations and specifications which are required at run time.

3. The entire entry corresponding to an identifier (or entity) in the identifier list may be assigned an *internal identifier* of standard format to represent it uniquely. This may be used to replace the identifier or entity throughout the process language program.

A detailed description of the identifier list as used in the formal translator is given in section 7.2.2.2.

3.2. Target language

At some point, elements of some language form derived from the process language must be converted into hardware instructions, or their equivalent, symbolic assembly instructions, for the machine for which the translator is designed. This process is a matter of simple substitution or transliteration if the set of elementary constituents of this derived language form is suitably chosen. With such a choice the target language, for the purpose of translation, is in effect precisely this set of elementary constituents which we may call *elementary language*. The ultimate target language is of course machine language. However, we shall not consider the final step from our target language to machine language. The elementary language used within the formal translator has been designed using the fact that most modern machines have *single-address*

2*

instructions. This is reflected in the fact that in elementary statements like

$$U := V + W,$$

representative of binary operations, the left part variable U and the first right-hand operand V are restricted to an exceptional variable AC. Similarly in

$$U := -V$$

U must, and V can be AC. In an implementation AC will generally be associated with the *accumulator* of the machine. The restriction in elementary constituents made here is not necessary in the case of a *three-address* machine.

The constituents of the elementary language for a single-address machine are listed below; we must first introduce notation for certain exceptional variables and for elements of certain generic classes.

Exceptional variables:

AC   to be interpreted as accumulator
ADDR AC   to be an exceptional register to take up the actual
                   address of a formal variable (address accumulator)
BFS   to be interpreted as a special register containing the
                   address of the beginning of free storage
SIR   to be interpreted as special index register of switch
                   components
$\Im\mathfrak{r}[\varkappa], \varkappa = 0, \dots, k$ to be interpreted as index registers.

Generic symbols:

ω unary or binary operations: $-_u$, $\daleth$, $+$, $+_i$, $+_{addr}$, $-$, $-_i$, $\times$, $\times_i$, $/$, $\div$, $\uparrow$, $\uparrow_{ii}$, $\uparrow_{ir}$, $\oplus$
ϱ relational operator: $<$, $\leq$, $\geq$, $>$, $=$, $\neq$.

The following names are repeatedly used with the meaning given below:

A	array
AP	actual parameter
FL	formal label
FP	formal parameter or formal procedure
GI	generated identifier
GL	generated label
GV	generated variable
I	identifier
L	label
N	formal not value listed variable

P	procedure
S	switch
V	simple or generated variable
VV	value listed formal variable.

Furthermore we introduce an operator to denote the content of an address variable V:

content of V.

This operator may sometimes be applied to a variable V; V must then have a value that can serve as an address. This occurs in connection with subscripted variables where a generated variable is used in order to store the address of a subscripted variable computed at run time.

In terms of the notation discussed above, elementary constituents of the elementary language include the following:

a) Binary operations:

$$\textbf{AC} := \textbf{AC} \, \omega \, V$$
$$\textbf{AC} := \textbf{AC} \, \omega \, \textbf{content of} \; V$$
$$\mathfrak{Ir}[\varkappa] := \mathfrak{Ir}[\varkappa] +_i V \, .$$

b) Unary operations:

$$\textbf{AC} := \omega \, V$$
$$\textbf{AC} := \omega \; \textbf{content of} \; V$$
$$\textbf{AC} := \omega \, \textbf{AC}$$
$$\textbf{AC} := \omega \; \textbf{content of ADDR AC} \, .$$

c) Assignments:

$$V := \textbf{AC}$$
$$\textbf{content of} \; V := \textbf{AC}$$
$$\textbf{AC} := V$$
$$\textbf{AC} := \textbf{content of} \; V$$
$$\textbf{ADDR AC} := \textbf{AC}$$
$$GV := \textbf{ADDR AC}$$
$$\textbf{AC} := \textbf{content of ADDR AC}$$
$$\mathfrak{Ir}[\varkappa] := V \, .$$

d) Jump elements. Three kinds of jump elements may be distinguished:

(1) Jumps to non-formal labels or switch designators:

$$\textbf{jump to} \; L$$
$$\textbf{SIR} := \textbf{AC}$$
$$\textbf{SIR} := V$$
$$\textbf{SIR} := \textbf{content of} \; V$$
$$\textbf{local jump to content of} \; S[\textbf{SIR}] \, .$$

(2) Jumps to formal labels:

formal procedure exit FL.

(3) Jumps generated by the translator:

local jump to L
if AC ϱ **0 then local jump to** L
if AC then local jump to L
if ¬ **AC then local jump to** L .

For the purpose of clarity, the element

$L:$

is introduced, whose machine equivalent target action is a "no operation".

e) Subroutine linkages:

subroutine jump L
save return address V
subroutine return V .

At times a part of an ALGOL program, such as in some instances the statement subject to a for clause, must be formulated as a closed subroutine, i.e., a generated procedure. The three linkage elements given above are used in this connection. The first, as its name indicates, is a jump to a subroutine beginning at label L; at the same time the return address is placed into a fixed register. If there are nested for statements, or if recursive procedures are implemented, the content of this register must be stored in an auxiliary cell, denoted by a generated variable V. The return address is then stored as the value of V using the second element above. The exit from the subroutine is effected by the third element, which involves a jump to the label (or address) stored in V.

f) Standard procedure calls:

(1) **transfer to type real**
transfer to type integer
transfer to address
possible transfer to real depending on GV
possible transfer to integer depending on GV
possible transfer to real depending on ADDR AC
possible transfer to integer depending on ADDR AC
possible transfer from real depending on GV
possible transfer from integer depending on GV
possible type transfer from ADDR AC to GV
compare real to GV

compare integer to GV
compare Boolean to GV
compare Boolean to ADDR AC
compare ADDR AC to GV

(2) **storage allocation** A
BFS $:= V$

(3) **procedure call** P
formal procedure call FP
real procedure call P
integer procedure call P
Boolean procedure call P
real formal procedure call FP
integer formal procedure call FP
Boolean formal procedure call FP
type formal procedure call FP
real name call N
integer name call N
Boolean name call N
name call N
name address N
value array A

(4) **normal procedure exit**
name procedure exit

(5) **absolute address** V
absolute subscripted variable address V.

The statements are calls of internal subroutines with which the target program is supplied. These deal with type handling (1), dynamic allocation of storage to arrays (2), the call of procedures and formal parameters (3), the body of ALGOL procedures and procedures generated from actual parameters (4), and the determination of the absolute memory location corresponding to a simple or subscripted variable (5).

The elements listed here may be considered to be macroinstructions written in machine code, since in general no simple standard machine code equivalent is available. A detailed description will be found in sections 4.3, 4.4, and 4.10, and in chapter 6.

The way in which these auxiliary programs are called and furnished with parameters is a question of machine code. The description given in the next chapters uses a generally accepted form of subroutine linkage that consists of a subroutine jump instruction followed by a list of parameters, which in general are internal identifiers.

g) A special standard procedure:

end of the ALGOL program

the meaning of which is obvious.

3.3. Meta-language for describing the translator

It is desirable to avoid designing a special new language for the formal description of the translation process; this can be done by modifying an existing algorithmic language as required for this purpose. For the formal translator described in this book, ALGOL itself will be used. However, it must be extended to include operations on sequences of symbols or, as they may properly be called, strings.

The additional elements have the following syntactic descriptions:

⟨elementary string expression⟩ ::= ⟨variable⟩|
    ⟨function designator⟩ | (⟨string expression⟩) | ⟨string⟩
⟨string operator⟩ ::= ⊕
⟨simple string expression⟩ ::= ⟨elementary string expression⟩ |
    ⟨simple string expression⟩ ⊕ ⟨elementary string expression⟩
⟨string expression⟩ ::= ⟨simple string expression⟩ |
    ⟨if clause⟩ ⟨simple string expression⟩ **else** ⟨string expression⟩
⟨string assignment statement⟩ ::= ⟨left part list⟩ ⟨string expression⟩
⟨string relation⟩ ::=
    ⟨simple string expression⟩ = ⟨simple string expression⟩ |
    ⟨simple string expression⟩ ≠ ⟨simple string expression⟩
⟨string type⟩ ::= **string**.

This means that

1. Variables, arrays, and functions over strings are introduced by means of the declarator **string**.

2. String expressions can be formed by means of concatenation ⊕ and of conditional clauses. A special convention in concatenation is that the sequence

'⟨open string⟩' ⊕ '⟨open string⟩'

is equivalent (equal) to

'⟨open string⟩ ⟨open string⟩'.

Basic symbols used to form proper strings are ALGOL basic symbols and some additional ones by which ALGOL is extended in order to describe the translator. The "delimiter" **empty** is only used as a more understandable denotation for the empty proper string. Thus, the empty string can be written in the form ' ' or **'empty'**.

3. The assignment statement is extended to include the assignment of the value of an arbitrary string expression to a string variable.

4. String comparison is permitted as a valid Boolean expression, i.e. strings or string variables are permitted to be operands of the relations $=$ and $\neq$. The equality in this case means strict identity of strings.

Within the model translator three different types of simple or subscripted variables occur: **Boolean, integer,** and **string**. The possible values of

Boolean variables are logical values (**true, false**),
integer variables are integer numbers (15, $+4$, -23, ...) or the "delimiter" **empty,**
string variables are strings, consisting of concatenated basic symbols.

In order to facilitate the readability of the model translator program we do not demand the types of the right and left hand expressions of binary operators (e.g. $+$, $=$, or $\oplus$) and of the assignment symbol $:=$ to be identical. Assignments of integer values to string variables and strings to integer variables, comparison, as well as concatenation of strings and integers may occur. In all cases where strings and integers interfere appropriate transfer functions are assumed to be called automatically, as e.g.

string (x), which transforms the value of the integer type expression x into the form of a string (*string* (15) is '15', *string* $(+4)$ is '4', *string* (-23) is '-23'),

and

value (y), which executes the transformation from strings to values (*value* ('15') is 15, *value* ('$+4$') is 4, *value* ('-23') is -23).

With these definitions, the relations

value $(string(x)) = x$,
string $(value\,(y)) = y$, and
value $(string\,(value\,(z))) = value\,(z)$

hold as long as x is an integer type expression, y is a string expression having a value of the form '⟨unsigned integer⟩' or '$-$ ⟨unsigned integer⟩', and z is a string expression having a value of the form '⟨integer⟩'.

The assignment statement requires the transfer of the right hand expression to the type of the left part variable. Operators like $+$, $-$, $\times$, $<$, $\leq$, $>$, $\geq$, and $\oplus$ require the values of the operands to be elements of the domain of integers or elements of the domain of strings exclusively. Therefore, occasionally a transfer of the operands to the appropriate type **integer** resp. **string** has to be executed. For the operators $=$ and $\neq$ the domain depends on the left and right hand types. If both operands are integer type expressions, the domain is the set of integers, otherwise the set of strings.

4. Correspondence between elements of source and target language

Before we enter into the detailed discussion of the translator program in chapters 7 and 8 we describe how a target program must be produced and how it must work at run time. This is the aim of chapters 4, 5, and 6.

Our task is to produce from any correct ALGOL program a program consisting of instructions of the elementary language described in 3.2 for our model machine, a target program which when executed does exactly what the ALGOL programmer has intended. Hence we must describe by which elementary language instructions each constituent of an ALGOL expression or statement is replaced and in what manner these instructions are interpreted.

The main object of the following chapters is to point out the problems of ALGOL translation and to develop a complete run time system which will do everything required and can also serve as a model for practical considerations. For a given machine certain simplifications (e.g. actual parameters) may be possible and useful. However, we here design a logically complete and understandable system which is relatively machine independent, but of course lacking certain optimizations possible for specific machines.

At this point a question of compiler economics arises. In general the aim of a compiler builder must be the production of a compiler which is as economical as possible under the conditions under which it is to be used. This economy is determined by several factors, among them, the speed of the translation process, the execution time of the target program, and the storage requirements of the latter. If the primary use is compile-and-run operation, translation time is as important as execution time, even more so if the average execution time of a program is short. On the other hand, if the primary use will be in the generation of programs which will be executed many times, translation time becomes of much less importance and more attention should be devoted to producing programs which are as efficient as possible, even at the expense of delegating much work to the translator which would take considerable time at translation time.

In that case, only what actually must be done dynamically is left to the execution of the target program. In a large computation center, both kinds of operation may be of equal importance, in which case it is often desirable to have two compilers, one which minimizes compile-and-run time, and one which minimizes run time alone. In our discussion here the latter type of compiler is primarily in view.

4.1. Declarations in general

Declarations have two functions since they may contain both directives to the translator, and information which must be transformed into instructions to be executed at run time.

Declarations of simple variables are directives only to the translator, which uses them for purposes of *storage allocation* and some form of *type handling*; after that they may be eliminated from the program.

Array declarations also determine type and storage requirements. The latter, in the general case, must be handled at run time when the block to which they belong is entered. Thus, the declarations must generate instructions to be executed immediately before execution of the first ALGOL statement of the block. Procedure and switch declarations in general must be transformed into subprograms to be executed at the moment of call or jump.

If the instructions corresponding to the declarations in the target language program are compiled in the sequence given by the ALGOL program, switch and procedure declarations must be separated from the normal sequential flow of execution by means of appropriate skip instructions.

Labels of statements may be considered as implicit declarations. They are recorded in the identifier list, and left in place in the program until definite program addresses can be associated with them.

4.2. Declaration of variables and arrays
and data storage allocation in the main program

The declarators **real, integer, Boolean,** and **array**[1] serve (besides indicating the type) to reserve storage places for the values of the variables and fields declared by them. Storage places must also be allotted to numbers, truth values, and strings, which are used in an ALGOL program and which must be considered to be declared by their occurrence, and to generated variables. For our purposes, all constants are treated like variables declared in the outermost block of the main program. Generated variables are treated as local variables declared in the innermost block containing them.

Complete data storage allocation cannot be done by the translator for the following reasons:

---

[1] The declarator **own** will not be handled in this book, because the concept of **own** becomes ambiguous in connection with procedures.

(1) A source of flexibility in ALGOL programs is the possibility of the program itself computing the storage needed for certain sets of data and assigning storage locations for the individual elements. This is done by means of the dynamic array declaration whose subscript bound list contains arbitrary arithmetic expressions. The number of storage cells required for such an array (which may be called a *dynamic array*) is known only at run time.

(2) Each procedure call requires storage places for the data local to the procedure body. If a procedure is called recursively, the number of storage places is unknown at translation time as it depends on the number of recursive calls. The translator, however, can undertake certain preparations for the sake of efficiency. The block structure in ALGOL gives a simple and obvious means of economizing storage which is shown immediately by the structure itself.

The following dynamic storage allocation scheme suggests itself: A sequence of consecutive storage locations beginning with the initial *"beginning of free storage"* is assumed to be available for data storage. Whenever execution of a new block is initiated, storage locations for all variables and arrays declared in the block head (including generated variables) are lined up in sequence from the momentary "beginning of free storage", and the end of this sequence determines the new "beginning of free storage" for the next step. The whole sequence of storage locations is available for reuse as soon as execution of the block is terminated. Procedure calls are treated like new blocks inserted at the place of call according to the so called copy rule.

This scheme of storage allocation automatically takes care of re-cursive procedures. However, it involves a fully interpretive handling of blocks. To avoid this we shall modify somewhat this scheme. Consider at first the main program apart from procedures. Each block introduces (1) a fixed number of variables and arrays with fixed bounds (*static arrays*) where the number of places required may be fixed by the trans-lator and (2) a number of arrays with variable bounds.

Thus, the data storage for the main program may consist of two parts, a *fixed storage* part and a *dynamic array storage* part. Each procedure has the same structure as the main program; its storage may also be divided into a fixed storage part of fixed length for the variables[1] and a dynamic array storage part. To each data storage location an address is attached. The addresses of the entities of a fixed storage part may only be determined by the translator relative to the beginning of

---

[1] Static arrays within procedures will be handled like dynamic arrays, although the number of components is known to the translator. The reason why this is done will be understood once we have defined the "information vector" and explained the storage allocation for procedures (see also section 4.2.2.1, p. 32).

this fixed storage part. If the beginning of the data storage is known to the translator the addresses of the fixed storage part of the main program can be made absolute by the translator itself. Therefore we shall call this part of the data storage the *static storage*. In addition we have the *dynamic storage* which consists of the fixed storage parts of the procedures and the dynamic array storage parts, the absolute addresses of which are determined at run time, more precisely, at time of the call of the procedure, or the entrance into the block.

If recursive procedures are not implemented and if it is not necessary to make efficient use of storage capacity, storage can be allocated quite simply. The whole data storage is divided into a static and a dynamic storage. The static storage contains a fixed and unique location for each variable and element of a static array of the main program and of each procedure body without any regard to the block and procedure structure. The dynamic storage contains all dynamic arrays. We shall not use this solution or similar solutions.

Initially, we restrict our discussions to the main program and disregard procedure declarations in order to avoid the complications arising from the structure of procedure data storage. Those storage allocation problems are deferred to section 4.9.

4.2.1. Simple variables

Simple variables of the main program may be assigned fixed locations in data storage (static storage) by the translator. Assignment of addresses is essentially a sequential numbering, and may be done in connection with the processing of the block begins in the decomposition pass. Fixed storage is assigned with regard to block structure. Local quantities become meaningless on leaving a block; practically, this means that variables declared in "parallel" blocks may be assigned the same storage space. Therefore the address counter used for the numbering must be set back on exit to its value at the beginning of the block.

Numbers, truth values, and strings, which are constant for the entire program, may be listed on the lowest level and before the variables declared in the outermost block.

4.2.2. Arrays and their information vectors

An array declaration

$$\textbf{array } A\,[a_1\!:\!b_1, \ldots, a_n\!:\!b_n]$$

serves to allocate storage places to the n-dimensional field of the variables $A\,[s_1, \ldots, s_n]$, $s_1 = a_1, \ldots, b_1; \ldots ; s_n = a_n, \ldots, b_n$. The sequential nature of present-day computers forces us to arrange these variables in a sequential manner. Hence multidimensional arrays are transformed into one-dimensional fields to which storage is more easily assigned. Usually the components of such an array $A[a_1:b_1, \ldots, a_n:b_n]$ are arranged lexicographically so that $A[s_1, \ldots, s_n]$ is before $A[t_1, \ldots, t_n]$ whenever a j exists with $1 \leq j \leq n$ and $s_i = t_i$, $i < j$ and $s_j < t_j$.

The multidimensional array A with elements $A[s_1, s_2, \ldots, s_n]$ may be made into a one-dimensional one W with corresponding elements $W[s]$ as follows. Suppose that A has been declared by the declaration

$$\textbf{array } A[a_1:b_1, \ldots, a_n:b_n]. \tag{1}$$

Let $K_i = b_i - a_i + 1$ $(i = 1, \ldots, n)$ be the dimensions of the array. Then we may associate with each n-tuple $(s_1, \ldots, s_n)$ the simple subscript s by the following relation:

$$\begin{aligned} s &= \underset{A}{\varphi}([s_1, \ldots, s_n]) \\ &= (\ldots((s_1 \times K_2 + s_2) \times K_3 + s_3) \ldots) \times K_n + s_n. \end{aligned} \tag{2}$$

$\underset{A}{\varphi}([s_1, \ldots, s_n])$ is understood to be an abbreviated notation for the expression on the right side. $[s_1, \ldots, s_n]$ is called the *index vector* belonging to $A[s_1, \ldots, s_n]$, φ is called the *distance function*. The relation (2) yields c and d in the equivalent one-dimensional declaration:

$$\textbf{array } W[c:d]$$

as $c = \underset{A}{\varphi}([a_1, a_2, \ldots, a_n])$ and $d = \underset{A}{\varphi}([b_1, b_2, \ldots, b_n])$ respectively.

By way of example, if A is declared by

$$\textbf{array } A[1:6, 0:3, 3:5],$$

then

$$\underset{A}{\varphi}([i, j, k]) = (i \times 4 + j) \times 3 + k,$$

and the general element $A[i, j, k]$ corresponds to the element

$$W[(i \times 4 + j) \times 3 + k]$$

of the one-dimensional array W with the declaration

$$\textbf{array } W[15:86].$$

The address of the possibly fictitious element $W[0] (\equiv A[0, \ldots, 0])$ will be called the *reduced initial address* $W_{red\ init} (\equiv A_{red\ init})$ of the field W. This address satisfies the equation

$$W_{red\ init} = address\ (W[0]) = address\ (W[c]) - c$$

or

$$\begin{aligned} A_{red\ init} &= address\ (A[0, \ldots, 0]) \\ &= address\ (A[a_1, \ldots, a_n]) - \underset{A}{\varphi}([a_1, \ldots, a_n]). \end{aligned}$$

The address of $W[0]$ depends only on the dimensions of the array A and the address of the first element, so that it may be calculated at the time of the processing of the array declaration.

Certain information concerning the dimensions and the initial address is needed for the computation of the address of a special element of an array. What is needed depends on the method of computing the address.

The information connected with each array is collected and stored in a vector which will be called the *information vector*.

If m arrays are declared by

$$\textbf{array } A_1, \dots, A_m[a_1:b_1, \dots, a_n:b_n],$$

the information vector of the array A_1 consists of the storage location address of $A_1[0, \dots, 0] = A_{1red\ init}$ (even though $A_1[0, \dots, 0]$ is not an actual element of the array) and the dimensions $K_2, \dots, K_n$ where $K_i = b_i - a_i + 1$ $(i = 2, \dots, n)$.

In order to be able to handle value listed formal arrays (see section 4.10.2.2) we include the total length $\prod_{i=1}^{n} K_i$ of the array and the value of the expression $\underset{A_1}{\varphi}([a_1, \dots, a_n])$.

At the time of allocation of storage to an array, a location is known at which the storage of the array may begin. If FFC denotes the first free cell, we may determine the address of $A_1[0, \dots, 0]$ from the difference

$$address\ (FFC) - \underset{A_1}{\varphi}([a_1, \dots, a_n]) = address\ (A_1[a_1, \dots, a_n]) - \underset{A_1}{\varphi}([a_1, \dots, a_n]).$$

The numbers $a_1, \dots, a_n$ are not otherwise needed in the target program[1].

If, as in our example, several array identifiers have the same list of subscript bounds, the entire information vector is listed for each: the information vector for A_2 behind the information vector for A_1 etc. The information vectors for A_j $(j = 2, \dots, m)$ differ from the information vector for A_1 by only one component. The reduced initial address of A_j is determined by the reduced initial address of A_1 increased by the product

$$(j-1) \times \prod_{i=1}^{n} K_i.$$

---

[1] For purposes of error checking it might be desirable to keep the values of all lower and upper bounds in the information vector. Such checks are not provided in the model translator since they increase the program and the execution time for determining addresses of subscripted variables considerably. The information given in the information vector is sufficient for a simple check on whether the location of the subscripted variable is within the associated array W or not.

It is also possible to list the dimensions only once and the reduced initial addresses separately. This may save some space, but on the whole introduces other obvious disadvantages.

4.2.2.1. Static arrays

For *static arrays*, that is, arrays with fixed lower and upper bounds for the subscripts, the dimensions may be calculated during translation. The total length of the array is therefore also fixed, and the translator can assign fixed locations in storage to the array components and determine the reduced initial address, both relative to the beginning of fixed storage (static storage). The state of the address counter at the beginning of translation (see 4.2.1, p. 29) raised by one, may for example, be used as the address of the beginning FFC of the free storage, and thus is also accessible to the translator. The information vectors associated with the static arrays contain only constants which are known at translation time and remain constant at execution time; therefore, they may be added to the list of constants in the identifier list (see 3.1, p. 19).

This is not true for arrays with fixed lower and upper bounds within procedure bodies. Since procedures may be called recursively the reduced initial addresses can not be calculated at translation time. Therefore, we handle these arrays as dynamic arrays. With each recursive call of the procedure a new additional specimen of the array is associated (see footnote 1 in section 4.2, p. 28).

4.2.2.2. Dynamic arrays

While for static arrays the translator may determine the dimensions and the total storage requirements, in the dynamic case these computations must be carried out during the execution of the program. The necessary instructions for the computation and storage allocation must be furnished to the generated machine program.

The only thing known at translation time is the number of subscript positions and therefore the length of the information vector. The translator can do no more than to reserve space in the fixed storage for information vectors.

As we have already pointed out the elements of a *dynamic array* declared in the main program must be placed in the dynamic array storage of the main program which is a part of the general dynamic storage (see 4.2, p. 28).

During the computation, the macro instruction which undertakes the evaluation of declarations of dynamic arrays must continually have at its disposal the beginning FFC of the free dynamic array storage. A mechanism must be provided which, during the execution of the program, keeps track of the status of the dynamic array storage and can furnish free storage. This mechanism is naturally associated with block structure. It may be provided by using an exceptional global variable **BFS** (beginning of free storage) containing as its value the address of the first free cell of the free dynamic array storage. If during the execution of the program a block containing dynamic arrays (an "*array block*") is entered, the value of **BFS** is increased by the total length of all dynamic arrays declared in this array block AB.

On the other hand, on leaving the array block AB the value of **BFS** must be decreased by the same amount.

We define the *level of an array block* by induction:

1. If AB is an array block not contained in a larger array block, then AB has the level 1.

2. If AB is an array block on the level l and if it is the smallest array block containing the array block AB', then AB' has the level $l+1$.

In addition, we associate the level 0 (zero) with the main program itself, which therefore is also considered to be an array block.

Suppose L is the maximum level. Then an array of generated variables

$$BFS_0, BFS_1, \ldots, BFS_L$$

is produced, whose corresponding declarations are regarded as in the (fictitious) outermost block containing the constants and standard procedure declarations (see section 3.1, p. 19).

Now, when the value of **BFS** is increased the new value of **BFS** is also assigned to BFS_l, if l is the level of the array block AB.

On leaving an array block AB' with level l' and reentering the array block AB with level $l < l'$, the value of **BFS** is lowered by assigning the value of BFS_l to the exceptional variable **BFS**.

If the block AB' is left in consequence of a go to statement, the decrease of **BFS** is handled by the corresponding macro instruction **jump to** (cf. 4.8, p. 59). The case of the natural exit from AB' is handled by an assignment statement **BFS** $:= BFS_l$ inserted where **end** originally occurred.

The permanent value of BFS_0 is the initial address of the dynamic array storage which is known at the beginning of the actual run.

If an array block AB has the form

> **begin**
>   **array** $A_1, \ldots, A_m[a_1:b_1, \ldots, a_n:b_n]; \ldots$
> **end**

the following macro instruction executes all necessary actions connected with the evaluation of dynamic arrays:

$$\textbf{storage allocation } A_1 \,;$$
$$a_1 \,;$$
$$b_1 \,;$$
$$\vdots$$
$$a_n \,;$$
$$b_n$$

Here A_1 indicates the internal identifier associated with the array identifier A_1 and $a_1, \ldots, b_n$ mean the internal identifiers of the (possibly generated[1]) variables for the subscript bound expressions $a_1, \ldots, b_n$.

It should be recalled here, that with the help of the internal identifier A_1 alone the macro instruction **storage allocation** A_1 can obtain the number, the length, and the initial addresses of the information vectors and the level of the associated array block (see section 3.1, p. 19). An internal identifier characterizes the whole associated entry in the identifier list unambiguously, and an extract of the identifier list is accessable also during run time.

4.3. Handling of types

The decomposition of expressions and statements involves the consideration of the types of variables and expressions. The type **Boolean** poses no problem if every variable and formal parameter is declared or specified. Therefore, our discussion at present is primarily concerned with the handling of the arithmetic types **real** and **integer**. A single arithmetic operation, such as $a+b$, represents a class of operations, each of which applies to a different combination of operand types. The declarators and specifiers **real** and **integer** have then the effect of determining one operation out of this class. Thus, $a+b$ may represent, depending on the types of a and b, the addition of two real values, the addition of two integer values, or the addition of an integer and a real value. A second effect is concerned with the assignment statement $a := b$ where, depending on the types of a and b, the execution of the assignment sign may include a *type transfer* and a *rounding* of the value of b. The target code must differ accordingly.

The implementation of the two types **real** and **integer** may actually be regarded as a machine-oriented feature of ALGOL. As is well known, in most present-day computers both fixed point and floating point arithmetic are available; binary fixed point instructions are usually

---

[1] If a subscript bound expression ab is not a single simple variable then the value of it is determined and assigned to a generated variable before entering the macro instruction **storage allocation** A_1.

more economical in time than the floating point instructions. If values of type **integer** can be identified with fixed point numbers and values of type **real** with floating point numbers the ALGOL programmer has a useful means to reduce the execution time of his program. Another advantage is that mathematically integral results which are falsified by certain machine operations may be rounded off automatically[1].

The interpretation of the arithmetic type declarations may be handled in several ways. Roughly, we may distinguish the following extreme methods: We speak of *static type handling* if the handling of types is done by the translator which produces different machine instructions depending on whether the type of the variable or expression concerned is **real** or **integer**. We speak of *dynamic type handling* if the translator does not differentiate types and leaves the questions arising to the target program execution.

Static type handling is only possible if the types of all operands and the types of left part variables are known. This means that the translator must associate a definite type with each arithmetic expression and variable[2], even if the type can not be determined during translation time (as for the power operation), even if the ALGOL 60 Report does not give any exact type definition (as in the case of conditional expressions), and even if formal parameters are not specified. The following type definitions which exceed the definitions in the ALGOL 60 Report (see [75], 3.3.4, 4.2.4) but are compatible with SUBSET ALGOL 60 as described in vol. 1a), seem plausible, and have in some cases been adopted:

1*. $a \uparrow n$ is replaced by $exp(n \times ln(a))$ of type **real**, wherever n is not an unsigned integer.

Of course, this type modification for the power operation may lead to an inaccurate result since the functions exp and ln may introduce inaccuracies, or even may not be defined for certain arguments.

---

[1] It is most useful if one has the possibility of identifying addresses with integer numbers. Otherwise a transfer instruction from integers to adresses is needed when computing the address of a subscripted variable. This transfer does not cause any trouble; it only requires changes in the target language program: for e.g. $A[S_{A1}, S_{A2}]$ of the example in section 4.4.3, p. 49 becomes

$$\mathbf{AC} := S_{A1} \ ;$$
$$\mathbf{AC} := \mathbf{AC} \times A K_2 \ ;$$
$$\mathbf{AC} := \mathbf{AC} + S_{A2} \ ;$$
$$\textbf{transfer to address} \ ;$$
$$\mathbf{AC} := \mathbf{AC} +_{addr} A_{red\ init} \ ;$$
$$GV1 := \mathbf{AC}$$

Other program parts have to be changed accordingly.

[2] We consider to be variables also all unspecified formal parameters which by context are shown to have the character of variables, and we shall use the term "*name variable*" for simple formal variables not value listed.

2*. The type of conditional arithmetic expressions is defined as

$$\text{type } (\textbf{if } B \textbf{ then } E_1 \textbf{ else } E_2)$$

$$:= \begin{cases} \textbf{integer}, \text{ if type } (E_1) = \textbf{integer} \\ \qquad\qquad \text{and type } (E_2) = \textbf{integer} \\ \textbf{real}, \qquad \text{otherwise}. \end{cases}$$

3*. Unspecified name variables and unspecified formal subscripted variables occurring in an arithmetic expression are handled as if their type were **real**[1].

4*. Function designators with unspecified type procedure identifiers which are parts of arithmetic expressions are assumed to be real type expressions[1].

5*. All left part variables of an assignment statement which are formal and unspecified are handled as if their type were the type of the right hand expression defined by the ALGOL 60 Report [75], 3.3.4, 4.2.4 and by the adjustments 1*.—4*.[2].

The adjustments 1*.—5*. permit implementing static type handling in a simple way if, with respect to arithmetic parameters, the ALGOL programmer obeys the following three conditions:

a) For each formal (value listed or not) arithmetic variable FV, the associated actual arithmetic expression must have the same type as FV.

b) For each arithmetic function designator with a formal type procedure identifier FP, the actual type procedure identifier must have the same type as FP.

c) For each formal (value listed or not) arithmetic array identifier FA, the actual array identifier must have the same type as FA.

These conditions permit the production of a very efficient target language program[3] since no tests, and no redundant type transfers are required in the system. But they are objectionable, because first, they impose restrictions on ALGOL 60, second, the ALGOL programmer must learn rules which deviate from the ALGOL 60 Report, and third, experience has shown these conditions have led to crucial errors which could not be detected at translation time. In reference to this third point, the average ALGOL programmer should not be required to check the type of arithmetic *expressions* if he is forced to employ some of the adjust-

---

[1] This requirement is superfluous if we demand that the ALGOL programmer specify each formal parameter.

[2] This statement can be relaxed if we demand that the ALGOL programmer specify each formal parameter.

[3] The ALGOL translator ALCOR MAINZ 2002 requires the conditions a), b), and c) to be observed.

ments $1^*.-5^*$. These adjustments should only be an internal affair of the ALGOL translator.

It is, of course, possible to alter the translator and the run time system to meet the above conditions, in such a way that the target program checks for deviations at run time, and then gives alarm. However, run time alarms are undesirable, and it seems more sensible, instead of making user, translator, and run time system "learn" and check additional conditions, to construct the system in such a way that type definitions do not go beyond the definitions in the ALGOL 60 Report, with the exception of that dealing with the power operation. This course is adopted here, and the model translator handles types, which are not statically recognizable, according to the following rules $1.-5.$:

1. For the power operation three cases are distinguished:

a) $a \uparrow i$ with an unsigned integer i is treated according to the ALGOL 60 Report,

b) $a \uparrow i$ with an expression i which is statically recognizable to be of type **integer** is evaluated according to the ALGOL 60 Report, but is defined to be always of type **real**,

c) in all other cases, $a \uparrow n$ is replaced by $exp(n \times ln(a))$.

If the value of a at run time is negative, $a \uparrow n$ is undefined even if n is integer valued, and though $a \uparrow n$ is otherwise well defined according to the rules of the ALGOL 60 Report.

2. The type of conditional expressions is defined as follows

$$\text{type (if } B \text{ then } E_1 \text{ else } E_2)$$

$$:= \begin{cases} \textbf{integer} & \text{if type } (E_1) = \textbf{integer} \text{ and type } (E_2) = \textbf{integer} \\ \textbf{real} & \text{if type } (E_1) = \textbf{real} \text{ or type } (E_2) = \textbf{real} \\ \textbf{Boolean} & \text{if type } (E_1) = \textbf{Boolean} \text{ or type } (E_2) = \textbf{Boolean} \\ \text{to be determined by context otherwise.} \end{cases}$$

Context also may later lead to a type definition overriding the one given in the first three cases above.

3. Unspecified name variables and unspecified formal subscripted variables are handled as required by context when these variables occur in an expression[1].

4. Function designators with unspecified type procedure identifiers are handled as required by context when they are parts of expressions[1].

5. If all left part variables of an assignment statement are formal but unspecified variables then the translator is forced to produce a conditional type transfer and store instruction[1]. At run time this in-

---

[1] This requirement can be relaxed if we demand that the ALGOL programmer specify each formal parameter.

struction determines the type of the inserted actual variables, checks the identity of types for all the left part variables[1], and depending on the type of the right hand expression executes the necessary type transfer. If we do not require that the ALGOL programmer specify each formal parameter then some type handling decisions must be left to run time instructions.

After the adjustments 1.—5. the ALGOL programmer needs not check the type of arithmetic expressions and actual parameters other than simple variables and he needs not learn type handling rules deviating from the ALGOL 60 Report. Only if a (simple or subscripted) name variable is used as an *output variable* (e.g. as a left part variable) and is specified, then the (possibly by successive substitutions) associated declared or value listed variable must have the same type. And only if a formal and not value listed array identifier is specified then the (possibly by successive substitutions) associated declared or value listed array identifier must have the same type.

In connection with the possibility of specifying formal parameters the position adopted here is as follows:

The model translator in chapter 7 and 8 does not deal with specifications as if they were bare comments. Specifications are not required, but if formal parameters are specified then these specifications will be used to produce better and more efficient target language programs. Accordingly, specifications must meet a certain condition: Originally, this condition says that the specification of a formal parameter must agree with the declaration of the associated actual parameter. By the adjustments 1.—5. this condition is weakened: If a formal and not value listed parameter FP is specified then the associated actual parameter AP (after execution of successive substitutions of name parameters) should look as follows[2]:

specification of FP: appearance and declaration of AP:

real AP must be an arithmetic expression. If FP is used as an output variable (e.g. as a left part variable) then AP must be a declared or value listed (simple or subscripted) real variable.

real array AP must be a declared or value listed real array.

real procedure AP must be declared an arithmetic procedure. If FP is used as a left part type procedure identifier then AP must be declared a real procedure.

---

[1] This check must also be made if some (but not all) left part variables are declared or specified.

[2] In the interest of completeness, the replacement rules for procedure, switch, label, and string parameters are included in the list.

Analogous rules hold for integer and Boolean quantities where we only have to replace "real" by "integer", or "real" and "arithmetic" by "Boolean" respectively.

procedure AP must be declared a proper procedure.

switch AP must be declared a switch.

label AP must be a designational expression.

string AP must be a string.

In contrast with static type handling, dynamic type handling means that no types are to be distinguished by the translator. The type declarators **real** and **integer** are put into the identifier list and preserved for run time. Certain instructions must then be generated to deal with the types at run time.

Intermediate forms of type handling are possible. The most efficient way may be to make use of type declarations and specifications by the translator as far as possible. Taking care of situations undefined at translation time is left to the run time system.

Every compiler builder tries to design the compiler in such a way that the translation as well as the target language programs produced by this translator will work rapidly and efficiently. Hence, dynamic handling should be avoided as far as possible if the additional execution time is not negligible compared with the proper execution time needed for the arithmetic operations themselves. Type handling thus is closely related to number representation and to the set of arithmetic instructions available in a given computer, and thus in general is machine dependent.

It may be useful to discuss some of the possibilities in greater detail, disregarding the question of error checks.

(1) In the trivial case a single standard number representation and only one sort of arithmetic machine instructions is used in target language programs. Consequently, values of type **integer** must be regarded as a subset of the set of real values. Type handling therefore is reduced to rounding-off operations when assigning a value to an integer variable. This may be done statically or dynamically.

a) dynamic

The assignment of the accumulator to a variable or to the content of a variable must include type transfers. This may be done by extending the action of the instructions

$$V := \textbf{AC}$$

and

$$\textbf{content of } V := \textbf{AC} .$$

The effect of these instructions beside assigning values at execution time is that if the type of V or **content of** V is recognized to be **integer** then the value of **AC** is rounded off. The type of V (or **content of** V) may be taken from the identifier list or from a marker connected directly with the actual address, that is the addresses of simple variables and the reduced initial addresses of arrays (cf. 4.2.2) are coupled with a *type marker*[1]. Generated variables used for intermediate results have an undefined type and at execution time no type transfer is performed.

b) static

Static type handling requires the translator to analyze each arithmetic operation to see whether the type of the result is real or integer. When a real result must be assigned to an integer variable, then a transfer to type **integer** will be generated by the translator which changes the type of the content of the accumulator **AC**.

If all formal variables or arrays are specified so that at translation time the type of each variable can be recognized, this type transfer is no problem. But if in an assignment statement all left part variables are formal and not specified and if the expression on the right hand side could possibly deliver a number of type **real**, then at execution time, the type of the actual variable of at least one left part variable must be determined (by definition the other actual variables must have the same type, see [74] and [75]). This can be done, as in case a), if the addresses of simple variables and the reduced initial addresses of arrays are accompanied by a type marker. After the evaluation of the expression on the right hand side, a conditional transfer to type **integer** must be executed. If the address of the left part variable is coupled with an integer type marker, the type transfer (a rounding-off process) takes place. Otherwise the content of the **AC** remains unchanged[2].

(2) The target language may admit only one representation of numbers, that is, integer values must be regarded as a subset of the real

---

[1] If one uses such a type marker, it is necessary that operations on the reduced initial address do not destroy the marker but take it over into the result (e.g. in the instruction $\mathbf{AC} := \mathbf{AC} + A_{red\ init}$ in section 4.4.3, p. 49).

[2] The case (1 b) is essentially realized in the ALGOL translator ALCOR ILLINOIS IBM 7090/94 although the machine would permit case (3). This translator does not demand specifications, but exploits them if they are available.

The instruction which executes a conditional transfer to type **integer** has not been implemented. Thus adjustment 5*. applies instead of adjustment 5. As a consequence, the ALGOL programmer has to obey the following condition: If all left part variables of an assignment statement are formal and unspecified and if the right hand expression is a real type expression then the associated actual variables must be real type variables.

If a formal parameter FP is specified then the conditions for appearance and declaration of the associated actual parameter AP can be weakened as compared values. But there may exist two sets of binary operations, one of which is faster but operates only on the subset mentioned above. As in case (1) the type transfer is essentially a rounding-off process and, concerning the assignment statements, the same as in (1) is valid.

Efficient handling of binary operation requires the use of different machine instructions depending on the types of the operands.

a) dynamic

Dynamic handling is only possible if the different types can be distinguished dynamically, that is if either the integral values can be recognized[1] or the integral and real numbers can be coupled with a type marker [see point (1), p. 40]. The binary operation instruction generated by the translator has the task of determining at execution time the types of the operands and selecting the suitable machine instruction. Operations involving values of type **integer** run faster only if the time needed for the test of the types is negligible.

b) static

The handling of assignments is the same as in case (1 b). The static utilization of "fast" and "slow" machine operations requires that the translator produces a "fast" binary instruction when both operands are of type **integer**. If real operands or operands of unknown types occur, then a "slow" binary instruction is produced, the result of which is again of type **real**.

---

with the model translator (see p. 38) since the model translator deals with the more complicated case of "fully static type handling" [case (3 b)]. Two entries in the list on p. 38 can be replaced:

specification of FP: appearance and declaration of AP:

real AP must be an arithmetic expression. If FP is used as an
                          output variable then AP must be a declared or value listed
                          real variable.

Analoguous rules hold for integer and Boolean name variables where we have to replace "real" by "arithmetic", or "real" and "arithmetic" by "Boolean" respectively.

real array AP must be the identifier of a declared or value listed arith-
                          metic array. If $FP[...]$ is used as an output variable then
                          AP must be the identifier of a declared or value listed real
                          array.

An analoguous rule holds for formal integer arrays where we have to replace "output" by "input" and "real" by "integer".

Boolean array AP must be the identifier of a declared or value listed
                          Boolean array.

[1] In this case operands of type **real** with accidentally integral values are handled like operands of type **integer**.

(3) The most complicated case is that where the target language has two different number representations and two kinds of arithmetic operations. In addition to the points mentioned already the type transfer must be taken into consideration whenever two entities of different types are connected[1]. Whereas in the former cases only transfers from real to integer were necessary now also transfers from integer to real must be made.

a) dynamic

Dynamic type handling is only possible if the types of numbers can be distinguished dynamically. Each assignment to a variable V or to content of V requires at first a possible transfer of the content of the accumulator depending on the type of V or on the type marker given by the content of V. At execution time the type coupled with the left part address [see (1 a), p. 40] and the type marker coupled with the right part value are tested and the type transfer is executed if they differ. The binary operations differ from (2a) in that if two operands of different types are found, the operand of integer type must be made real[2, 3].

b) static[4]

In the former case (1 b) and (2 b) of static type handling the translator did not have to produce any type transfer instruction, when processing an arithmetic (right hand) expression. Now we have to observe the following:

If a binary operator with two integer operands occurs, then the integer machine instruction can be generated. Otherwise the real machine instruction must be used. If the two operands have different types, the translator must arrange for the necessary type transfer at execution time by generating the instructions

$$\textbf{AC} := V;$$
transfer to type real

where V is the operand of type **integer**.

---

[1] In most machines the range of admissible integer numbers is smaller than that for real numbers. This requires an error device when transferring a real value which has no integer equivalent.

[2] Certain operators $(/, \uparrow_{rr})$ work only with real operands. Then both operands require the type transfer from integer to real.

[3] The case (3a) is essentially realized in the ALGOL translator ALCOR MAINZ Z 22. This translator disregards real and integer specifications. With respect to assignment statements the ALGOL programmer must obey the following condition: If a left part variable is a declared or value listed integer variable or if a left part variable is a name variable and associated with an actual integer variable then, at run time, the right hand expression must deliver an integer value.

[4] In this case we shall speak of "*fully static type handling*". This case will be realized in the model translator, chapters 7 and 8. All other cases can be obtained by specializing.

The handling of assignments is also somewhat more complicated than in the cases previously discussed (1 b), p. 40 and (2 b), p. 41. In addition, the following must be done by the translator: If an integer value is to be assigned to a real variable V or content of V then a transfer to type **real** must be generated. If the type of the left part variable is unknown (unspecified name or formal subscripted variable) then the translator generates a conditional transfer to type **real** [compare (1 b), p. 40][1].

4.4. Assignment statements

The principles for handling expressions have been discussed in sections 2.4, 2.5, and 2.6. Now these principles will be applied to arithmetic expressions disregarding conditions.

In the next three sections, we handle the possible constituents of these expressions, which may be (1) numbers and simple variables, (2) name variables and functions, and (3) subscripted variables. Conditional expressions will be treated in connection with conditional statements (see section 4.6.3, p. 55). Boolean expressions will be discussed in section 4.5, p. 51.

4.4.1. Simple arithmetic variables

For an assignment statement with a simple arithmetic expression (not containing conditions) as right part, the operands of which are numbers or simple variables which, if formal, are value listed (*"value*

---

[1] The case (3 b) is essentially realized in the ALGOL translator ALCOR MAINZ 2002. This translator demands each formal parameter to be specified by the ALGOL programmer. The list of adjustments 1*. — 5*. reduces to:

1**. The power operator ↑ is not implemented as such.
2**. The type of a conditional arithmetic expression is defined to be **real**.

The conditions for the appearance and declaration of an actual parameter AP associated with a formal parameter FP are stronger than in the model translator (see p. 38): ALCOR MAINZ 2002 requires the conditions a), b), c) on p. 36 to hold.

The case (3 b) is also realized in the ALGOL translator ALCOR MUENCHEN 2002. This translator demands that each formal parameter be specified by the ALGOL programmer and the remarks in the footnote on p. 37 apply. The list of adjustments 1. — 5. reduces to 1., 2. The conditions for the appearance and declaration of an actual parameter AP associated with a formal parameter FP are slightly stronger than those for the model translator. The following point should be added to the list on p. 38:

specification of FP: appearance and declaration of AP:

real procedure   If AP is the identifier of a declared type procedure with parameters then AP must be declared to be a real procedure.

integer procedure   Replace "real" by "integer" in the text above.

variables"), the target language equivalent is a sequence of statements of the kinds listed in 3.2a), b), c), and f1):

$$\mathbf{AC} := \mathbf{AC}\,\omega\,V$$
$$\mathbf{AC} := \omega\,V$$
$$\mathbf{AC} := \omega\,\mathbf{AC}$$
$$V := \mathbf{AC}$$
$$\mathbf{AC} := V$$

transfer to type real
transfer to type integer .

We assume that the result of each operation will be left in the accumulator **AC**; if the accumulator is needed for further operations this result must be put into intermediate storage. This consists of generated variables $GV1, ..., GVn$ which are treated as if they were declared within the innermost, possibly fictitious, block including the expression just handled.

The assignment of the content of the accumulator **AC** to a generated variable GVi is necessary when this value is not an operand of the operation immediately following. The accumulator must then be cleared to make room for one operand of the following operation. The translator then generates $GVi := \mathbf{AC}$. The variable GVi is free for further auxiliary storage once GVi has appeared at the right of $:=$ in an elementary language statement.

The storage places reserved for the generated variables act as a push down (called *numbers push down*), since the variable which has been generated last will always be the first to be free again, see [8].

4.4.2. Name variables, function designators, and type procedure identifiers

Clearly the decomposition of the bracket structure of an expression does not change in principle if the expression contains *name variables*. The target language program for name variables will be different from the program for declared and value listed variables, e.g. the binary operation instructions are restricted to simple non-name variables as operands. As will be shown in section 4.10.1 the name variables and function designators will be translated into special elementary language instructions which assign the value in question to the exceptional variable **AC**, and its type to the exceptional variable **ADDR AC** (address accumulator). The *call* of the name variable N will have the form

name call N.

When a name variable N occurs in a left part list, the address of the memory cell of the respective actual variable must be determined. For this purpose we use the instruction

name address N

which assigns the actual address and type to the exceptional variable **ADDR AC**. This is normally followed by an instruction

$$GV_N := \textbf{ADDR AC}.$$

If type handling [case (3 b) of 4.3, p. 42] is implemented statically, a certain amount of dynamic type checking of the name variables is necessary. This is provided in the following way:

a) For a name variable N specified to have a given type we introduce instead of the general name call N a type name call N. If, e.g., N is specified to be integer, N will be called by

integer name call N

and analogously for other types. This instruction will deliver the value of the actual parameter corresponding to N, transferred to the type indicated if this transfer is admissible, and will deliver an alarm if the type transfer is not admissible (for instance **Boolean** to **integer**). The information on the type of the actual parameter is available in the address accumulator (see section 4.10.1.1):

If context requires a type transfer immediately following a type name call, the transfer can be incorporated into the call. This means, for instance, that instead of the sequence

integer name call N;
transfer to type real

the simple instruction

real name call N

is generated.

b) An unspecified name variable N is treated like a specified name variable if from immediate context a required type can be deduced (e.g. N is necessarily of type **integer**, if appearing as a subscript expression, or as an operand of integer division, N is of type **real** as the numerator of a real division). If context shows only that N is arithmetic, the type is defined to be **real**. If immediate context gives no information about the type, then the simple

name call N

is used. This will be followed by a *conditional type transfer* or type check for Boolean before the next operation on **AC** takes place.

If the translator from a wider context can deduce the required type, then an instruction

possible transfer to ⟨type⟩ **depending on ADDR AC**

where $\langle$type$\rangle$ may be **integer** or **real** or

compare Boolean to ADDR AC

is generated.

The remaining cases are actual parameter expressions and assignment statements which contain no information on types.

For actual parameters, the conditional type transfer will be handled by the corresponding name calls.

In assignment statements of this class, in particular, no left part variable can have a specification or declaration since otherwise the type of the right hand side expression is already established. Since left part variables play the deciding part here, we include this case in the discussion of assignments to name variables in general which now follows.

Here we must again distinguish the two cases of specified and un-specified left part name variables N.

a) If N is specified to have a given type, the model translator requires all corresponding actual variables to have the same type. In this case, the translator can in all circumstances associate a type with the right hand side expression, and types are handled in the same way as for declared variables.

b) If N is unspecified but a type is defined for the right hand side expression, a conditional transfer

possible transfer from $\langle$type$\rangle$ **depending on** GV_N

where $\langle$type$\rangle$ again may be **integer** or **real** or

compare Boolean to GV_N

is generated.

c) In the case mentioned above, where the left part is unspecified and no type can be associated with the right hand side by the translator a conditional transfer instruction

possible type transfer from ADDR AC to GV_N

is generated. This instruction will execute a type transfer if the types of the actual value of the right hand side expression, and of the actual variable corresponding to the left part name variable differ but are compatible, and will give an alarm otherwise. The necessary information is given by type markers in **ADDR AC** for the former, GV_N for the latter constituent.

If N is accompanied by other left part variables, the identity of the type of the actual corresponding to N with the types of the companions is checked by an instruction

compare $\langle$type$\rangle$ **to** GV_N.

If one of the left part variables is specified or declared its type determines the comparison, and the translator inserts **real**, or **integer**, or **Boolean**, as the case may be, in the place of ⟨type⟩. Otherwise, ⟨type⟩ is replaced by **ADDR AC**, which then contains the information on the type of the companion left part variables.

For example consider: $N_1 := N_2 := N_3$; N_1, N_2 are unspecified name variables, N_3 is a name variable specified real.

This is translated into

> **name address** N_1 ;
> $GV1 :=$ **ADDR AC** ;
> **name address** N_2 ;
> $GV2 :=$ **ADDR AC** ;
> **compare ADDR AC to** $GV1$;
> **real name call** N_3 ;
> **possible transfer from real depending on** $GV2$;
> **content of** $GV2 :=$ **AC** ;
> **content of** $GV1 :=$ **AC** .

In this target program N_1, N_2, N_3 represent the internal identifiers of the name variables N_1, N_2, N_3. This principle of using the same symbol to represent ALGOL variables and their internal identifiers is generally applied below.

Function designators are handled in connection with *procedure calls* in section 4.10.1. At this stage we can restrict ourselves to *function calls* whose actual parameters are only procedure identifiers or array identifiers. With regard to type handling, function calls behave exactly like name calls, and are treated in the same way. A normal or formal function call $F(AP_1, \ldots, AP_r)$ then can be translated into

> ⟨type⟩ **procedure call** F;
> $\qquad AP_1$;
> $\qquad \vdots$
> $\qquad AP_r$

or ⟨type⟩ **formal procedure call** F ;
> $\qquad AP_1$;
> $\qquad \vdots$
> $\qquad AP_r$

respectively. As noted above, in these target language instructions, F, $AP_1, \ldots, AP_r$ refer to the internal identifiers associated with the function identifier and the actual parameters. ⟨type⟩ may be **real**, **integer**, or **Boolean**; in the second case **type** is also possible if no information on the required type is available.

To produce this simple form of function designator, an actual parameter which is not yet a (generated) procedure or array identifier is extracted and transformed into a special kind of procedure; the generated name of this procedure is used to replace the actual parameter. The manner in which this is done is described in section 4.10.1.1.

With each *function identifier* P a simple generated variable GVP is associated which serves as storage for the value assigned to the function identifier in the body. Thus, a statement $P := V$ may be translated into:

$$\textbf{AC} := V ;$$
$$GVP := \textbf{AC}$$

(see section 4.10.2.3, p. 75).

A formal function identifier in left part position is handled like a name variable.

The execution of the instructions **name call, name address, procedure call,** and **formal procedure call** requires the exceptional variable **AC** to be free. If necessary its value must be saved by assigning the **AC** to a generated variable as described in 4.4.1 for intermediate results.

4.4.3. Subscripted variables

According to the discussion in section 4.2.2 we assume that the elements of the array A are stored sequentially in the machine; the address of the arbitrary element $A[s_1, \ldots, s_n]$ is then given by

(1) $address\,(A[s_1, \ldots, s_n]) = address\,(A[0, \ldots, 0]) + \varphi_{A}([s_1, \ldots, s_n]).$

If the reduction to $address(A[0, \ldots, 0])$ is not carried out, it is necessary to use

(2) $address\,(A[s_1, \ldots, s_n]) = address\,(A[a_1, \ldots, a_n]) - \varphi_{A}([a_1, \ldots, a_n])$

$\qquad + \varphi_{A}([s_1, \ldots, s_n])$

$\qquad = address\,(A[a_1, \ldots, a_n]) + \varphi_{A}([s_1 - a_1, \ldots, s_n - a_n]).$

Whenever a subscripted variable has a subscript expression containing a variable, the location of the array element must be determined dynamically, that is, at the time the target program is executed; it is then necessary to incorporate an algorithm equivalent to (1) or (2) into the target program.

The simplest, though spacewise and timewise the poorest, way of determining the location is to explicitly insert the Horner algorithm (1) or (2) itself in order to determine the values of the distance function. Clearly this computation (as well as the computation of the expressions s_i) may be decomposed into elementary constituents of types listed in section 4.4.1. Particularly in the execution of the for statement, this involves the repeated execution of many unproductive instructions. A mechanism for reducing the number of operations dynamically is that of *recursive address calculation* [48, 94, 20] which will be discussed in detail in chapter 5. However, recursive address calculation cannot be carried out for all cases of subscripted variables. In chapter 5 the full conditions are listed. Therefore the explicit Horner algorithm, or its equivalent, is in general needed in the over-all system.

If an expression contains subscripted variables, the list of target language statements of 4.4.1 must be extended by the instructions:

a)   **ADDR AC := AC**

which is used when the address of a subscripted variable has been determined and stored in the accumulator **AC** and when the value of this subscripted variable is needed immediately. Otherwise the address is saved by being assigned to a generated variable: $GV :=$ **AC**.

b)   **AC := AC** ω **content of** V
                      AC := ω **content of** V
                      AC := ω **content of ADDR AC**
                      content of $V :=$ **AC**
                      AC := content of V

which contain the element

content of V.

V means the internal identifier of an address variable. The operand of the instruction is not the value of the variable V itself but the content of that memory cell whose address is given by the value of V.

c) Certain type transfer instructions which, with a few exceptions to be treated at the end of this section, have already been discussed in connection with name variables.

In order to show how subscripted variables may be translated into target language instructions we give an example.

$$A\,[S_{A1}, S_{A2}] := B\,[S_{B1}, S_{B2}] + C\,[S_{C1}, S_{C2}]\,;$$

$S_{A1}, \ldots, S_{C2}$ are assumed to be simple variables of integer type. The dimensions of A, B, and C are denoted by

$$AK_1, AK_2, BK_1, BK_2, CK_1, CK_2 \quad \text{respectively.}$$

The target language program may then be:

AC $:= S_{A1}$;
AC $:=$ **AC** $\times AK_2$;
AC $:=$ **AC** $+ S_{A2}$;
AC $:=$ **AC** $+ A_{red\ init}$;
$GV1 :=$ **AC** ;

} calculate the address of $A\,[S_{A1}, S_{A2}]$ and store it in $GV1$

$$\mathbf{AC} := S_{B1} \,;$$
$$\mathbf{AC} := \mathbf{AC} \times BK_2 \,;$$
$$\mathbf{AC} := \mathbf{AC} + S_{B2} \,;$$
$$\mathbf{AC} := \mathbf{AC} + B_{red\ init} \,;$$
$$GV2 := \mathbf{AC} \,;$$

address of $B\,[S_{B1},\,S_{B2}]$ into $GV2$

$$\mathbf{AC} := S_{C1} \,;$$
$$\mathbf{AC} := \mathbf{AC} \times CK_2 \,;$$
$$\mathbf{AC} := \mathbf{AC} + S_{C2} \,;$$
$$\mathbf{AC} := \mathbf{AC} + C_{red\ init} \,;$$
$$\mathbf{ADDR\ AC} := \mathbf{AC} \,;$$

address of $C\,[S_{C1},\,S_{C2}]$ into **ADDR AC**

AC := **content of ADDR AC** ;
AC := **AC** + **content of** $GV2$;
content of $GV1$:= **AC**

$C\,[S_{C1},\,S_{C2}]$ in **AC**,
AC $+\,B\,[S_{B1},\,S_{B2}]$ in **AC**,
AC stored into $A\,[S_{A1},\,S_{A2}]$

Another way of determining the location of an array element is given by computing the necessary multiplications in advance during the processing of the array declaration and listing them in a table. The scheme may be reduced to the point where, if the values of $s_1, \ldots, s_n$ are known, a look into the table is sufficient to gain the desired element $A\,[s_1, \ldots, s_n]$ of the array A. This requires that the table referred to be stored in a special way (Leitzellentechnik). To illustrate, consider the n-dimensional array $A = A_n$ as a set of vectors each of which corresponds to a certain set of values of the first $n-1$ subscripts and whose components correspond to the possible values of the last. The reduced initial addresses of these vectors form an array A_{n-1} with $n-1$ dimensions. In this array we proceed in the same way. Finally we reach a vector A_1 whose reduced initial address is then the value of the variable A_0. Thus the arrays A_i are assigned values as follows:

for $\quad i = n-1\,(-1)\,1, \quad j = 1\,(1)\,i, \quad$ and $\quad v_j = a_j(1)\,b_j,$

$$A_i\,[v_1, \ldots, v_i] := address\,(A_{i+1}\,[v_1, \ldots, v_i,\, 0]).$$

That is,

$$A_i\,[v_1, \ldots, v_i] := address\,(A_{i+1}\,[0, \ldots, 0]) + \underset{A_i}{\varphi}\,([v_1, \ldots, v_i]) \times K_{i+1}$$

and

$$A_0 := address\,(A_1\,[0]).$$

To obtain the element $A\,[s_1, \ldots, s_n]$, we obtain first from A_0 the location of the vector A_1; the value of s_1, the first subscript, then determines a component of A_1. This is continued until a component of A_{n-1} determines the location of a definite vector of A_n, from which the component $A\,[s_1, \ldots, s_n]$ is then obtained using the value of s_n.

This method is most useful if (1) sufficiently many index registers are available and (2) indexing has precedence over the **content of** operation. If one assigns to $A_i\,[v_1, \ldots, v_i]$ the value

content of $(address\ (A_{i+1}\,[v_1, \ldots, v_i,\, 0]) \oplus \mathfrak{Jr}\,[i+1])$, for $i = 0, 1, \ldots, n-2$, and, in the last step, to $A_{n-1}\,[v_1, \ldots, v_{n-1}]$ the value

$$address\,(A_n\,[v_1, \ldots, v_{n-1},\, 0]) \oplus \mathfrak{Jr}\,[n]$$

and if the index registers $\mathfrak{Ir}[1], \ldots, \mathfrak{Ir}[n]$ are loaded with the values of $s_1, \ldots, s_n$ respectively then

content of A_0

delivers immediately (is equivalent to) the address of $A[s_1, \ldots, s_n]$, which means, e.g. that the target language instruction

AC := **content of** A_0

assigns immediately the value of $A[s_1, \ldots, s_n]$ to the accumulator **AC**.

Formal subscripted variables are handled in the same way as declared ones as long as they are specified since the information vector associated with the actual array is available to the procedure at the moment of the procedure call.

The situation is different in the case of unspecified formal subscripted variables if the translator works with static type handling.

In arithmetic expressions, the translator tries to determine from context a required type (e.g. a subscript position is required to be of type **integer**). If the translator succeeds, it inserts a conditional type transfer. Depending on the situation two different instructions

possible transfer to ⟨type⟩ **depending on ADDR AC**

or

possible transfer to ⟨type⟩ **depending on** GV

occur. The former one can be used if the operation on the subscripted variable immediately follows evaluation of its address, the latter one in other cases.

In the remaining cases, that is in expressions without any information on type, and in left parts, unspecified subscripted variables $A[s_1, \ldots, s_n]$ are handled in the same way as unspecified name variables. The only formal difference is that **name address** is replaced by

$$\mathbf{AC} := A_{red\ init} + \varphi([s_1, \ldots, s_n]) \; ; $$
$$\quad\quad\quad\quad\quad\quad\quad {}_{A}$$
$$\mathbf{ADDR\ AC} := \mathbf{AC} ,$$

for **name call** the instruction

AC := **content of ADDR AC**

is added to this.

4.5. Boolean expressions

4.5.1. Truth values

There is no fundamental difference between arithmetic variables and Boolean variables, or arithmetic and Boolean function designators with the exception that the Boolean quantities assume one of the two values

true and **false**. The instructions used for storing or for bringing a truth value into the accumulator are the same as for arithmetic values.

Truth values will usually be "arithmetized" internally. Thus, Boolean values may be represented by integer or real numbers such as 1 and 0 or 1 and -1 for **true** and **false**, respectively.

4.5.2. Relations

A relation is a binary operation with two arithmetic operands that yields a Boolean result. If ϱ denotes one of the operators $<$, $\leqq$, $=$, $\neq$, $\geqq$, $>$, $E_1 \varrho E_2$ is equivalent to $(E_1)-(E_2) \varrho 0$ and also equivalent to the conditional expression

$$\textbf{if } (E_1)-(E_2) \varrho 0 \textbf{ then true else false}.$$

The relation may therefore be replaced by a standard form

$$\textbf{AC} := (E_1)-(E_2) \text{ ;}$$
$$\textbf{if AC} \varrho 0 \textbf{ then true else false}.$$

The handling of relations is thus in principle reduced to the handling of conditional expressions that will follow in section 4.6.3, p. 55.

4.5.3. Boolean operators

If there are no machine instructions that correspond to the Boolean operators of ALGOL, various possibilities exist for evaluating Boolean expressions. If Boolean values are internally represented by the numbers 1 and 0, the operations may be replaced by arithmetic ones in a field of characteristic 2. It is possible to carry out these computations, not during the execution of the program, but already during translation, with a resulting saving of execution time. The translator must then establish the truth value table for each Boolean expression.

Another possibility is to replace the Boolean operations by conditional Boolean expressions (cf. 4.6.3). This is permissible if we rule out side effects of type procedures, and the possibility of "generating" undefined values of expressions by undefined values of variables. In the opinion of the authors this is of no real value in programming anyway. For example we may handle

$a \wedge b$	like	**if** a **then** b **else false**
$a \vee b$	like	**if** a **then true else** b
$a \supset b$	like	**if** a **then** b **else true**
$a \equiv b$	like	**if** a **then** b **else if** b **then false else true**
$\neg a$	like	**if** a **then false else true**.

4.6. Conditional statements and expressions

4.6.1. The if clause

The elementary language instructions

if ¬ AC then local jump to L
if AC ϱ 0 **then local jump to** L [1]
local jump to L

permit the translation of the if clause and the decomposition of conditional statements and expressions.

An if clause may contain a simple relation or a Boolean expression.

a) The if clause

if $E_1 \varrho E_2$ **then** ...

is equivalent to

AC $:= (E_1) - (E_2)$;
if AC ϱ 0 **then** ... .

b) The if clause

if B **then** ... ,

where B is any Boolean expression is equivalent to

AC $:= B$;
if AC then ... .

Thus it is only necessary to give the target language equivalent for **if AC** ϱ 0 **then** ... and **if AC then** .... Each if clause is essentially a conditional go to statement that causes the expression or statement following **then** to be skipped in the case **false**.

a) Thus **if AC** ϱ 0 **then** ... is translated into

if AC $\bar{\varrho}$ 0 **then local jump to** L ; ...
$L:$

where L is associated with the closing bracket that corresponds to **then** (**else** or ; etc.) and $\bar{\varrho}$ is defined by

$$E_1 \bar{\varrho} E_2 \text{ equivalent } \neg (E_1 \varrho E_2).$$

_____

[1] The splitting of the elementary conditional go to statement into two instructions is not essential. It is suggested by practical reasons, since machine languages may include equivalents. If truth values are arithmetized then the instruction

if ¬ AC then local jump to L

may obviously be replaced by an instruction

if AC ϱ 0 **then local jump to** L.

On the other hand, if machine operations exist for **AC** ϱ 0 which assign the value **true** or **false** to the accumulator in dependence on its value then the second instruction **if AC** ϱ 0 **then local jump to** L may be replaced by **AC** $:=$ **AC** ϱ 0 and **if AC then local jump to** L.

b) The target equivalent of **if AC then** ... is

> **if ¬ AC then local jump to** L ; ...
> $L:$

4.6.2. Conditional statements

The conditional statement

> **if** $E_1 \varrho E_2$ **then** S_1 **else** S_2 or   **if** B **then** S_1 **else** S_2

where S_1 is an unconditional and S_2 any statement may simply be transformed into:

> $\mathfrak{T}(\textbf{AC} := (E_1) - (E_2))$;
> **if AC** $\bar{\varrho}$ **0 then local jump to** L_1 ;
> $\mathfrak{T}(S_1)$;
> **local jump to** L_2 ;
> $L_1:$ $\mathfrak{T}(S_2)$;
> $L_2:$

or

> $\mathfrak{T}(\textbf{AC} := B)$;
> **if ¬ AC then local jump to** L_1 ;
> $\mathfrak{T}(S_1)$;
> **local jump to** L_2 ;
> $L_1:$ $\mathfrak{T}(S_2)$;
> $L_2:$

respectively.

Here $\mathfrak{T}(x)$ is used as an abbreviation for the target language equivalent of the expression or statement x.

The special case

> **if** $E_1 \varrho E_2$ **then** S_1

or

> **if** B **then** S_1

may be handled likewise:

> $\mathfrak{T}(\textbf{AC} := (E_1) - (E_2))$;
> **if AC** $\bar{\varrho}$ **0 then local jump to** L ;
> $\mathfrak{T}(S_1)$;
> $L:$

or

> $\mathfrak{T}(\textbf{AC} := B)$;
> **if ¬ AC then local jump to** L ;
> $\mathfrak{T}(S_1)$;
> $L:$

respectively.

4.6.3. Conditional expressions

Each assignment statement

$$a := \textbf{if } RB \textbf{ then } E_1 \textbf{ else } E_2$$

where RB is a relation or a Boolean expression and E_1 an unconditional and E_2 any expression may be replaced by the statements

$$\textbf{if } RB \textbf{ then } \textbf{AC} := E_1 \textbf{ else } \textbf{AC} := E_2 ;$$

$$a := \textbf{AC} .$$

Thus the translation of conditional expressions can be made according to the model for conditional statements.

If we consider an expression to be internally an assignment statement that assigns a value to the exceptional variable **AC** then we may show: The target equivalent of

$$\textbf{if } E_1 \varrho E_2 \textbf{ then } E_3 \textbf{ else } E_4$$

or

$$\textbf{if } B \textbf{ then } E_1 \textbf{ else } E_2$$

is

$$\mathfrak{T}(\textbf{AC} := (E_1) - (E_2)) ;$$
$$\textbf{if AC } \bar{\varrho} \ 0 \textbf{ then local jump to } L_1 ;$$
$$\mathfrak{T}(\textbf{AC} := E_3) ;$$
$$\textbf{local jump to } L_2 ;$$
$$L_1 : \quad \mathfrak{T}(\textbf{AC} := E_4) ;$$
$$L_2 :$$

or

$$\mathfrak{T}(\textbf{AC} := B) ;$$
$$\textbf{if } \neg \ \textbf{AC then local jump to } L_1 ;$$
$$\mathfrak{T}(\textbf{AC} := E_1) ;$$
$$\textbf{local jump to } L_2 ;$$
$$L_1 : \quad \mathfrak{T}(\textbf{AC} := E_2) ;$$
$$L_2 :$$

respectively.

As an example we give the translated program for an assignment statement with a relation (4.5.2, p. 52) and a Boolean operation (4.5.3, p. 52) as right part[1].

a) $a := b > c$; will yield

$$\textbf{AC} := b ;$$
$$\textbf{AC} := \textbf{AC} - c ;$$

---

[1] *true* and *false* are used as internal identifiers for the truth values **true** and **false**.

$$\textbf{if AC} \leq 0 \textbf{ then local jump to } L_1 ;$$
$$\textbf{AC} := true ;$$
$$\textbf{local jump to } L_2 ;$$
$$L_1 : \quad \textbf{AC} := false ;$$
$$L_2 : \quad a := \textbf{AC}.$$

b) $a := b \wedge c$; yields

$$\textbf{AC} := b ;$$
$$\textbf{if } \neg \textbf{ AC then local jump to } L_1 ;$$
$$\textbf{AC} := c ;$$

$$\textbf{local jump to } L_2 ;$$
$$L_1 : \quad \textbf{AC} := false ;$$
$$L_2 : \quad a := \textbf{AC}$$

The two instructions set in ⬚ are of course superfluous. This shows that one should consider possible optimizations when handling a Boolean operation like a conditional expression.

4.7. For statements

The decomposition rules for for statements with a single list element are given in the ALGOL Report [75]. Elementary constituents already introduced so far suffice for the decomposition. When several for list elements occur, an obvious expedient is that of writing the statement subject to the for clause after every list element. This is, however, practically undesirable. An alternative which appears more acceptable is that of using pseudo-procedures or generated procedures.

Generated procedures, as the name implies, are constructed during the translation process for specific purposes; whether the procedure is actually generated explicitly, or whether for the purpose of translation a certain structure is treated as if a procedure were present, is for our present purposes immaterial. The logic in any case is the same.

A for statement with several for list elements LE_1, LE_2, LE_3 and a do statement[1] S:

$$\textbf{for } V := LE_1, LE_2, LE_3 \textbf{ do } S$$

may be transformed using a generated procedure into

$$\textbf{begin}$$
$$\textbf{procedure } GPS; S ;$$
$$\textbf{for } V := LE_1 \textbf{ do } GPS ;$$
$$\textbf{for } V := LE_2 \textbf{ do } GPS ;$$
$$\textbf{for } V := LE_3 \textbf{ do } GPS$$
$$\textbf{end}$$

where GPS is the name of the generated procedure.

---

[1] We call the statement after the delimiter **do** in a for statement a *do statement.*

By using this generated procedure the handling of a for statement with more than one for list element reduces to the handling of one with only one for list element.

The decomposition of such a for statement

$$\textbf{for } V := LE \textbf{ do } S$$

is defined in the ALGOL Report by the following.

a) LE is an arithmetic expression A. Then we have to handle the for statement like

$$V := A \ ; \ S \ .$$

b) LE is a step until element. Then

$$\textbf{for } V := A \textbf{ step } B \textbf{ until } C \textbf{ do } S$$

is handled like

$$
\begin{aligned}
&V := A \ ; \\
L: \quad &\textbf{if } (V - C) \times sign(B) > 0 \textbf{ then go to } NEXT \ ; \\
&S \ ; \\
&V := V + B \ ; \\
&\textbf{go to } L \ ; \\
NEXT: \quad &
\end{aligned}
$$

or

$$
\begin{aligned}
&V := A \ ; \\
&\textbf{go to } L_2 \ ; \\
L_1: \quad &V := V + B \ ; \\
L_2: \quad &\textbf{if } (V - C) \times sign(B) > 0 \textbf{ then go to } NEXT \ ; \\
&S \ ; \\
&\textbf{go to } L_1 \ ; \\
NEXT: \quad &
\end{aligned}
$$

The first version is the one given in the ALGOL 60 Report. The second version is equivalent but simpler to translate.

c) LE is a while element:

$$\textbf{for } V := A \textbf{ while } B \textbf{ do } S \ .$$

This is to be handled like

$$
\begin{aligned}
L: \quad &V := A \ ; \\
&\textbf{if } \neg B \textbf{ then go to } NEXT \ ; \\
&S \ ; \\
&\textbf{go to } L \ ; \\
NEXT: \quad &
\end{aligned}
$$

The translation of these ALGOL program parts into elementary language instructions needs no further explanation since only statements occur which are treated in other sections.

In the target language program the generated procedures and generated procedure calls can be handled more simply than the ALGOL procedures and procedure calls, since they have a simpler structure. We use the *subroutine elements* listed in 3.2e), p. 22. The generated procedure

$$\textbf{procedure } GPS \; ; \; S \; ;$$

may, thus, be translated into

$$L: \quad \textbf{save return address } GV \; ;$$
$$\mathfrak{T}(S) \; ;$$
$$\textbf{subroutine return } GV$$

where $\mathfrak{T}(S)$ denotes the target language equivalent of the statement S. The call GPS of this generated procedure must be translated into the special elementary language instruction

$$\textbf{subroutine jump } L \; .$$

A simple example follows:

$$\textbf{for } v := a \textbf{ step } b \textbf{ until } c, \, d \textbf{ do } S$$

where v, a, b, c, d are simple variables, is translated into the following program[1]. GV is a generated variable. The call of the standard function *sign* is avoided by using conditional jump instructions. It is assumed that no type transfers are necessary.

```
          AC := a ;
          local jump to L2 ;
L1:       AC := v ;
          AC := AC + b ;
L2:        v := AC ;
          AC := AC − c ;
          ┄┄┄┄┄┄┄┄┄┄┄┄┄┄┄┄┄┄┄┄┄┄┄
          GV := AC ;
          AC := b ;
          if AC = 0 then local jump to L5 ;       step until element
          if AC > 0 then local jump to L3 ;
          AC := −ᵤ GV ;
          local jump to L4 ;
L3:       AC := GV ;
L4:       if AC > 0 then local jump to NEXT ;
          ┄┄┄┄┄┄┄┄┄┄┄┄┄┄┄┄┄┄┄┄┄┄┄
L5:       subroutine jump L ;
          local jump to L1 ;
```

---

[1] In order to simplify the translation process, the generated subroutine body is left at the place of the original do statement S, and is skipped after execution of the for statement by a generated **local jump to** $OVER$.

$NEXT:$ **AC** $:= d$;
      $v := $ **AC** ;
      subroutine jump L ;
      local jump to $OVER$;   simple list element
  $L:$ **save return address** GV ;
      $\mathfrak{T}(S)$;   do statement
      subroutine return GV ;
$OVER:$

In most cases occurring in practice, the step expression reduces to a possibly signed literal (number). Thus, the step value is known to the translator. This allows considerable reduction in the target program. The entire sequence of statements enclosed by the broken lines is replaced by

$$\textbf{if AC} \, \varrho \, 0 \textbf{ then local jump to } NEXT$$

where ϱ is $>$ for a positive, and $<$ for a negative step. The unlikely case of a literal zero step allows a further simplification.

4.8. Go to statement and switch declaration

A go to statement is of the form

$$\textbf{go to } D$$

where D is a designational expression which may be

(1) a simple identifier
(2) a switch designator
(3) a conditional designational expression
(4) ($\langle$designational expression$\rangle$).

The first three cases are handled independently in the next three paragraphs. The last case, **go to** (D), can be handled like **go to** D.

Case (1) has the form **go to** L where L is an identifier, which can be either a) a label or b) a formal parameter.

A particularly simple subcase of (1a) is the statement with a label L local to the innermost array block or a procedure body containing the statement. This statement can be translated into a simple jump instruction

$$\textbf{local jump to } L$$

available in all machines.

All other cases of (1a) require storage allocation operations. These involve a generated macro instruction of the form

$$\textbf{jump to } L.$$

In case (1 b) additional operations must be carried out: the actual label corresponding to the formal parameter L and the correct "procedure state" associated with this label must be determined. Here a target language instruction

formal procedure exit L

is generated.

Case (2): the ALGOL statement **go to** $S[I]$ requires first the assignment of the index I to the exceptional index register **SIR**. Then a jump instruction is executed as in the case (1). The same subcases may be distinguished. Thus **go to** $S[I]$ is translated as follows:

(2a) S is a declared switch:

$$\text{\textbf{SIR}}:= I ; \qquad \text{or} \qquad \text{\textbf{SIR}}:= I ;$$
$$\text{\textbf{local jump to}}\ S \qquad\qquad \text{\textbf{jump to}}\ S$$

(2b) S is a formal parameter:

$$\text{\textbf{SIR}}:= I ;$$
$$\text{\textbf{formal procedure exit}}\ S.$$

The final compilation associates with each label or switch an address of the target program. In manipulating designational expressions, therefore, the translator must provide for that fact.

Case (3): a go to statement with a conditional designational expression:

go to if B **then** $D1$ **else** $D2$

may be handled like conditional statements and thus transformed first into

if B **then go to** $D1$ **else go to** $D2$.

go to $D1$ and **go to** $D2$ are to be recursively decomposed in the same manner.

The handling of a switch designator involves a suitable treatment of the corresponding switch declaration. Consider

switch $S := D1, \ldots, Dn$

where Di are arbitrary designational expressions. The switch declaration is transformed into the following piece of program:

$$\text{\textbf{local jump to}}\ L0 ;$$
$$Li_1:\quad \mathfrak{T}\ (\text{\textbf{go to}}\ Di_1) ;$$
$$\cdots$$
$$Li_K:\quad \mathfrak{T}\ (\text{\textbf{go to}}\ Di_K) ;$$

S: **local jump to content of** $S[\textbf{SIR}]$;
    $L1$;
    $\cdots$
    Li_1 ;
    $\cdots$
    Li_K ;
    $\cdots$
    Ln ;
  $L0$:

In the list $L1$;
    $\cdots$
    Ln ;

Li is identical to Di if Di is a simple label which causes no global jump. If Di is not such a simple label, then Di is extracted out of the list of designationals and there replaced by the generated label Li. This label Li is used to mark the sequence of instructions corresponding to **go to** Di which is stored before the beginning S of the switch declaration.

The following example illustrates this:

    $\cdots$
  $L2$: $\cdots$
    begin
      switch $S := L1$, **if** B **then** $L1$ **else** $L2$; $\ldots$
    $L1$: $\ldots$
    end ;
    $\cdots$

is translated into:

    $\cdots$
  $L2$: $\cdots$
    local jump to $L0$;
  $L3$: $\textbf{AC} := B$;
    if $\neg$ $\textbf{AC}$ **then local jump to** $L4$;
    local jump to $L1$;
  $L4$: **jump to** $L2$;
  S: **local jump to content of** S $[\textbf{SIR}]$;
    $L1$;
    $L3$;
  $L0$: $\ldots$
  $L1$: $\ldots$

The macro instruction

    local jump to content of $S[\textbf{SIR}]$

acts as follows. The values of **SIR** may be integer numbers $1, ..., n$, where n is the number of designational expressions in the switch list associated with S. If the value of **SIR** is m then the m-th label of the list following

local jump to content of $S[\textbf{SIR}]$

is selected and a jump to this label is executed.

4.9. Procedures and dynamic storage

We shall now consider the problem of storage allocation with respect to procedures, whose detailed discussion we have deferred up to this point. The handling of data storage for procedures depends on whether the recursive use of procedures is permitted in full generality or whether calls are limited to the simple kind.

In the simple case, where recursive calls of procedures of every kind are forbidden, the run time organization needed for recursive procedures becomes superfluous. As was mentioned in 4.2, p. 28 a procedure declaration may be handled up to a point like a block which would occur in its place at the same point in the ALGOL program. Simple variables, arrays with fixed subscript bounds, and the information vectors of dynamic arrays may be placed in the fixed data storage; the handling of dynamic arrays is of course like that in the main program. The only, though essential, difference from the handling of the block lies in the fact, that "parallel" procedure declarations cannot in general share the use of the same fixed storage cells. The space reserved for variables and arrays local to a procedure declared in a block A must not conflict with that reserved for those local to other (non-recursive) procedures in A or blocks contained in A which call the procedure. On the other hand, two blocks of level l which are not procedures but which are both contained in a block of level $l-1$, may share storage. This difference affects therefore even the static storage allocation for simple variables and arrays with fixed subscript bounds. Apart from this restriction, no new difficulties arise with respect to procedures.

Below, we discuss what is required if the recursive use of procedures is allowed. The essential property is that they may, directly or indirectly, call themselves. Thus, the same procedure may simultaneously have several *dynamic levels*. There will thus be a simultaneous use of as many sets of variables local to the procedure[1] as there are calls of it in

---

[1] It must be said there remain some interesting theoretical questions regarding the variables of such procedures. For example:

Variables which are local to a procedure but which are not considered different in the various calls of a nesting are desirable. There seems no provision for such variables in ALGOL. A flexible apparatus for the use of recursive procedures will require it.

the nesting. Each set must have its own reserved storage which is allotted at the time of call.

4.9.1. Variables in procedures

As stated in 4.2, p. 28, a procedure P may, in a certain sense, be handled like a simple main program, if procedures declared within the former are ignored. The declaration of P defines a fixed storage part containing simple and generated variables, information vectors of P, and the set $BFSP_0, \ldots, BFSP_{LP}$ of "initial free storage locations" of P (where LP means the maximum array block level within the body of P) as described in 4.2.2.2, p. 33.

Each *dynamic call level* of P requires a new specimen of this fixed storage part. Consequently, the translator which necessarily has no knowledge of dynamic call levels, can assign addresses for the fixed storage part only relative to an origin which, for each specimen, is determined at run time by the dynamic call level.

Appropriate means of keeping these origins, and executing the transformation from relative to absolute addresses of the fixed storage components at run time are *index registers*. If the use of index registers for locating these components in procedures is adopted throughout, the translator has the tasks of

a) assigning these registers to the procedures, and

b) associating components of fixed storage parts of procedures with the proper index registers in the instructions generated.

In order to make efficient the run time mechanism of locating free storage components we require that reloading of index registers takes place at most only when a procedure call is initiated or terminated in the natural way (via the end), or when a formal parameter is actualized.

These requirements are met by the following scheme for assigning index registers to procedure declarations which is used in the model translator:

To each *static procedure level* (which is defined by inclusion of procedure declarations) a unique index register is assigned. For details see section 6.2, p. 104.

In this way, the translator associates an index register sP with the fixed storage of every ALGOL procedure P[1]. At run time, this index register for each call of P is loaded with the momentary "beginning of

---

[1] With respect to storage allocation, generated procedures [i.e. generated procedures representing do statements and actual parameters, see sections 4.7, p. 56 and 4.10.1.1, c), p. 67] are handled like blocks. Index registers need not be associated with them.

free storage" which constitutes the origin of the associated specimen of the fixed storage part of P.

The set of available index registers can be considered to be a new type of exceptional global array defined by a pseudo declaration

$$\textbf{address array } \Im r\,[0:S]\; ;$$

generated in the outermost fictitious block containing ALGOL programs, standard procedure declarations, and constants (see section 3.1, p. 19).

The translator assigns a relative address $RADV$ to every component V of the fixed storage of every procedure P. The pair $(RADV, sP)$ which at run time determines the absolute address $ADV = RADV + \Im r\,[sP]$, is the main constituent of the entry in the identifier list associated with the component V (see section 3.1, p. 19).

To obtain a uniform storage allocation scheme for the whole ALGOL program it is convenient to consider the main program also as a procedure with the associated static level and index register number 0 (zero).

Basically the following may be said about data storage for procedures: The choice of time at which the determination of absolute addresses is undertaken depends on the facilities of the machine and on the required facilities of the programming system. If the determination of absolute addresses is undesirable during the execution of the target program, it must be undertaken by the translator; in that case, however, the mechanism which makes recursion possible becomes extremely inconvenient, and it is often better to avoid the use of recursive calls. The mechanism is much simpler if index registers are available for setting absolute addresses so that the process can be carried out conveniently in the running program.

If there are not enough index registers in a computer, it is possible (in a present-day machine) to simulate them. This way of implementing recursive procedures sometimes is economical, if arithmetic operations are executed by coded subroutines. If, on the other hand, arithmetic operations are wired in the machine, the simulation of index registers may take as much time as the arithmetic operations themselves with the result that simulation is quite definitely uneconomical.

4.9.2. Dynamic storage

The usual machine memory may be considered as a one dimensional array $\mathfrak{M}$ of memory cells (words, or storage locations, or bytes) where the subscripts on $\mathfrak{M}$ are the addresses belonging to these memory cells. Those parts of data storage which have been considered separately before can now be arranged in a sequence as outlined in 4.2, p. 27.

fixed storage of the main program including that for arithmetic constants, truth values, and strings occurring in the ALGOL program	static storage of ALGOL program
dynamic array storage of main program	
fixed storage of the first procedure called and not yet completed	
dynamic array storage of first procedure called and not yet completed	
. . .	dynamic storage of ALGOL program
fixed storage of momentarily last procedure called and not yet completed	
dynamic array storage of momentarily last procedure called and not yet completed	
free storage	

Such a scheme is always dependent on a definite point in time in the execution of the program; it contains the currently relevant dynamic arrays and other data of procedures whose calls are still current in dynamic sequence.

The "*fixed storage*" of a procedure P may be divided into six sections consisting of the storage locations for

1. the *procedure linkage* necessary for the organization of calls and exits,

2. the array block levels

$$BFSP_0, \ldots, BFSP_{LP}$$

of those array blocks contained in the body P, LP meaning the maximum array block level in the body of P,

3. the "*parameter block*" associated with the formal parameters and containing proper information on the corresponding actual parameters,

4. the information vectors of formal arrays,

5. the value listed non-array parameters,

6. the simple and generated variables and information vectors of declared arrays local to the procedure body P.

4.10. Procedure calls and declarations

The handling of procedure statements and function designators[1] and the handling of procedure declarations are related, since any decision made regarding one will be reflected in the other. We begin with the discussion of procedure calls.

In this connection we introduce certain generated procedures called *name procedures* with the syntax

⟨name procedure declaration⟩ ::= **name procedure**

⟨generated procedure identifier⟩ ;

⟨name procedure body⟩ ;

Here the delimiter **name** must be thought to be an additional type declarator and the procedure itself must be regarded as a special type procedure. The designation "name procedure" has been chosen, as name procedures are exactly those procedures called by name variables.

4.10.1. Procedure calls

In order to handle *procedure calls*

$$P(AP_1, \ldots, AP_r),$$

or

$$FP(AP_1, \ldots, AP_r),$$

respectively, (where FP is a formal procedure identifier) in a simple systematic way, we extract certain actual parameters out of the calls and replace them by generated identifiers. The following sections explain which parameters are concerned.

Evidently, it is necessary at translation time to handle actual and formal parameters in such a way that at run time they correspond with each other in an appropriate manner. This handling does not involve great difficulty if the declarations and specifications of all entities within an actual parameter and the specifications of the associated formal parameters of an actual parameter are known at translation time.

The actual situation is, however, not this simple. In ALGOL, specifications for name parameters may be omitted (see [75], 5.4.5). Even if we demand that each formal parameter be specified by the programmer we have the problem of correct correspondence when handling actual parameters in formal procedure calls. Here it is in general not feasible to determine the specifications of the associated formal parameters at translation time. One may try to find them by pursuing all syntactically

---

[1] We shall say "*procedure calls*" for both concepts.

admissible replacements of formal parameters by actual parameters, but there contradictions between the specifications of the different possible associated formal parameters may arise, which will not actually occur at run time.

As a consequence, we do not assume that specifications are given. If specifications are available we shall use them since the translation into the target language often will be more efficient and less complicated (see [75], 4.7.5.5, compare also p. 38 of section 4.3).

4.10.1.1. Actual parameters in procedure calls

This section identifies precisely the different forms which an actual parameter AP may take and explains how these forms are handled. The following pages define when an actual parameter remains unchanged or not. In the latter case, a description of the rearranged actual parameter is given.

a) The actual parameter AP is a single identifier. AP remains unchanged whenever (exactly) one of the following conditions is fulfilled:

a1) The actual parameter AP is a formal not value listed (name) identifier.

a2) The actual parameter AP is a value listed array identifier.

a3) The actual parameter AP is a declared array, label, switch, or procedure identifier.

b) The actual parameter AP is of the form

$$\underbrace{(\ldots(}_{m \geq 1}\; I\; \underbrace{)\ldots)}_{m \geq 1}$$
opening parentheses   closing parentheses

where I is a single identifier. Here the actual parameter AP is replaced by I if one of the following conditions is fulfilled:

b1) I is a name identifier.

b2) I is a declared label or type procedure identifier (function designator without any parameter).

c) The actual parameter AP is a string or an arithmetic or Boolean expression different from a form mentioned under a) or b). Here the actual parameter AP is replaced by a generated identifier $GIAP$ of a generated procedure. The expression AP itself will be transformed into a *"name procedure"* $GIAP$. Such a generated procedure is a special function procedure (see section 4.10, p. 66), which

(1) has no parameters,

5*

(2) assigns a storage location address coupled with a type marker to the exceptional internal address variable **ADDR AC** (address accumulator)[1],

(3) assigns the value of the actual expression AP to the exceptional variable **AC**, and

(4) its declaration is added to the declarations of the smallest block or procedure body B_P containing the procedure call

$$P(AP_1, \ldots, AP_r),$$

or

$$FP(AP_1, \ldots, AP_r).$$

We distinguish five cases c 1) to c 5) of procedures to be constructed.

c1) The actual parameter AP is a string with an internal identifier STR.

The name procedure has a form that can be described in the following "ALGOL-like" way:

> **name procedure** $GIAP$;
> **begin**
>   **absolute address** STR
> **end** ;

absolute address STR is understood to be a target language instruction. Its parameter is the internal identifier STR of the string AP, and the instruction assigns the absolute initial storage location address of the actual string AP to the exceptional variable **ADDR AC**.

c2) The actual parameter AP is a declared or value listed simple variable V.

The name procedure has the following ALGOL-like form:

> **name procedure** $GIAP$;
> **begin**
>   **absolute address** V ;
>   **AC** $:= V$
> **end** ;

The target language instruction **absolute address** V assigns the absolute storage location address of the actual variable V coupled with the appropriate type marker of V to the exceptional address variable **ADDR AC**.

---

[1] The action (2) of a name procedure is the action of a "thunk" as given in [55] and [56].

c3) The actual parameter AP is a subscripted variable

$$A\,[s_1, \ldots, s_n]$$

which is not handled by recursive address calculation.

The name procedure $GIAP$ looks as follows:

name procedure $GIAP$;
    begin
        AC $:= A_{red\ init} + \underset{A}{\varphi}\,([s_1, \ldots, s_n])$;
        ADDR AC $:=$ **AC** ;
        AC $:=$ **content of ADDR AC**
    end ;

The definitions of the reduced initial address $A_{red\ init}$ and of the notation

$$\underset{A}{\varphi}\,([s_1, \ldots, s_n])$$

have been given in section 4.2.2, p. 30.

It is to be remembered that the address in the exceptional variable **ADDR AC** is coupled with a type marker representing the type of the declared array A or of that actual array associated with the formal array A (see section 4.3, p. 40). The marker is transferred from the reduced initial address $A_{red\ init}$ to **AC** through the operation

$$\textbf{AC} := A_{red\ init} + \underset{A}{\varphi}\,([s_1, \ldots, s_n])$$

which is an address addition [compare footnote 1 in section 4.3, (1 a), p. 40].

c4) The actual parameter AP is a subscripted variable

$$A\,[s_1, \ldots, s_n]$$

handled by recursive address calculation. Here we distinguish two cases depending on whether $A\,[s_1, \ldots, s_n]$ is associated with an index register or not. Suppose, $A\,[s_1, \ldots, s_n]$ is connected with a generated address variable $GVADDR$ (see chapter 5). If no index register is involved the name procedure $GIAP$ simply looks as follows:

name procedure $GIAP$;
    begin
        ADDR AC $:= GVADDR$;
        AC $:=$ **content of ADDR AC**
    end ;

If an index register is involved, then we have

> **name procedure** $GIAP$;
>   **begin**
>   **absolute subscripted variable address** $GVADDR$;
>   **AC**:= **content of ADDR AC**
>   **end** ;

absolute subscripted variable address $GVADDR$ is understood to be a target language instruction finding out the absolute storage location address of $A[s_1, \ldots, s_n]$ defined by the content of $GVADDR$ and by the content of the involved index register, and its type, and assigning both to **ADDR AC**.

c5) The actual parameter AP is an arithmetic or Boolean expression which does not consist of a single variable.

The name procedure looks like:

> **name procedure** $GIAP$;
>   **begin**
>   **AC**:= AP ;
>   **ADDR AC**:= '⟨type⟩'
>   **end** ;

⟨type⟩ means the type **integer**, **real**, or **Boolean** of the expression AP if this type can be determined during translation. Otherwise the statement

$$\textbf{ADDR AC}:= \text{'⟨type⟩'}$$

is omitted as then **ADDR AC** already contains the correct type.

d) If an actual parameter AP is a designational expression different from a form mentioned under a) and b) the actual parameter is replaced by a generated label $GLAP$. The designational expression itself is transformed into a generated goto statement **go to** AP, which is labelled by $GLAP$:

$$GLAP: \textbf{go to } AP ;$$

This statement is added to the statements of the block or procedure body B_p mentioned above under point c), p. 68 in the section 4.10.1.1.

e) A situation may occur where it is impossible to decide at translation time whether the actual parameter AP is a designational expression to be subsumed under d), p. 70 or an arithmetic or Boolean expression to be subsumed under c3), c4), or c5), p. 69. Here the actual parameter AP is replaced by a generated identifier $GIAP$. The expression itself is

α) transformed into a name procedure called $GIAP$ [see section 4.10.1.1, c3), c4), c5), p. 69], and

β) changed into a generated go to statement labelled by $GLAP$ [see section 4.10.1.1, d), p. 70].

The run time system (see section 6.6, p. 117) must be able to determine the label $GLAP$ when given the identifier $GIAP$.

Below we shall explain how the main constituents of an actual parameter are to be handled both in the name procedure $GIAP$ and in the designational expression $GLAP$. Apart from if clauses and delimiters **then** and **else**, the innermost actual simple expressions have the form

$$\underbrace{(\ldots((E))\ldots)}$$
$$\underset{m\geq 0}{}\quad \underset{m\geq 0}{}$$
$$\text{parentheses} \quad \text{parentheses}$$

where E is either

  e1) an unsigned integer, e.g. 15^1, or

  e2) an unspecified formal parameter N, or

  e3) an unspecified formal subscripted parameter $A[s_1]$.

Evidently the parentheses may be deleted during translation.

  e1. α) Within the name procedure $GIAP$ the expression AP can only be interpreted to be an arithmetic expression and is translated e.g. by

$$\textbf{AC} := 15 ;$$
$$\textbf{ADDR AC} := \text{'integer'}$$

[see 4.10.1.1, c5), p. 70].

  e1. β) Within the designational expression $GLAP$ the unsigned integer number e.g. 15 is an integer label. It is translated by the target language instruction

$$\textbf{jump to } I_{15}$$

where I_{15} means the internal identifier for the integer label 15 (see section 3.2, p. 21 and 4.8, p. 59).

  e2. α) Within the name procedure $GIAP$, N can be a name variable or a formal function designator without parameters. In both cases N can be translated correctly by

$$\textbf{name call } N$$

(see section 4.4.2, p. 44, 4.10.1.2, p. 72, 4.10.1.3, p. 73).

  e2. β) Within the designational expression $GLAP$, N can only be interpreted as a formal label. N is translated by

$$\textbf{formal procedure exit } N$$

(see section 4.8, p. 60).

  e3. α) Within the name procedure $GIAP$, $A[s_1]$ can be interpreted as a formal integer, real, or Boolean subscripted variable. The target

---

[1] This case e1) of an undetermined actual parameter cannot be avoided even if we demand that each formal parameter must be specified, whereas case e2) and case e3) do not apply then.

language instructions look as follows:

$$\mathbf{AC} := A_{red\ init} + \varphi([s_1])\ ;$$
$$\mathbf{ADDR\ AC} := \mathbf{AC}\ ;$$
$$\mathbf{AC} := \mathbf{content\ of\ ADDR\ AC}$$

e3. β) Within the designational expression $GLAP$, $A[s_1]$ can only be a formal switch component. The associated target language instructions are

$$\mathbf{AC} := s_1\ ;$$
$$\mathbf{SIR} := \mathbf{AC}\ ;$$
$$\mathbf{formal\ procedure\ exit}\ A$$

(see section 4.8, p. 60).

4.10.1.2. Procedure calls after transformations of actual parameters

The transformations of actual parameters AP described in section 4.10.1.1 transform the procedure call into a sequence of generated procedures and go to statements followed by the procedure call proper which now has only (ALGOL or generated) identifiers IAP as parameters.

The procedures and statements generated from the actual parameters are executed at run time only upon calls from within the body of the procedure called. In the flow of execution of the program leading to the call proper, they must be skipped like switch and procedure declarations in block heads (c.f. 4.1, p. 27). The sequence of generated procedures must therefore be headed by a jump instruction leading to the call instruction.

The procedure calls

$$P(AP_1, ..., AP_r) \quad \text{and} \quad FP(AP_1, ..., AP_r)\,, \quad (r \geq 0)$$

are transformed as follows:

local jump to $OVER$;

$\cdots$

name procedure IAP_{g_1} ;
   begin
      $\cdots AP_{g_1} \cdots$
   end ;
   $\cdots$

$IAP_{l_1}:$ **go to** AP_{l_1} ;
   $\cdots$

$OVER:$ **procedure call** P ; (or **formal procedure call** FP ; resp.)
      IAP_1 ;
      $\vdots$
      IAP_r

where IAP_ϱ $(\varrho=1,\dots,r)$ is (1) the internal identifier of the generated name procedure into which AP_ϱ may be transformed or (2) the internal identifier of the generated label in the case of a designational expression AP_ϱ or (3) the internal identifier of the actual parameter AP_ϱ itself in the cases listed in section 4.10.1.1, case a) and b). **procedure call** P and **formal procedure call** FP are understood to be macro instructions of the target language program. Their parameters P, FP, IAP_1, $\dots$, IAP_r are the internal identifiers associated with the corresponding ALGOL identifiers. The operation of these macro instructions is discussed under run time organization (see section 6.2, p. 104 and 6.3, p. 111).

If
$$P(AP_1, \dots, AP_r)$$
or
$$FP(AP_1, \dots, AP_r)$$
is a function designator the macro instructions

$$\langle\text{type}\rangle \text{ \textbf{procedure call}} \ \ P$$

or

$$\langle\text{type}\rangle \text{ \textbf{formal procedure call}} \ \ FP$$

assign the value and type of the called function procedure to the exceptional variables **AC** and **ADDR AC**, respectively (see section 4.10.2.3, p. 75).

We subsume the handling of standard functions under that of non-standard functions. However, in most cases standard function calls may be executed in a simpler way, since it is known that the parameters are to be handled like value listed parameters and that therefore no recursive calls can occur.

4.10.1.3. Name calls

A *name call* (call of a formal, not value listed variable) is a special sort of function procedure call without any parameter. The name procedures introduced in section 4.10.1.1, c), p. 67 are just those procedures which are called by name variables.

Since a name variable N may occur both on the left side and on the right side of the assignment symbol $:=$ we distinguish two different calls:

$$\text{\textbf{name address}} \ \ N$$

and

$$\text{\textbf{name call}} \ \ N \ \ (\text{or} \ \langle\text{type}\rangle \ \textbf{name call} \ N)$$

name address N assigns the storage location address and the type of an actual variable to the exceptional address variable **ADDR AC**. **name call** N assigns the value of an actual expression to the exceptional variable **AC** (see also section 4.4.2, p. 44) and the type of that value to the exceptional address variable **ADDR AC**.

Like **procedure call** P and **formal procedure call** FP in section 4.10.1.2, p. 72 **name address** N and **name call** N are understood to be target language macro instructions the exact description of which we shall give when speaking about run time organization (see section 6.7, p. 119).

4.10.2. Procedure declarations

In section 4.10.1 we have discussed the translation of a procedure call which in particular involves generating name procedures from the actual parameters. Correspondingly, in the body of the procedure declaration, formal parameters are replaced by the associated name calls. Apart from this, the procedure body is treated largely like the main program. In particular, the storage allocation for the individual procedure body is handled in the same way as for the main program as described in 4.2 and 4.9.

The information conveyed by means of the procedure heading requires the generation of some instructions, or sequences of instructions, peculiar to procedures. These are described in the following.

4.10.2.1. Value listed formal variables

The value listing for a formal simple variable VV is replaced by the appropriate type declaration of a generated variable GVV at the beginning of the procedure body. A type specification must be given in the procedure heading (see [75], 5.4.5, p. 447). Before the first statement of the procedure body, assignment statements assigning values to these variables of the ALGOL-like type

$$GVV := VV$$

are inserted. In target language, these statements will be of the form

$$\langle\text{type}\rangle \ \textbf{name call} \ VV \ ;$$
$$GVV := \textbf{AC}$$

(see section 4.10.1.3, p. 73).

Throughout the whole body each occurrence of VV is replaced by the generated variable GVV.

4.10.2.2. Formal arrays

If by a prepass the number n of subscripts is made part of the specification[1] then the treatment of the specifications of formal arrays in several respects corresponds to that of value listed variables.

---

[1] No new difficulties arise, if the number of subscripts of a formal array A can be determined only during execution of the program. See section 6.2.

An array declaration for the array A is generated at the beginning of the procedure body. This means that storage for the array information vector[1], the length of which depends on the number of subscripts, is allocated by generating variables local to the body. The elements of the information vector are then parameters of the procedure which are treated like value listed simple variables.

The assigning of values to the generated variables is left to the macro instructions **procedure call** P or **formal procedure call** FP and is simply a copying of the information vector of the actual array, as will be seen in the discussion of run time organization (see section 6.2, action 2, p. 107). As a consequence, with respect to address calculation the translation of a formal subscripted variable $A[s_1, \ldots, s_n]$ by no means differs from the translation of one that is not formal.

A macro instruction for a value listed formal array A

value array A

is also set up. At execution time this macro instruction allocates storage to the *value array* A and then assigns the values of all components of the actual array to the corresponding components of the value array A. The macro instruction will be described formally in chapter 8.

Since for value arrays a type specification must be given, the actions of the instruction **value array** include a type transfer of the actual array components to the type given by the specification, when required. A translator handling type statically can generate, in place of the above instruction, different instructions such as **integer value array** and **real value array**. These must test only for the types of the actual arrays[2].

4.10.2.3. Function procedures

Since a function procedure yields a value, additional features have to be provided. For each function procedure P a new variable GVP is generated whose appropriate type declaration is considered to be at the beginning of the body. Throughout the body, each occurrence of the identifier P on the left hand of the assignment symbol $:=$ is replaced by GVP. Immediately before the end of the body the value of GVP and its type are assigned to the exceptional variables **AC**, and **ADDR AC** respectively (see section 4.4.2 and 4.10.1.2).

---

[1] Section 4.2.2 gives an exact description of the concept "information vector of an array".

[2] Situations may occur where certain components of the actual array have no significant value at execution time. In such cases type transfers may lead to inappropriate run time alarms. These may be avoided, at some cost in efficiency, by suitable initialization of the values of the elements of declared arrays.

4.10.2.4. Normal procedure exit

The normal exit out of a procedure or function procedure body is organized by the target language macro instruction

normal procedure exit.

In the case of name procedures this task is done by

name procedure exit.

These macro instructions will be explained in sections 6.4 and 6.6.

5. Recursive address calculation

5.1. Introduction

Recursive address calculation concerns optimization of the calculation of the addresses of subscripted variables ocurring in the statement following do, which we call the *do statement*.

The discussion here is restricted to recursive evaluation of subscript functions which in part are given only implicitly by storage mapping functions. It is obvious that the same methods could be used for a more general optimization of expressions and statements depending on the loop variable. The simplest case is the removal, from loops, of expressions which have constant value throughout the loop, and of multiple assignments of a constant value to the same variable. Since these features, however, can be handled if desired by the ALGOL programmer and the translator would here assume what amounts to an editorial function, we restrict ourselves here to the case of subscripted variables which are not very amenable to manipulation by the programmer.

Consider the for statement

for $i := a$ **step** b **until** c **do** ... $A[s_1, ..., s_n]$ ... ;

The $s_1, ..., s_n$ can be functions (arithmetic expressions) of the loop variable i. The distance function $\varphi([s_1, ..., s_n]) = \varphi_A^*(i)$ (see sections 4.2.2 and 4.4.3) which is needed for the computation of the address of $A[s_1, ..., s_n]$ is then also a function of i.

Consider the transition from i to $i+b$. $\varphi_A^*(i)$ changes into $\varphi_A^*(i+b)$, which may be represented in the form

$$\varphi_A^*(i+b) = \varphi_A^*(i) + \left(\varphi_A^*(i+b) - \varphi_A^*(i)\right)$$
$$= \varphi_A^*(i) + \Delta \varphi_A^*(i, b).$$

If the computation of the value $\Delta \varphi_A^*(i, b)$ needs less operations (especially multiplications) than the computation of $\varphi_A^*(i+b)$, then it is obviously worthwhile to compute $\varphi_A^*(a)$ at the beginning of the for statement as the initial value and to evaluate and add $\Delta \varphi_A^*(i, b)$ for each new step[1].

We assume here that

1) no assignments are made in the do statement to i or the variables occurring in the expressions $s_1, \ldots, s_n$ (explicitly or implicitly);

2) b does not change during the execution of the entire loop.

We restrict ourselves to the special case where all s_ν's $(\nu = 1, \ldots, n)$ are polynomials of degree $\leq k$ in i. Then φ_A^* is also a polynomial of degree $\leq k$ in i. If φ_A^* is of degree k then $\Delta \varphi_A^*$ is a polynomial of degree $k-1$ in i. Now $\Delta \varphi_A^*$ can be determined recursively in the following way: When changing i to $i+b$, $\Delta \varphi_A^*(i, b)$ changes to

$$\Delta \varphi_A^*(i+b, b) = \Delta \varphi_A^*(i, b) + \Delta^2 \varphi_A^*(i, b).$$

Here, $\Delta^2 \varphi_A^*$ is a polynomial of degree $k-2$ in i and so on: $\Delta^\varkappa \varphi_A^*(i, b,)$ $(\varkappa = 1, 2, \ldots)$ changes to

$$\Delta^\varkappa \varphi_A^*(i+b, b) = \Delta^\varkappa \varphi_A^*(i, b) + \Delta^{\varkappa+1} \varphi_A^*(i, b)$$

with $\Delta^k \varphi_A^*(i, b) = \text{constant}$ and $\Delta^\varkappa \varphi_A^*(i, b) = 0$ for $\varkappa > k$.

This leads to the following scheme:

Before beginning the first step of the loop one has to compute:

$$\varphi_A^*(i)$$
$$\Delta \varphi_A^*(i, b) := \varphi_A^*(i+b) - \varphi_A^*(i)$$
$$\vdots$$
$$\Delta^k \varphi_A^*(i, b) := \Delta^{k-1} \varphi_A^*(i+b, b) - \Delta^{k-1} \varphi_A^*(i, b)$$

for the initial value $i = a$.

Before each new step one has to compute:

$$\varphi_A^*(i+b) := \varphi_A^*(i) + \Delta \varphi_A^*(i, b)$$
$$\Delta \varphi_A^*(i+b, b) := \Delta \varphi_A^*(i, b) + \Delta^2 \varphi_A^*(i, b)$$
$$\vdots$$
$$\Delta^{k-1} \varphi_A^*(i+b, b) := \Delta^{k-1} \varphi_A^*(i, b) + \Delta^k \varphi_A^*(i, b)$$
$$\vdots$$
$$i := i+b.$$

Most programs occurring in practice contain only subscript expressions which are linear in the loop variable $(k=1)$. In this case $\Delta^\varkappa \varphi_A^* = 0$ for $\varkappa \geq 2$ and $\Delta \varphi_A^*$ does not depend on i. Therefore $\Delta \varphi_A^*$

---

[1] We assume that the subscripted variable is used in the do statement and that therefore all s_j's have significant values at the beginning of the loop. Otherwise we are calculating with undefined entities.

remains constant throughout the loop and can be evaluated once before entering. The former scheme reduces to the simple statements:

Before beginning the first step compute:

$$\varphi_A^*(i)$$
$$\varDelta\,\varphi_A^*(i,\,b) := \varphi_A^*(i+b) - \varphi_A^*(i) \quad \text{for} \quad i=a$$

and before the next steps compute:

$$\varphi_A^*(i+b) := \varphi_A^*(i) + \varDelta\,\varphi_A^*(i,\,b)$$
$$i := i+b.$$

We shall restrict our discussions to this linear case, since the non-linear case is too involved to be of practical importance in translation. If necessary, the linear recursive calculation description may be extended to the non-linear case.

As may be seen from the above remarks, the simpler the expressions in the index positions are, the more efficient is the recursive address calculation (rec addr calc). Since a general rec addr calc can be carried out only at great expense of time during translation, it is reasonable to limit the implementation to the most used cases. These are precisely also the simplest.

Everything necessary to insert recursive address calculation can be done at translation time. This includes the investigation of the conditions under which it is admissible; some form of prepass is required. Recognizing identical subscript expressions and evaluating them only once are possible further optimizations.

In section 4.2.2 the *index vector* $[s_1, \ldots, s_n]_A$ and the distance function φ were introduced. We state some relations satisfied by the index vector to be used below.

$$address\,(A\,[s_1, \ldots, s_n]) = A_{red\ init} + \varphi_A([s_1, \ldots, s_n]).$$

The following binary operations and equivalence relations may be established for index vectors:

$$[t_1, \ldots, t_n]_A + [u_1, \ldots, u_n]_A = [t_1+u_1, \ldots, t_n+u_n]_A,$$
$$\alpha \times [t_1, \ldots, t_n]_A = [\alpha \times t_1, \ldots, \alpha \times t_n]_A,$$
$$[s_1, \ldots, s_n]_A \equiv [s_1, \ldots, s_n]_B \text{ if and only if the lengths } K_2, \ldots, K_n$$

of the 2-nd, 3-rd, ..., n-th dimensions of array A are the same as the lengths of the corresponding dimensions of array B.

Since φ is a *linear function* of the components of the argument vectors, the following relations are true:

$$\varphi\underset{A}{([s_1, \ldots, s_n]} + [t_1, \ldots, t_n]) = \varphi\underset{A}{([s_1, \ldots, s_n])} + \varphi\underset{A}{([t_1, \ldots, t_n])},$$

$$\varphi\underset{A}{(\alpha \times [s_1, \ldots, s_n])} = \alpha \times \varphi\underset{A}{([s_1, \ldots, s_n])},$$

$$\text{if } \underset{A}{[s_1, \ldots, s_n]} \equiv \underset{B}{[s_1, \ldots, s_n]} \text{ then } \varphi\underset{A}{([s_1, \ldots, s_n])} = \varphi\underset{B}{([s_1, \ldots, s_n])}.$$

Furthermore, $\varphi\underset{A}{([0, 0, \ldots, 0, s_n])} = s_n$.

5.2. Assumptions necessary for the use of recursive address calculation

We implement recursive address calculation only for subscripted variables with subscript expressions which are *constant* or *linear in the loop variable* with coefficients which are not changed in the do statement in loops with constant step elements and a loop variable which is not changed in the do statement.

Under these conditions we compute

$$\varphi_A^*(i) \quad \text{and} \quad \varDelta \varphi_A^*(i, b) \equiv B \quad \text{before the first step.}$$

Then, for each new pass through the do statement only the addition of B to the address $A_{red\ init} + \varphi_A^*(i)$ of the subscripted variable $A[s_1, \ldots, s_n]$ computed in the foregoing step is necessary.

The compiler must first determine if it is possible to carry out recursive address calculation. To that end the following questions must be answered for each loop:

(1) Is rec addr calc possible for the for statement?

(2) Does the do statement contain at least one subscripted variable whose address can be calculated recursively?

The following restrictions are made:

R1: The for clause must contain at least one step until or while element[1].

R2: The loop variable must be a simple variable of type **integer**, and, if a formal parameter, listed by value.

---

[1] With while elements, the rec addr calc reduces to a simple precalculation of addresses which remain constant. Rec addr calc has no advantages for simple list elements.

If rec addr calc is to be used for one list element in a for statement with more than one list element, it must be implemented for all list elements, including the simple and the while elements. The do statement should then be implemented as a generated procedure, called by each list element.

R3: rec addr calc will only be implemented for subscripted variables whose index expressions are linear functions of the loop variable consisting of integer constants, simple variables of type **integer**, (which, if formal, are value listed), the delimiters (,), $+$, $-$, $\times$, and the standard functions *entier* and *sign*.

Point (1) may now be more concretely expressed:

For a given loop (do statement) the compiler must determine that the following conditions are met:

(1) a) the loop variable is a simple variable of type **integer** and not a name variable.

b) no assignment is made within the do statement to the loop variable, to the variables appearing in the step element, or those in the expression before **while**. All assignment statements and procedure calls must be investigated for this.

c) the for clause contains at least one step until or while element.

d) the do statement contains at least one subscripted variable.

If this is the case, the loop is *suitable* for rec addr calc. Otherwise, the for clause and the subscripted variables appearing in the do statement will be translated as described earlier.

In a loop which is suitable for rec addr calc the distance function φ for a subscripted variable $A[s_1, \ldots, s_n]$ can be evaluated recursively if

(2) a) the array name A is declared outside the for loop (and not within the do statement),

b) the do statement is free from assignments to the variables appearing in $s_1, \ldots, s_n$,

c) the condition R3 holds.

If we define declared or value listed variables of type **integer** which remain constant during execution of the loop to be "*admissible variables*" then (2b) reads:

the subscript expressions $s_1, \ldots, s_n$ contain only admissible variables or the loop variable.

The necessary tests are easy to implement in a compiler with enough passes. First, the for clause and do statement can be checked for the suitability of rec addr calc [points (1a) through (1d)]. Point (1b) [and also (2b)] requires checks of left part variables. A list of all simple left part variables in a do statement must be collected, since these variables are inadmissible for the surrounding loops. Name variables occuring as left parts and procedure statements cause all surrounding loops to be unsuitable for rec addr calc, since admissibility of variables can not be established. Calls of name variables and function designators can be disregarded here, since for us side effects do not occur. However, such calls affect the use of index registers in loops which will be discussed later (see p. 98).

A following pass can transform the for clause into a piece of program adapted to recursive address calculation. At the same time also the subscripted variables in the do statement can be checked and the associated distance functions translated into appropriate instructions.

A true one pass compiler must follow both branches simultaneously; the produced program should allow for both rec addr calc and normal address calculation, since only at the end of a do statement it is known whether rec addr calc is possible. At the end of the do statements, parts of the produced program are corrected or discarded.

The following compromise solution may be preferred. Outside of for statements the compiler acts as a *one-pass* translator. When a **for** is encountered, another pass takes over in order to check the for clause and do statement for rec addr calc possibilities. Control returns then to the main pass starting at the **for** again.

The model translator decribed here is a *multi-pass* translator, as was already mentioned, so that there is a pass for checking and one for actual translating.

5.3. Use of recursive address calculation for one loop

This section is an explanation of the target language for a single for list element and a subscripted variable in the do statement of a loop in which rec addr calc is possible. From this example it will be easy for the reader to extend it to the general case.

Let

$$\textbf{for } i := a_i \textbf{ step } b_i \textbf{ until } c_i \textbf{ do } \dots A[s_1, \dots, s_n] \dots ;$$

be a for loop. Then

$$address\,(A\,[s_1, \dots, s_n]) = A_{red\ init} + \varphi_A([s_1, \dots, s_n]).$$

Let the $s_\nu\,(\nu = 1, \dots, n)$ be linear functions of i:

$$s_\nu = t_\nu + i \times u_\nu, \text{ with } t_\nu \text{ and } u_\nu \text{ independent of } i \text{ for all } \nu.$$

Then

$$address\,(A\,[s_1, \dots, s_n]) = A_{red\ init} + \varphi_A([t_1 + i \times u_1, \dots, t_n + i \times u_n]).$$

Define $\varphi_A^*(i) = \varphi_A([t_1 + i \times u_1, \dots, t_n + i \times u_n]).$

Call $\varphi_A^*(a_i) = \varphi_A([t_1 + a_i \times u_1, \dots, t_n + a_i \times u_n])$ the *"initial value"* and

$$\Delta\varphi_A^*(i, b_i) = \varphi_A^*(i + b_i) - \varphi_A^*(i) \equiv B \text{ the *"increment"*}.$$

6   Grau · Hill · Langmaack, Translation of ALGOL 60

For the general case it is necessary to generate a list of variables which can be thought of as being declared in the smallest block containing the for statement[1]. In certain limited cases they may not be needed. See 5.6.2.

The following variables are necessary:

$GV\ ADDR$ for the address of a variable to be calculated recursively
$GV\Delta i$ for the increment relative to the loop variable i
GVa_i for the initial value of the loop variable i
GVb_i for the value of the step element
GVc_i for the end value, if it is constant.

We can now form the following program scheme in the produced program for a for loop with rec addr calc with a subscripted variable $A\,[s_1, \ldots, s_n]$:

```
                    begin
                    procedure DO STATEMENT i ;
                        begin
                            ... content of GV ADDR ...
                        end ;
                    procedure INITIAL VALUE i ;
                        begin
                            GV ADDR := A_red init + φ*_A (GVa_i) ;
                            GVΔi := Δ φ*_A (i, GVb_i)
                        end ;
                    procedure INCREMENT i ;
                        begin
                            GV ADDR := GV ADDR + GVΔi ;
                            i := i + GVb_i
                        end ;
    FOR CLAUSE :     GVa_i := i := a_i ; GVb_i := b_i ; GVc_i := c_i ;
                     INITIAL VALUE i ;
                     go to TEST i ;
          Li :       INCREMENT i ;
        TEST i :     if (i - GVc_i) × sign (GVb_i) > 0 then
                        go to NEXT i ;
                     DO STATEMENT i ;
                     go to Li ;
        NEXT i :
                    end
```

[1] Or the for statement F can be replaced by the block

begin Δ ; F **end** ,

where Δ represents the declaration of the necessary variables.

In case c_i changes during execution of the do statement, the statement $GVc_i := c_i$ should be eliminated and, in the statement at $TEST\ i$, GVc_i should be replaced by c_i.

The execution of the loop proceeds as follows:

First, i, GVa_i, GVb_i, and GVc_i are given the values a_i, a_i, b_i, and c_i respectively. This is followed by initialization of the initial address value and the increment. The generated variable $GV\ ADDR$ receives the value of the initial address value plus the reduced initial address. This is actually the address of the element $A\,[s_1, \ldots, s_n]$ with $i = a_i$.

Next comes the usual test to see whether the do statement should be executed. In the do statement, the variable $A\,[s_1, \ldots, s_n]$ is replaced by **content of** $GV\ ADDR$, since $GV\ ADDR$ contains the address of $A\,[s_1, \ldots, s_n]$. After the do statement has been executed, i is increased by the step value b_i and $GV\ ADDR$ is increased by the increment — that is by the value with which the address of $A\,[s_1, \ldots, s_n]$ changes when i is changed by b_i. The test for continuing the do loop is then repeated.

It still remains to show how the expressions $\varphi_A^*(a_i)$ and $\varDelta\,\varphi_A^*(i, b_i)$ are to be calculated.

First of all, it is clear that $\varphi_A^*(i)$ and $\varphi_A^*(GVa_i)$ are the same as $\varphi_A^*(a_i)$, since $i = GVa_i = a_i$.

Of the many ways in which the expressions can be implemented, we will describe two. Method I is simpler for the compiler, and produces an object program which, in the usual cases, is almost as efficient as that obtained by method II.

If the index vector $[s_1, \ldots, s_n]$ is independent of the loop variable, then $GV\varDelta i \equiv 0$ and the calculation of $GV\varDelta i$ and incrementing of $GV\ ADDR$ can be omitted.

5.3.1. Method I: Difference method

In the procedure $INITIAL\ VALUE\ i$, the Horner scheme

$$\varphi_A([s_1, \ldots, s_n])\big|_{i=GVa_i}$$

will be explicitly written. $i = GVa_i$ means that i is replaced by GVa_i in the expressions s_ν.

The increment is determined by

$$\varphi_A([s_1, \ldots, s_n])\big|_{i=GVa_i+GVb_i} - \varphi_A([s_1, \ldots, s_n])\big|_{i=GVa_i}.\text{[1]}$$

Of course, it will be better first to calculate the value $GVa_i + GVb_i$, and assign it to a generated variable.

---

[1] This may be extended to the non linear case without any problem.

The procedure $INITIAL\ VALUE\ i$ can then be written as:

> **procedure** $INITIAL\ VALUE\ i$;
> **begin**
> $$GV0 := \underset{A}{\varphi}([s_1, \ldots, s_n])|_{i=GVa_i} ;$$
> $$GV\ ADDR := A_{red\ init} + GV0 ;$$
> $$GV1 := GVa_i + GVb_i ;$$
> $$GV\varDelta i := \underset{A}{\varphi}([s_1, \ldots, s_n])|_{i=GV1} - GV0$$
> **end**

The increment procedure is written:

> **procedure** $INCREMENT\ i$;
> **begin**
> $$GV\ ADDR := GV\ ADDR + GV\varDelta i ;$$
> $$i := i + GVb_i$$
> **end**

The Horner scheme must therefore be calculated twice. Since $s_\nu = t_\nu + i \times u_\nu$, not counting the calculation of t_ν and u_ν, $2(n+n-1) = 4n-2$ multiplications, $2(n+n-1)+2 = 4n$ additions, and 1 subtraction are necessary.

The following simple rules may be given for the initial value and increment calculations:

a) $\underset{A}{[s_1, \ldots, s_n]} = \underset{A}{[0, \ldots, 0]}$:

Then no instructions are necessary. The subscripted variable $A[0, \ldots, 0]$ in the do statement may be replaced by **content of** $A_{red\ init}$.

b) $\underset{A}{[s_1, \ldots, s_n]} = \underset{A}{[t_1, \ldots, t_n]}$ and t_ν, $\nu = 1, \ldots, n$ are independent of the loop variable:

A variable $GV\ ADDR$ is generated and the statement

$$GV\ ADDR := A_{red\ init} + \underset{A}{\varphi}([s_1, \ldots, s_n])$$

is inserted in the procedure $INITIAL\ VALUE\ i$. No corresponding incrementing is necessary. In the do statement, $A[s_1, \ldots, s_n]$ is replaced by **content of** $GV\ ADDR$.

c) $\underset{A}{[s_1, \ldots, s_n]}$ depends on the loop variable:

In this case the variables $GV\ ADDR$, $GV\varDelta i$, $GV0$, and $GV1$ are generated and the instructions are produced as explained in the above scheme.

For further optimizations see section 5.6. The precise rules under which the compiler works are given in chapter 8.

5.3.2. Method II: Decomposition method

The distance function φ is a linear function of its arguments. Therefore the following decomposition is valid:

$$\varphi\left([t_1+i\times u_1, \ldots, t_n+i\times u_n]\right) = \varphi\left([t_1, \ldots, t_n]\right) + i\times\varphi\left([u_1, \ldots, u_n]\right)$$

where t_ν, u_ν $(\nu=1, \ldots, n)$ are independent of i. This corresponds to a decomposition of the index vector:

$$[s_1, \ldots, s_n] = [t_1, \ldots, t_n] + i\times[u_1, \ldots, u_n].$$

Because of requirement (2b) in section 5.2, p. 80, the expressions t_ν and u_ν remain constant during the execution of the for statement (since they do not depend on i). Therefore the *constant part* and the *factor*

$$\varphi_A\left([t_1, \ldots, t_n]\right) \equiv C \quad \text{and} \quad \varphi_A\left([u_1, \ldots, u_n]\right) \equiv D$$

remain constant relative to the loop variable i. The following relation is true for the increment:

$$\begin{aligned}
\Delta\,\varphi_A^*(i, b_i) \equiv B &= \varphi_A^*(i+b_i) - \varphi_A^*(i) \\
&= \varphi_A\left([t_1, \ldots, t_n]\right) + (i+b_i)\times\varphi_A\left([u_1, \ldots, u_n]\right) \\
&\quad - \varphi_A\left([t_1, \ldots, t_n]\right) - i\times\varphi_A\left([u_1, \ldots, u_n]\right) \\
&= b_i\times\varphi_A\left([u_1, \ldots, u_n]\right) \\
&= b_i\times D.
\end{aligned}$$

If the s_ν are decomposed into the form $t_\nu+i\times u_\nu$, then the procedure *INITIAL VALUE i* may be written as follows:

> **procedure** *INITIAL VALUE i* ;
>   **begin**
> $GVD := \varphi_A\left([u_1, \ldots, u_n]\right)$;
> $GV\ ADDR := A_{red\ init} + \varphi_A\left([t_1, \ldots, t_n]\right) + i\times GVD;$
> $GV\Delta i := GVb_i\times GVD$
>   **end**

where GVD is another generated variable.

Procedure *INCREMENT i* is the same as in the difference method.

In the general case, not counting the operations for calculating the t_ν and u_ν, two Horner scheme calculations are necessary for initialization, with a total of $2(n-1)+2=2n$ multiplications and $2(n-1)+2=2n$ additions.

The rules are similar to those used in the difference method for rec addr calc:

a) see 5.3.1 a)

b) see 5.3.1 b)

c) $[s_1, \ldots, s_n]_A$ depends on the loop variable i:

Variables GVD, $GV\Delta i$ and $GV\ ADDR$ will be generated.

Instructions will be generated according to the above scheme, and subscripted variables in the do statement will be replaced by **content of** $GV\ ADDR$. Compare also with 5.6.

The decomposition method means in effect that one has to represent

$$\varphi\left([s_1, \ldots, s_n]\right)_A$$

as a polynomial of degree 1 in i with constant index vectors in the coefficients. This may be simply extended to the *non-linear* case as follows (see 5.1):

Represent $\varphi\left([s_1, \ldots, s_n]\right)_A$ as the polynomial

$$\sum_{\mu=0}^{k} \varphi\left(\alpha_\mu\right) i^\mu$$

with α_μ = constant index vectors. Represent $\Delta^\varkappa \varphi_A^*(i, b) = \Delta^{\varkappa-1} \varphi_A^*(i + b, b)$ $- \Delta^{\varkappa-1}\varphi_A^*(i, b)$ as the polynomial

$$\sum_{\mu=0}^{k-\varkappa} \varphi\left(\alpha_\mu^\varkappa\right) i^\mu \qquad (\varkappa = 1, \ldots, k)$$

where $\alpha_\mu^\varkappa$ are constant index vectors.

In restrictions R2 and R3 we required in section 5.2 for the model translator that the loop variable be of type **integer** and that index positions contain only integer valued entities and not the operators $/$, $\div$, and $\uparrow$. These restrictions insure that the constant part and the factor of the loop variable deliver integral values of the distance function. This is necessary, since in the ALGOL Report ([75], section 3.1.4.2) it is required that a subscript expression E is to be replaced internally by *entier* $(E + .5)$ and since the *entier*-function is not a linear operator.

Because of this, the variation of an address when changing the loop variable by the step value is in effect not equal to

$$\varphi_A^*(i+b) - \varphi_A^*(i)$$

when the loop variable, or any factor of the loop variable is not integral. The difference then does not remain constant.

Therefore, restrictions are necessary for both the difference method and for the decomposition method.

When using the decomposition method for an index vector we require the equality

entier (*constant*+LV×*factor*+.5)=
entier (*constant*+.5)+LV×*entier* (*factor*+.5)

to hold for each subscript. With our restrictions, of course,

entier (*constant*+LV×*factor*+.5)=*constant*+LV×*factor*.

The following example makes this clear. The index vector

$$[2/3+1/3\times LV, LV]_A$$

yields for $LV=3$ and the number of columns 100 as value of the distance function:

$$\varphi=entier\,(2/3+1/3\times3+.5)\times100+3=203\,.$$

The decomposed vector (not regarding the restrictions) would yield:

$$[2/3+1/3\times LV, LV]_A=[2/3, 0]_A+LV\times[1/3, 1]_A,$$

$$\varphi'=entier\,(2/3+.5)\times100+0+3\times(entier\,(1/3+.5)\times100+1)$$
$$=103\,.$$

This discrepancy, of course, would not occur if ALGOL required that index expressions be mathematically integral valued. Then it would suffice to apply the *entier*-function to the value of φ in order to correct rounding errors and it would not be necessary to apply it to the single subscript expressions.

5.4. Nested loops

The calculation of the initial address and the increment for a subscripted variable in initializing a loop is only a first step in rec addr calc. Further optimizations may be attempted, such as moving address calculation as much as possible outside nested loops, and, if possible, into the initialization of the outermost loop. In order to explain this procedure, let us consider an example.

for $k:=a_k$ **step** b_k **until** c_k **do**
    begin ...
            for $i:=a_i$ **step** b_i **until** c_i **do** ... $A\,[s_1, ..., s_n]$ ... ;
            ... $B\,[r_1, ..., r_m]$ ...
    end

Since the rec addr calc is to be pushed from the "inside" to the "outside", we investigate the loops in the same manner. The inside loop is therefore first translated according to the rules given in section 5.3, whereby the program is formed into the following:

$$\textbf{for } k := a_k \textbf{ step } b_k \textbf{ until } c_k \textbf{ do}$$

        begin ...

            begin

                procedure $DO\ STATEMENT\ i$;

                    begin

                        ... **content of** $GV\ ADDR$ ...

                    end ;

                procedure $INITIAL\ VALUE\ i$;

                    begin

$$GV\ ADDR := A_{red\ init} + \varphi_A^*(GVa_i)\ ;$$
$$GV\Delta i := \Delta\,\varphi_A^*(i, GVb_i)$$

                    end ;

                procedure $INCREMENT\ i$;

                    begin

$$GV\ ADDR := GV\ ADDR + GV\Delta i\ ;$$
$$i := i + GVb_i$$

                    end ;

$FOR\ CLAUSE\ i$:

            ...

            end i ;

            ... $B[r_1, \ldots, r_m]$ ...

        end k

Now the outer loop (with loop variable k) is handled according to the rules of section 5.3 in the same way. The variable $B[r_1, \ldots, r_m]$ is also translated likewise. If the do statement with loop variable i contains a subscripted variable whose address cannot be calculated recursively with respect to i, then the subscripted variable is also not recursively addressable with respect to k; for (1) if the s_ν contain variables which are not admissible for the i-loop then they are not admissible with respect to the k-loop, (2) an array identifier which is local to the i-loop is local to the k-loop, and (3) the loop variable i is not admissible with respect to the k-loop.

For additional optimization, the procedure $INITIAL\ VALUE\ i$ may be investigated. The expressions $\varphi_A^*(GVa_i)$ and $\Delta\,\varphi_A^*(i, GVb_i)$, which may also be represented by index vectors

$$[s_1, \ldots, s_n]\big|_{i=GVa_i} \quad \text{and} \quad [s_1, \ldots, s_n]\big|_{i=GVa_i+GVb_i} - [s_1, \ldots, s_n]\big|_{i=GVa_i}$$
$$_A \qquad\qquad\qquad\qquad\quad {}_A \qquad\qquad\qquad\qquad\qquad {}_A$$

respectively, are investigated for recursive address calculation with respect to the k-loop in the same way as the subscripted variable $B[r_1, \ldots, r_m]$. What is actually done depends on whether $INITIAL\ VALUE\ i$ is produced by method I or method II.

a) In the difference method, the index vectors

$$[s_1, \ldots, s_n]\big|_{i=GVa_i}^{A} \quad \text{and} \quad [s_1, \ldots, s_n]\big|_{i=GV1}^{A} \quad \text{with} \quad GV1 = GVa_i + GVb_i$$

must be tested to see whether they may be calculated recursively. It must be ascertained whether or not the values of a_i and b_i, as well as the values of the other variables appearing in the s_ν remain constant with respect to the k-loop, and whether or not $\varphi_A^*(GVa_i)$ is linear in k. In case they are, $\varphi_A^*(GVa_i)$ and $\Delta\varphi_A^*(i, GVb_i)$ may be replaced in the procedure $INITIAL\ VALUE\ i$ by two new generated variables, whose values will be calculated before the execution of the k-loop in the procedures $INITIAL\ VALUE\ k$ and $INCREMENT\ k$. For a restriction with loops containing more than one for list element, see section 5.5.

b) In the decomposition method, the index vectors $[t_1, \ldots, t_n]$ and A
$[u_1, \ldots, u_n]$ must be checked. Thus the vectors produced by the de-A
composition are again decomposed if possible. If all variables appearing in the $t_\nu(u_\nu)$, with the exception of k itself, are constant relative to the k-loop and if the $t_\nu(u_\nu)$ are linear in k, then the vector $[t_1, \ldots, t_n]$ A
$([u_1, \ldots, u_n])$ can be replaced in $INITIAL\ VALUE\ i$ by a generated A
variable which will receive its value in $INITIAL\ VALUE\ k$ and $INCREMENT\ k$.

Here, the advantage of the decomposition method is obvious. The subscript expressions appearing in the vectors produced by decomposition become simpler as they are pushed from inner to outer loops. This does not happen in the difference method. Moreover, fewer assumptions are necessary. For example, in case a) t_ν is not constant or linear with respect to the k-loop; then no further optimization is possible with the difference method, while with the decomposition method it may still be possible to optimize the vector $[u_1, \ldots, u_n]$.[1]
A

The procedures $INITIAL\ VALUE\ k$ and $INCREMENT\ k$ are generated in the same manner as the corresponding i-loop procedures. Rules for the optimization of the index vectors are similar to those for subscripted variables (section 5.3.1 and 5.3.2):

a) $[s_1, \ldots, s_n] = [0, \ldots, 0]$:
$\quad A \qquad\qquad A$
The value $\varphi([0, \ldots, 0])$ is replaced by 0. No variable is generated.
$\qquad\qquad A$

b) $[s_1, \ldots, s_n]$ is independent of the loop variable:
$\quad A$

---

[1] In practice, the constant part vector $[t_1, \ldots, t_n]$ is still coupled to the reduced
A
initial address $A_{red\ init}$ and this address will be moved together with the index vector in the admissible case.

A variable $GV0$ is generated and the instruction $GV0 := \varphi([s_1, \ldots, s_n])$
is inserted in $INITIAL\ VALUE\ k$. In the do statement, $\underset{A}{\varphi}([s_1, \ldots, s_n])$
is replaced by $GV0$.

c) general case: see the tables in 5.4.1 and 5.4.2.

For further optimization see section 5.6 and chapter 8.

If an index vector is pushed to an outer loop, the necessary generated variables must be declared in the block in which this outer loop lies.

The following sections treat an example using both the difference and the decomposition methods.

5.4.1. Method I: Difference method

On page 91 the possible optimization steps are given for the example:

> **for** $k := a_k$ **step** b_k **until** c_k **do**
>   **for** $i := a_i$ **step** b_i **until** c_i **do** $\ldots\ A[s_1, \ldots, s_n]\ \ldots$;

We give here once more the necessary assumptions.

step 1: a) A is declared outside the i-loop.

b) no assignment occurs to i and to the variables in b_i and in $s_1, \ldots, s_n$ (check procedure calls!).

c) $s_1, \ldots, s_n$ are linear in i.

d) i is a simple variable (not called by name).

step 2: a) A is declared outside the k-loop.

b) no assignment occurs to k, to the variables in a_i, b_i, b_k, and to the variables different from i in $s_1, \ldots, s_n$.

c) $s_1, \ldots, s_n$ are linear in k.

d) k is a simple variable (not called by name).

e) The for clause of the loop i has only one list element (see 5.5).

$GV\ ADDR$, $GV\Delta i$, $GV0, \ldots, GV5$ are generated variables, whose declarations are introduced in the smallest block which contains the k-loop.

5.4.2. Method II: Decomposition method

On page 92 the possible optimization steps for the example

> **for** $k := a_k$ **step** b_k **until** c_k **do**
>   **for** $i := a_i$ **step** b_i **until** c_i **do** $\ldots\ A[s_1, \ldots, s_n]\ \ldots$;

have been diagrammed. The necessary assumption for the steps are:

step 1: see 5.4.1.

step 2: a) A is declared outside the k-loop.

Optimization scheme (difference method)

ALGOL program	step 1	step 2

ALGOL program

FOR CLAUSE k
  ...
DO STATEMENT k
  FOR CLAUSE i
    ...
  DO STATEMENT i
    ... $A[s_1, ..., s_n]$ ...

step 1

FOR CLAUSE k
  ...
DO STATEMENT k
  DO STATEMENT i
    ... **content of** $GV\,ADDR$ ...
  INITIAL VALUE i
    $GV0 := \varphi([s_1, ..., s_n])|_A\, i = GVa_k$;
    $GV1 := GVa_i + GVb_i$;
    $GV\,ADDR := A_{red\,init} + GV0$;
    $GV\Delta i := \varphi([s_1, ..., s_n])|_A\, i = GV1 - GV0$;
  INCREMENT i
    $GV\,ADDR := GV\,ADDR + GV\Delta i$;
    $i := i + GVb_i$;
  FOR CLAUSE i
    $GVa_i := i := a_i$; ...

step 2

DO STATEMENT k
  DO STATEMENT i
    ... **content of** $GV\,ADDR$ ...
  INITIAL VALUE i
    $GV\,ADDR := A_{red\,init} + GV0$;
    $GV\Delta i := GV4 - GV0$;
  INCREMENT i
    $GV\,ADDR := GV\,ADDR + GV\Delta i$;
    $i := i + GVb_i$;
  FOR CLAUSE i
    $i := GVa_i$; ...
INITIAL VALUE k
  $GV0 := \varphi([s_1, ..., s_n])|_A\, i = GVa_k, k = GVa_k$;
  $GV2 := GVa_k + GVb_k$;
  $GV3 := \varphi([s_1, ..., s_n])|_A\, i = GVa_k, k = GV2 - GV0$;
  $GV1 := GVa_i + GVb_i$;
  $GV4 := \varphi([s_1, ..., s_n])|_A\, i = GV1, k = GVa_k$;
  $GV5 := \varphi([s_1, ..., s_n])|_A\, i = GV1, k = GV2 - GV4$;
INCREMENT k
  $GV0 := GV0 + GV3$;
  $GV4 := GV4 + GV5$;
  $k := k + GVb_k$;
FOR CLAUSE k
  $GVa_i := a_i$; $GVa_k := k := a_k$; ...

Optimization scheme (decomposition method)

ALGOL program	step 1	step 2
FOR CLAUSE k $\quad$ ... *DO STATEMENT* k *FOR CLAUSE* i $\quad$ ... *DO STATEMENT* i $\quad$ ... $A[s_1, \ldots, s_n]$ ...	*FOR CLAUSE* k $\quad$ ... *DO STATEMENT* k *DO STATEMENT* i $\quad$ ... **content of** $GV\ ADDR$ ... *INITIAL VALUE* i $\quad GV1 := \varphi_A([u_1, \ldots, u_n])\ ;$ $\quad GV\ ADDR := A_{redinit} + \varphi_A([t_1, \ldots, t_n])$ $\qquad\qquad\qquad\qquad + i\times GV1\ ;$ $\quad GV\varDelta i := GVb_i \times GV1\ ;$ *INCREMENT* i $\quad GV\ ADDR := GV\ ADDR + GV\varDelta i\ ;$ $\quad i := i + GVb_i\ ;$ *FOR CLAUSE* i $\quad i := a_i\ ;\ \ldots$	*DO STATEMENT* k $\quad$ *DO STATEMENT* i $\qquad$ ... **content of** $GV\ ADDR$ ... $\quad$ *INITIAL VALUE* i $\qquad GV\ ADDR := GV5 + i\times GV1\ ;$ $\qquad GV\varDelta i := GVb_i \times GV1\ ;$ $\quad$ *INCREMENT* i $\qquad GV\ ADDR := GV\ ADDR + GV\varDelta i\ ;$ $\qquad i := i + GVb_i\ ;$ $\quad$ *FOR CLAUSE* i $\qquad i := a_i\ ;\ \ldots$ *INITIAL VALUE* k $\quad GV2 := \varphi_A([u_1'', \ldots, u_n''])\ ;$ $\quad GV1 := \varphi_A([t_1', \ldots, t_n']) + k\times GV2\ ;$ $\quad GV3 := GVb_k \times GV2\ ;$ $\quad GV4 := \varphi_A([u_1', \ldots, u_n'])\ ;$ $\quad GV5 := A_{redinit} + \varphi_A([t_1, \ldots, t_n]) + k\times GV4\ ;$ $\quad GV6 := GVb_k \times GV4\ ;$ *INCREMENT* k $\quad GV1 := GV1 + GV3\ ;$ $\quad GV5 := GV5 + GV6\ ;$ $\quad k := k + GVb_k\ ;$ *FOR CLAUSE* k $\quad k := a_k\ ;\ \ldots$

b) no assignment is made within the k-loop to k and the variables occurring in b_k, t_ν', t_ν'', u_ν', u_ν'', $\nu = 1, \ldots, n$.

c) the following decompositions are possible:

$$\underset{A}{\varphi}([t_1, \ldots, t_n]) = \underset{A}{\varphi}([t_1', \ldots, t_n']) + k \times \underset{A}{\varphi}([u_1', \ldots, u_n'])$$

$$\underset{A}{\varphi}([u_1, \ldots, u_n]) = \underset{A}{\varphi}([t_1'', \ldots, t_n'']) + k \times \underset{A}{\varphi}([u_1'', \ldots, u_n''])$$

where t_ν', t_ν'', u_ν', u_ν'', $\nu = 1, \ldots, n$ are independent of k.

d) k is a simple variable (not called by name).

5.4.3. An example

The target language program for the example

> **begin array** $A[1:n, 1:m]$; ...
>   **for** $k := 1$ **step** 1 **until** n **do**
>   **for** $i := 1$ **step** 1 **until** m **do** ... $A[k, i]$ ... ; ...
> **end**

using the optimization schemes given in 5.4.1 and 5.4.2 will now be given.

5.4.3.1. Difference method

The following expressions are calculated:

$$\underset{A}{\varphi}([s_1, \ldots, s_n]) = \underset{A}{\varphi}([k, i])$$

$$\underset{A}{\varphi}([s_1, \ldots, s_n])|_{i = GVa_i, \, k = GVa_k} = 1 \times m + 1$$

$$\underset{A}{\varphi}([s_1, \ldots, s_n])|_{i = GVa_i, \, k = GVa_k + GVb_k = GV2} = GV2 \times m + 1, \, GV2 = GVa_k + GVb_k$$

$$\underset{A}{\varphi}([s_1, \ldots, s_n])|_{i = GVa_i + GVb_i = GV1, \, k = GVa_k} = 1 \times m + GV1, \, GV1 = GVa_i + GVb_i$$

$$\underset{A}{\varphi}([s_1, \ldots, s_n])|_{i = GVa_i + GVb_i, \, k = GVa_k + GVb_k} = GV2 \times m + GV1.$$

The following program is produced:

> **begin** ...
>   **procedure** $DO\ STATEMENT\ k$;
>   **begin**
>   **procedure** $DO\ STATEMENT\ i$;
>   **begin**
>   ... **content of** $GV\ ADDR$ ...
>   **end** ;
>   **procedure** $INITIAL\ VALUE\ i$;
>   **begin**
> $GV\ ADDR := A_{red\ init} + GV0$;
> $GV\Delta i := GV4 - GV0$
>   **end** ;

```
            procedure INCREMENT i ;
              begin
                GV ADDR := GV ADDR+GVΔi ;
                i := i+1
              end ;
FOR CLAUSE i:
                i := 1 ;
                INITIAL VALUE i ;
                go to TEST i ;
RETURN i:       INCREMENT i ;
TEST i:         if i−m>0 then go to NEXT i ;
                DO STATEMENT i ;
                go to RETURN i ;
NEXT i:
              end do statement k ;
            procedure INITIAL VALUE k ;
              begin
                GV0 := m+1;   GV2 := 1+1 ;
                GV3 := GV2×m+1−GV0 ;   GV1 := 1+1 ;
                GV4 := m+GV1 ;
                GV5 := GV2×m+GV1−GV4
              end ;
            procedure INCREMENT k ;
              begin
                GV0 := GV0+GV3 ;
                GV4 := GV4+GV5 ;
                k := k+1
              end ;
FOR CLAUSE k:
                k := 1 ;
                INITIAL VALUE k ;
                go to TEST k ;
RETURN k:
                INCREMENT k ;
TEST k:
                if k−n>0 then go to NEXT k ;
                DO STATEMENT k ;
                go to RETURN k ;
NEXT k:    ...
          end
```

5.4.3.2. Decomposition method

The following expressions are calculated:

$$\varphi_A([u_1, \ldots, u_n]) = \varphi_A([0, 1]) = 1$$
$$\varphi_A([t_1, \ldots, t_n]) = \varphi_A([k, 0]) = k \times m$$
$$\varphi_A([u_1'', \ldots, u_n'']) = \varphi_A([0, 0]) = 0$$
$$\varphi_A([t_1'', \ldots, t_n'']) = \varphi_A([0, 1]) = 1$$
$$\varphi_A([u_1', \ldots, u_n']) = \varphi_A([1, 0]) = 1 \times m$$
$$\varphi_A([t_1', \ldots, t_n']) = \varphi_A([0, 0]) = 0.$$

The program is

```
        begin ...
          procedure DO STATEMENT k ;
            begin
              procedure DO STATEMENT i ;
                begin
                    ... content of GV ADDR ...
                end ;
              procedure INITIAL VALUE i ;
                GV ADDR := GV5 + i × GV1 ;
              procedure INCREMENT i ;
                begin
                  GV ADDR := GV ADDR + GV1 ;
                  i := i + 1
                end ;
FOR CLAUSE i :
                i := 1 ;
                INITIAL VALUE i ;
                go to TEST i ;
RETURN i :     INCREMENT i ;
TEST i :       if i − m > 0 then go to NEXT i ;
                DO STATEMENT i ;
                go to RETURN i ;
NEXT i :
            end do statement k ;
          procedure INITIAL VALUE k ;
            begin
              GV1 := 1 ;  GV4 := m ;
              GV5 := A_{red init} + k × GV4
            end ;
```

procedure *INCREMENT k* ;
  begin
   $GV5 := GV5 + GV4$;
   $k := k + 1$;
  end ;
FOR CLAUSE k :
  $k := 1$;
  INITIAL VALUE k ;
  go to *TEST k* ;
RETURN k :
  INCREMENT k ;
TEST k : **if** $k - n > 0$ **then go to** *NEXT k* ;
  DO STATEMENT k ;
  go to *RETURN k* ;
NEXT k : ...
  end

5.5. Loops with more than one list element

To complete the discussion of for loops, a scheme for handling loops with more than one list element may be described by means of the following example, which covers the different cases.

for $i := a_i$ **step** b_i **until** c_i, d_i, e_i **while** f_i
   do ... $A [s_1, \ldots, s_n]$ ... ;

is replaced by the following equivalent program:

begin
 procedure *DO STATEMENT i* ;
  begin
   ... **content of** *GV ADDR* ...
  end ;
 procedure *INITIAL VALUE i* ;
  begin
   $GV\ ADDR := A_{red\ init} + \varphi_A^* (GVa_i)$;
   $GV\Delta i := \Delta \varphi_A^* (i, b_i)$
  end ;
 procedure *INCREMENT i* ;
  begin
   $GV\ ADDR := GV\ ADDR + GV\Delta i$;
   $i := i + GVb_i$
  end ;

LE 1: $GVa_i := i := a_i$; $GVb_i := b_i$; $GVc_i := c_i$; [1]
    $INITIAL\ VALUE\ i$;
    go to $TEST\ i$;
LE 1 i: $INCREMENT\ i$;
TEST i: **if** $(i - GVc_i) \times sign\,(GVb_i) > 0$ **then go to** $LE\ 2$;
    $DO\ STATEMENT\ i$;
    go to $LE\ 1\,i$;
LE 2: $GVa_i := i := d_i$;
    $INITIAL\ VALUE\ i$;
    $DO\ STATEMENT\ i$;
LE 3: $GVa_i := i := e_i$;
    $INITIAL\ VALUE\ i$;
LE 3 i: **if** $\neg\ f_i$ **then go to** $NEXT\ i$;
    $DO\ STATEMENT\ i$;
    go to $LE\ 3\,i$;
NEXT i:
    end

In this program, unnecessary calculation of the increment is performed in procedure $INITIAL\ VALUE\ i$ for simple list elements and
while elements. Also, if the first list element is not a step-until element, a
value must be given to GVb_i (perhaps with the statement $GVb_i := 0$)
before the first call of $INITIAL\ VALUE\ i$. Of course $INITIAL$
$VALUE\ i$ may be split into two procedures, whereby the latter of
which, which serves to calculate the increment, will only be called by a
step-until element.

f_i cannot be calculated only once before the loop, since it must be changed
during the execution of the do statement in order to indicate the end of the loop.
The assumption is, of course, that e_i does not change during the execution of the
loop.

For the sake of simplicity a loop is considered to be not suitable for
recursive address calculation, if a single step element or value of a while
element is changed during the execution of the loop, although this is
not strictly necessary.

The optimization in case of nested for loops cannot be made for the
difference method when the inner loop has more than one list element.
A look at the $INITIAL\ VALUE\ k$ procedure, step 2 of the optimization
scheme given in section 5.4.1 will explain why. The initial value of the
inner loop variable i is already used in the initial value part of the outer
loop and therefore must not change in the whole outer loop: the statement

$$GV0 := \underset{A}{\varphi}([s_1, \ldots, s_n])\big|_{i=GVa_i,\ k=GVa_k}$$

---

[1] Here c_i is assumed to remain constant during the execution of the loop.
Compare with 5.3, p. 83.

assumes that the initial value GVa_i of the loop variable i is always the same, which may not be the case when the i-loop has more than one list element. This illustrates a further advantage of the decomposition method, since for this a similar restriction is not necessary.

5.6. Further optimization possibilities

The *identification* of the same subscripted variable used more than once or of identical Horner scheme expressions, the saving of generated variables, and the use of *index registers* are further optimization possibilities.

5.6.1. Identification of subscripted variables

The addresses of identical subscripted variables can be replaced by the same address variable. If the information vectors of two subscripted variables with identical subscripts differ only in the reduced initial address, then the Horner scheme $\varphi_A^*(a_i)$ and the increment are the same for both variables and need only be calculated once; the same auxiliary variables will be used for both variables. Moreover, if method II is used, the index vectors produced by decomposition can be checked for identity in order to save more operations in the target program. Subscripted variables, which appear in the c_i of a step until element or in the Boolean expression following a while, can be calculated recursively exactly as other subscripted variables in the do statement.

5.6.2. Generated variables

In a for loop with one list element, the variable $GVa_i(GVb_i)$ may be omitted if $a_i(b_i)$ is either a simple variable or a constant, or if $GVa_i(GVb_i)$ appears only once in the generated procedures. $GVa_i(GVb_i)$ will then be replaced by $a_i(b_i)$ in the procedures. The same applies to GVc_i, which in any case is generated only if c_i is constant. Similar remarks apply to all generated variables. In case an index vector appears only once and can not be calculated recursively, a generated variable is not needed for it.

5.6.3. Use of index registers

If the computer has index registers[1], these can well be used for rec addr calc, if the loop variable, initial, step, and end values are integers. A remark rather often made is that index registers should always be used for loop variables. But this actually adds relatively little to efficiency. The program is more efficient when address variables and the generated variables for the Horner scheme are replaced or modified by index registers.

---

[1] It must be observed that when index registers are used, the range of integers which are allowed as loop variable or subscript values may be restricted.

a) An address variable $GV\ ADDR$ can be implemented by an index register. Wherever $GV\ ADDR$ appears in the do statement, in $INITIAL\ VALUE$, and in $INCREMENT$, it is to be replaced by $\Im\mathfrak{r}\,[\varkappa_{GV\ ADDR}]$. Similarly, the storing of and addition to $GV\ ADDR$ are replaced by index register loading and incrementing.

b) Suppose the increment $GV\varDelta$ serves for the recursive calculation of the address variables $GV\ ADDR1, \ldots, GV\ ADDRn$. Then the initial addresses for $GV\ ADDR1, \ldots, GV\ ADDRn$ calculated in procedure $INITIAL\ VALUE$ may be modified by an index register by concatenating them with the index register. The instructions in $INITIAL\ VALUE$ are changed to

$$GV\ ADDR1 := \left\{ A1_{red\ init} + \varphi^*_{A1}(GVa) \right\} \oplus \Im\mathfrak{r}\,[\varkappa]$$

$$\ldots$$

$$GV\ ADDRn := \left\{ An_{red\ init} + \varphi^*_{An}(GVa) \right\} \oplus \Im\mathfrak{r}\,[\varkappa]$$

This means that when **content of** $GV\ ADDR\ \nu$ is executed in the do statement procedure, the contents of $\Im\mathfrak{r}\,[\varkappa]$ will be added to the address in the variable $GV\ ADDR\ \nu$ in order to get the correct array element address. The index register $\Im\mathfrak{r}\,[\varkappa]$ should be set to 0 in the beginning by a suitable instruction in procedure $INITIAL\ VALUE$.

Procedure $INCREMENT$ can now be simplified by replacing the n instructions

$$GV\ ADDR\ \nu := GV\ ADDR\ \nu + GV\varDelta\ \nu = 1, \ldots, n$$

by the single instruction

$$\Im\mathfrak{r}\,[\varkappa] := \Im\mathfrak{r}\,[\varkappa] + GV\varDelta .$$

Suppose now that the values $\varphi^*_{A1}(GVa), \ldots, \varphi^*_{An}(GVa)$ are also identical. This is so if the subscripted variables have identical subscripts and the arrays the same bound pair list. Procedure $INITIAL\ VALUE$ can be simplified by changing the sequence of instructions to

$$\Im\mathfrak{r}\,[\varkappa] := \varphi^*_{A1}(GVa) \ ;$$

$$GV\ ADDR1 := A1_{red\ init} \oplus \Im\mathfrak{r}\,[\varkappa]$$

$$\ldots$$

$$GV\ ADDRn := An_{red\ init} \oplus \Im\mathfrak{r}\,[\varkappa] .$$

c) In case an index register $\Im\mathfrak{r}\,[\varkappa_{LV}]$ is used for the loop variable LV, the locations reserved for LV and $\Im\mathfrak{r}\,[\varkappa_{LV}]$ must also be initialized and incremented each time through the loop. The only advantages obtained by this are in some cases simpler and faster instructions for index register incrementing and testing.

7*

d) Of practical interest is the case of one or more subscripted variables with the same increment as LV. Because of the row-wise storing of arrays this is the case if the loop variable LV appears only in the last index position with a factor 1:

$$A[t_1, \ldots, t_{n-1}, t_n + LV], t_1, \ldots, t_n \text{ independent of } LV.$$

For such a variable

$$\varphi_A^*(LV) = \varphi_A^*(0) + LV = \underset{A}{\varphi}([t_1, \ldots, t_n]) + LV.$$

Therefore, if the index register $\Im\mathfrak{r}[\varkappa_{LV}]$ contains the value of the loop variable, the address variable corresponding to $A[t_1, \ldots, t_{n-1}, t_n + LV]$ can be modified by the index register. The instructions

$$GV\ ADDR := \{A_{red\ init} + \varphi_A^*(0)\} \oplus \Im\mathfrak{r}[\varkappa_{LV}]$$
$$LV := \Im\mathfrak{r}[\varkappa_{LV}] := GVa_{LV}$$

are inserted into the procedure $INITIAL\ VALUE$.

It suffices then to increment the loop variable and the address variable in the single instruction

$$LV := \Im\mathfrak{r}[\varkappa_{LV}] := \Im\mathfrak{r}[\varkappa_{LV}] + GV\Delta LV.$$

e) Index registers can naturally be used for other generated variables as well as for address variables.

In practice the index registers are assigned according to the static loop structure such that "parallel" loops use the same index registers, but without regarding the procedure structure. Thus, one must make provision for the case that within a loop using an index register a function (or name variable) is called which might use the same index registers again. Some of the several possibilities are (1) to save index registers in function bodies by pushing them into the number cellar. Since it seems to be rather laborious to find out exactly which index registers require such saving, one might be forced to push down all index registers, which action should be fairly time-consuming at run time. (2) A quite restrictive method would be to declare all loops with name and function calls in the do statement to be "unsuitable". (3) We have decided for the model version to mark all those loops to be "unsuitable for index register use", but suitable for recursive address calculation.

5.6.4. Program organization

In the case of a for statement with only one list element, it is not necessary to implement the do statement, $INITIAL\ VALUE$, and $INCREMENT$ as procedures; the bodies of these procedures can replace what would otherwise be the sole call of the procedure.

5.6.5. The example of 5.4.3

To complete the discussion, the example given in section 5.4.3 will be used once more to indicate the use of index registers and other simplifications.

5.6.5.1. Difference method

begin ...
FOR CLAUSE k:

$\qquad k := \Im\mathfrak{r}[\varkappa_k] := 1$;

INITIAL VALUE k:

$\qquad GV0 := m+1$; $GV2 := 1+1$;

$\qquad GV3 := GV2 \times m + 1 - GV0$;

$\qquad$ **go to** *TEST* k ;

INCREMENT k:

$\qquad GV0 := GV0 + GV3$;

$\qquad k := \Im\mathfrak{r}[\varkappa_k] := \Im\mathfrak{r}[\varkappa_k] + 1$;

TEST k: $\quad$ **if** $\Im\mathfrak{r}[\varkappa_k] > n$ **then go to** *NEXT* k ;

DO STATEMENT k:

FOR CLAUSE i:

$\qquad i := \Im\mathfrak{r}[\varkappa_i] := 1$;

INITIAL VALUE i:

$\qquad GV\ ADDR := \{A_{red\ init} + GV0 - 1\} \oplus \Im\mathfrak{r}[\varkappa_i]$;

$\qquad$ **go to** *TEST* i ;

INCREMENT i:

$\qquad i := \Im\mathfrak{r}[\varkappa_i] := \Im\mathfrak{r}[\varkappa_i] + 1$;

TEST i: $\quad$ **if** $\Im\mathfrak{r}[\varkappa_i] > m$ **then go to** *NEXT* i ;

DO STATEMENT i:

$\qquad\qquad$ ... **content of** *GV ADDR* ... ;

$\qquad$ **go to** *INCREMENT* i ;

NEXT i: $\quad$ **go to** *INCREMENT* k ;

NEXT k: $\quad$ ...

$\qquad$ **end**

Remark: $GV0$ could be replaced by an index register $\Im\mathfrak{r}[\varkappa_{GV0}]$. When calculating *GV ADDR* and incrementing $GV0$ the value must then be taken from this index register. We assume that the translator can determine that the last index position of $A[k, i]$ is just the inner loop variable i, with the result that it is unnecessary to calculate $GV\Delta i$ since its value is then 1.

5.6.5.2. Decomposition method

begin ...
FOR CLAUSE k:

$\qquad k := 1$;

INITIAL VALUE k :
$$GV\ ADDR := \{A_{red\ init} + m + 1\} \oplus \Im\mathfrak{r}[\varkappa_i]\ ;$$
go to *TEST k ;*
INCREMENT k :
$$GV\ ADDR := GV\ ADDR + m\ ;$$
$$k := k + 1\ ;$$
TEST k :   **if** $k > n$ **then go to** *NEXT k ;*
DO STATEMENT k :
FOR CLAUSE i :
$$i := 1\ ;$$
INITIAL VALUE i :
$$\Im\mathfrak{r}[\varkappa_i] := 0\ ;$$
go to *TEST i ;*
INCREMENT i :
$$i := i + 1\ ;\quad \Im\mathfrak{r}[\varkappa_i] := \Im\mathfrak{r}[\varkappa_i] + 1\ ;$$
TEST i :   **if** $i > m$ **then go to** *NEXT i ;*
DO STATEMENT i :
... **content of** *GV ADDR* ... ;
go to *INCREMENT i ;*
NEXT i :   **go to** *INCREMENT k ;*
NEXT k :   ...
end

Remark: In this example we have avoided using index registers for the loop variables in order to show exactly what would be produced by the model translator pass for recursive address calculation.

6. Run time organization

Having discussed the structure of the target language programs and the data storage allocation, we turn to the manner in which the instructions produced by the translator act in order to execute the program.

The machine storage used for the target program at run time contains (1) the target program, (2) the identifier list, and (3) the data storage together with the index registers and exceptional variables. Moreover, the organizational system including standard functions must be available. The target program is composed of instructions from the elementary language (section 3.2, p. 19). This language consists on the one hand of instructions which are in reality basic ALGOL statements, whose actions are quite clear [instructions 3.2a)—d) except the **jump to** instruction]. On the other hand, we use instructions which have no direct equivalent in ALGOL [3.2e)—f) and the **jump to** instruction]. Their actions are more complex since they are concerned with declarations, procedure and subroutine calls. These instructions may be formulated as subroutines in ALGOL-like form; a programmer can code them in a machine language.

The simple as well as the more complicated elementary language instructions work with internal identifiers as operands. All essential information concerning these internal identifiers is contained in the identifier list (see section 3.1).

As a matter of fact the simple instructions make use only of the data storage address and the static procedure level. Thus, a possible optimization at run time is to replace each internal identifier by this information only once, before execution. The same is true for each local jump instruction (which does not lead out of an array block or procedure); the internal identifier may be replaced by the corresponding program address. For most machines these basic instructions may each be represented by a single machine order containing e.g. an address and an index register.

Furthermore it is possible to extract constants and strings from the identifier list and to condense them into a constant list which becomes the initial part of the data storage.

In the following sections we shall describe explicitly the macro instructions which deal with dynamic storage allocation, the call of procedures and formal parameters, and certain jump instructions. The subroutine orders 3.2e) require no further discussion as most machines contain equivalent machine instructions. The description of the other macro instructions can conveniently be given in the form of ALGOL-like subprograms allowing immediate translation into a machine language subroutine. This is not attempted here, since it is not essential to our discussion and it is for the most part a well-known procedure.

6.1. The instruction storage allocation

As described in 4.2.2.2, p. 33 the target equivalent of the array declaration

$$\textbf{array } A_1, \ldots, A_m [a_1 : b_1, \ldots, a_n : b_n] \ ;$$

is

$$\textbf{storage allocation } A_1 \ ;$$
$$a_1 \ ;$$
$$b_1 \ ;$$
$$\vdots$$
$$a_n \ ;$$
$$b_n$$

Explicitly the actions of the macro instruction **storage allocation** A_1 are:

1. The reduced initial address of A_1 is calculated by evaluating the expression

$$\textbf{BFS} - \underset{A_1}{\varphi([a_1, \ldots, a_n])}.$$

It is coupled with the type of A_1 and is assigned to the proper component $A_{1\,red\,init}$ of the information vector for A_1.

2. The dimensions

$$K_i = b_i - a_i + 1 \qquad (i = 2, \ldots, n)$$

are evaluated and assigned to the proper components of the information vector for A_1.

3. The total length

$$\prod_{i=1}^{n} K_i$$

is evaluated and assigned to the proper component of the information vector for A_1. This will be used only if the array is called by value.

4. Analoguously the expression

$$\underset{A_1}{\varphi([a_1, \ldots, a_n])}$$

is treated. This will be used only if the array is called by value.

5. The information vectors for $A_j (j = 2, \ldots, m)$ are copied from the information vector for A_1 with the difference mentioned in section 4.2.2. The reduced initial address of A_j is the value of the expression

$$A_{1\,red\,init} + (j - 1) \times \prod_{i=1}^{n} K_i$$

and is assigned to the proper component $A_{j\,red\,init}$ of the information vector for A_j.

6. The address

$$\textbf{BFS} + m \times \prod_{i=1}^{n} K_i$$

of the new first free cell is evaluated and assigned to both the exceptional variable **BFS** and the generated address variable BFS_l, l being the level of the array block AB.

6.2. The instruction procedure call

Before the organization of procedure calls and exits is described explicitly, some concepts are explained. The *static predecessor* of a procedure P is understood to be the innermost procedure whose body contains the declaration of P. The *static procedure level* is defined by induction for declared procedures:

1. The main program MP, which is also considered as a procedure, has the level 0 (zero).

2. If P is a procedure on the level s and if it is the static predecessor of a procedure P' then P' has the level $s+1$.

The static level of a generated procedure representing a do statement and of a name procedure is undefined since with respect to storage allocation name procedures are handled like blocks (see footnote 1 in section 4.9.1, p. 63). The static level of a procedure can be identified with the number of the associated index register. This identification is due to the following consideration: If s is a static procedure level, then at each stage of program execution non-formal variables of at most one procedure on level s are accessible. If S is the maximum static procedure level, $S+1$ index registers

$$\mathfrak{Ir}[0], \ldots, \mathfrak{Ir}[S]$$

are needed.

The *dynamic call level* of a called procedure is given by the memory address of the beginning of the associated fixed storage. Now, the procedure linkage of the ν-th procedure P called and not yet completed has five entries:

1. the return address,

2. the dynamic level of the $(\nu-1)$-th procedure called and not yet completed,

3. the static level of the called procedure P,

4. a dynamic call level of the static predecessor of the called procedure P[1],

5. a type marker indicating the type required by the call.

The determination of the fourth entry will be given in this section, p. 107, and section 6.3, p. 111. With the help of this fourth entry in each procedure linkage the execution of formal procedure calls, of name calls, and formal procedure exits becomes simple.

In addition, a new exceptional variable MDL (*momentary dynamic level*) is introduced containing as its value the dynamic call level of the momentarily last procedure called and not yet completed.

The actions of the macro instruction

$$\textbf{procedure call } P \; ;$$
$$AP_1 \; ;$$
$$\vdots$$
$$AP_r$$

(see section 4.10.1.2, p. 72) are as follows:

---

[1] In case of recursiveness there may be several dynamic call levels for the same procedure declaration.

1. First, proper information on each actual parameter AP_ϱ ($\varrho=1$, $\ldots, r$) is obtained and assigned to the parameter block storage location of the associated formal parameter. Its memory address is determined unambiguously by the sum

$$\mathbf{BFS}+5+LP+\varrho.$$

Here LP is the maximum array block level in the procedure P. The proper information on the actual parameter AP_ϱ consists either of one of the chains

$$AP \oplus DL, \quad \text{or} \quad AP \oplus DL \oplus \text{'\textbf{RI}'}, \quad \text{or} \quad AP \oplus DL \oplus DL1, \quad \text{or}$$
$$AP \oplus DL \oplus DL1 \oplus \text{'\textbf{RI}'}$$

if AP_ϱ is a non array identifier or of a pair

$$A \oplus ADDR$$

if AP_ϱ is an array identifier.

The first component is the internal identifier of the (nonformal) actual parameter itself. $ADDR$ is the absolute initial address of the information vector associated with the non-formal array A. DL is the dynamic level of a procedure call. '$\textbf{RI}$' is a marker indicating the necessity of a round off. In this case AP is the generated identifier of a name procedure the values of which cannot statically be determined to be of type **integer** or **Boolean** and therefore possibly could be of type **real**, or a real function identifier. If AP is a (nonformal) type procedure identifier, AP may be sometimes also concatenated with a second dynamic level $DL1$. This is necessary whenever there are formal function identifiers on the left part of an assignment statement (see section 4.10.1.3, p. 73).

The construction of the proper information on AP_ϱ can be combined with a compatibility check and leads to three different cases:

a) If AP_ϱ is not a formal and not an array identifier, then the actual parameter AP_ϱ is paired with the value of the index register $\mathfrak{Ir}[sAP_\varrho]$:

$$AP_\varrho \oplus \mathfrak{Ir}[sAP_\varrho].$$

sAP_ϱ designates the static level of the smallest procedure Psm containing the declaration of AP_ϱ. If AP_ϱ is a procedure identifier, then $\mathfrak{Ir}[sAP_\varrho]$ is the momentary dynamic call level DL of Psm, the static predecessor of AP_ϱ (see this section, p. 104). The importance of the value of $\mathfrak{Ir}[sAP_\varrho]$ will become clear (see section 6.3, action 7) when the formal procedure call mechanism is explained.

If AP_ϱ is a function identifier, $DL1$ is given by the value of the index register $\Im r[sAP_\varrho+1]$, $sAP_\varrho+1$ being the static procedure level of AP_ϱ itself, and sometimes $\Im r[sAP_\varrho+1]$ is concatenated with $AP_\varrho \oplus \Im r[sAP_\varrho]$:

$$AP_\varrho \oplus \Im r[sAP_\varrho] \oplus \Im r[sAP_\varrho+1].$$

If the ϱ-th formal parameter of the procedure declaration P is specified to be integer then the chain produced above is concatenated with the round off marker '**RI**'.

b) If AP_ϱ is a nonformal array identifier then the actual parameter AP_ϱ is coupled with the absolute initial address of the associated information vector. This address $ADDR$ is given by the sum

$$\Im r[sAP_\varrho]+RADAP_\varrho,$$

$RADAP_\varrho$ designating the relative initial address of the associated information vector. If the associated formal array parameter has been supplied with its number n of subscripts by the prepass then the actual information vector must be copied. If the number of subscripts could not be found by the prepass, then the copying is not necessary.

c) If AP_ϱ is a formal identifier, the proper information about AP_ϱ is contained in the parameter block storage location associated with AP_ϱ. Its memory address is determined by the sum

$$\Im r[sAP_\varrho]+RADAP_\varrho.$$

$RADAP_\varrho$ designates the relative address of the parameter block storage location associated with AP_ϱ. If AP_ϱ is specified to be integer the round off indicator '**RI**' is joined to the afore mentioned information unless already present there.

2. The return address of the macro instruction is saved in the memory cell whose address is given by the value of **BFS**.

3. The momentary dynamic call level in MDL is stored in the memory cell addressed by **BFS**+1. In an ALGOL-like form we may write:

$$\mathfrak{M}[\textbf{BFS}+1]:=MDL.$$

4. The static level $sP+1$ of the called procedure P is stored in the memory cell addressed by **BFS**+2:

$$\mathfrak{M}[\textbf{BFS}+2]:=sP+1,$$

where sP is the static level of the static predecessor of P.

5. The value of $\Im r[sP]$, that is the instantly accessible dynamic level of the static predecessor of P, is stored in the memory cell addressed by **BFS**+3:

$$\mathfrak{M}[\textbf{BFS}+3]:=\Im r[sP].$$

6. In case of a ⟨type⟩ **procedure call**, the type marker corresponding to the call is stored in the memory cell addressed by **BFS**+4:

$$\mathfrak{M}[\mathbf{BFS}+4] := \text{'}⟨\text{type}⟩\text{'}.$$

In case of proper procedures ⟨type⟩ is empty.

7. The value of **BFS** is the new momentary call level. It is transferred
to MDL and to the index register $\mathfrak{Ir}[sP+1]$ associated with P:

$$MDL := \mathfrak{Ir}[sP+1] := \mathbf{BFS}.$$

8. The value of **BFS** enlarged by the length $FIXP$ of the fixed
storage belonging to P is the initial address of the new free storage and
is assigned to the generated variable $BFSP_0$ with the absolute storage
address **BFS**+5 and to the exceptional variable **BFS**:

$$\mathbf{BFS} := \mathfrak{M}[\mathbf{BFS}+5] := \mathbf{BFS}+FIXP.$$

9. The last action of the macro instruction is a jump to the program
address marking the beginning of the translated body of P.

Although the main program MP is also considered to be a procedure
(see section 4.9.1, p. 64) the execution of the translated ALGOL program
can not be started by the macro instruction

procedure call $MP_{internal}$

as is done with a normal procedure P. Only the actions 4, 6, 7, 8, and 9
have to be performed.

The following example of an ALGOL program illustrates how the
macro instruction **procedure call** works:

```
begin integer N ;
   procedure P(F, G) ;  procedure F, G ;
   begin procedure Q(R, S) ;  procedure R, S ;
      begin N := N+1 ;
         if N=4 then begin R(P, Q) ;  L4: end ;
         N := N+1 ;
         if N=7 then go to END P ;
      end Q ;
      N := N+1 ;
      if N=2 then begin P(Q, F) ;  L2: end ;
      if N=3 then begin F(Q, F) ;  L3: end ;
   END P:
   end P ;
```

$N := 1$;
   if $N=1$ **then begin** $P(P, P)$; $L1$: **end** ;
end

The maximum static procedure level is 2, and thus 3 index registers

$$\Im r[0], \ \Im r[1], \ \Im r[2]$$

have to be reserved. The fixed storage lengths of the main program and
of the procedures P and Q are, respectively, 7, 8, and 8.

The following diagram is the storage scheme shortly after the start
of the execution of the target language program (α designates the
initial memory address of the fixed storage of the main program):

BFS:	$\alpha + 7$
MDL:	α
$\Im r[0]$:	α
$\Im r[1]$:	empty
$\Im r[2]$:	empty

α:	empty
	empty
	0
	empty
	empty
	$\alpha + 7$
	empty

$\alpha + 7$:

$P(P, P)$ is the first procedure call and leads to the following new values
of the exceptional variables and additional entries in the free storage:

BFS:	$\alpha + 15$
MDL:	$\alpha + 7$
$\Im r[0]$:	α
$\Im r[1]$:	$\alpha + 7$
$\Im r[2]$:	empty

$\alpha+6$:	1[1]
$\alpha+7$:	$L1$[2]
	α
	1
	α
	empty
	$\alpha+15$
	$P \oplus \alpha$
	$P \oplus \alpha$
$\alpha+15$:	

The second procedure call is $P(Q,F)$:

BFS:	$\alpha+23$
MDL:	$\alpha+15$
$\Im\mathfrak{r}[0]$:	α
$\Im\mathfrak{r}[1]$:	$\alpha+15$
$\Im\mathfrak{r}[2]$:	empty
$\alpha+6$:	2
$\alpha+15$:	$L2$
	$\alpha+7$
	1
	α
	empty
	$\alpha+23$
	$Q \oplus \alpha+7$
	$P \oplus \alpha$
$\alpha+23$:	

---

[1] $\alpha+6$ is the memory address associated with the simple integer variable N.

[2] $L1$ indicates the internal label associated with the label $L1$ and also represents the return address of the procedure statement $P(P, P)$.

$P \oplus \alpha$ has been transferred from the memory cell addressed by $\alpha + 13$ to the cell addressed by $\alpha + 22$.

6.3. The instruction formal procedure call

The actions of the macro instruction

formal procedure call FP ;
$$AP_1 ;$$
$$\vdots$$
$$AP_r$$

(see section 4.10.1.2, p. 72) differ in the following respects from those actions described in the section above:

1. The actually called procedure is given by content Γ of the memory cell associated with FP and addressed by

$$\Im r[sFP] + RADFP.$$

sFP is the static level of that procedure whose formal parameter is FP. $RADFP$ designates the relative address of the parameter block storage location associated with FP.

Obviously content Γ has a structure

$$P \oplus DL,$$

where P is the internal identifier of the actually called procedure P, and DL is a certain dynamic call level of the static predecessor of P [see action 1 a) in section 6.2, p. 106].

2. As action 1 in section 6.2, p. 106.

3. As action 2 in section 6.2, p. 107.

4. As action 3 in section 6.2, p. 107.

5. As action 4 in section 6.2, p. 107.

6. DL is stored in the memory cell addressed by **BFS**$+3$:

$$\mathfrak{M}[\textbf{BFS}+3] := DL.$$

7. The main task and difficulty in connection with a formal procedure call is the reloading of certain index registers with their correct dynamic call levels.

The index registers in question are precisely those associated with procedures which contain the declaration of the called procedure P. Their numbers are
$$sP, \dots, 0$$

if $sP+1$ is the static procedure level of P (see action 5 of the present section or action 4 of the above section).

The correct values now to be assigned to the index registers $\mathfrak{Ir}[s]$ $(s=sP,\ldots,0)$ are precisely the same values which the index registers contained at that moment when the actual parameter P and the call level DL were coupled

$$P \oplus DL$$

and put into the proper parameter block storage location [see action 1 a) in section 6.2, p. 106]. Thus DL is the correct dynamic call level for $\mathfrak{Ir}[sP]$ associated with the static predecessor of P. The values for

$$\mathfrak{Ir}[sP-1], \ldots, \mathfrak{Ir}[0]$$

can be found with the help of the fourth entry in the procedure linkages: These reloadings can be done, for instance, by a subroutine call

RELOAD INDEX REGISTERS .

The appropriate and simple "procedure declaration" can be given in the following ALGOL-like form:

procedure *RELOAD INDEX REGISTERS* ;
   begin integer I ;   **address** B ;
     $B := DL$;
     $\mathfrak{Ir}[\mathfrak{M}[B+2]] := B$;
     for $I := \mathfrak{M}[B+2]-1$ **step** -1 **until** 1 **do** [1]
       $B := \mathfrak{Ir}[I] := \mathfrak{M}[B+3]$
  end ;

This subroutine is also used in connection with name calls and procedure exits.

It can well be optimized, since there exists a largest static level σ of a called procedure whose variables are accessible both immediately before and immediately after the execution of the instruction

formal procedure call FP .

As a consequence only the index registers with numbers between $\sigma+1$ and $\mathfrak{M}[DL+2]$ need to be reloaded. σ is determined by the largest index register number having the properties:

a) $\sigma \leq \mathfrak{M}[DL+2]$,

b) $\sigma \leq \mathfrak{M}[MDL+2]$, and

c) the value of $\mathfrak{Ir}[\sigma]$ is already correct and needs not be changed.

A further practically important optimization may be mentioned. If there are available as many index registers as procedure declarations in the ALGOL program and they are associated in a $1-1$ manner, then the

---
[1] The value of $\mathfrak{Ir}[0]$ associated with the main program never changes and therefore needs not be reloaded.

call of the subroutine $RELOAD\ INDEX\ REGISTERS$ can be avoided as long as no procedure has been called recursively or as soon as all recursive calls have been annulled again.

Special attention is called to the following fact:

If S is the maximum static procedure level, then at most S reloadings are needed at each formal procedure call, name call, or procedure exit. Thus, it is by no means necessary to step backwards from call to call in the dynamic data storage until the discovery of that procedure call whose actual parameter is the identifier P now in question.

8. As action 6 in section 6.2, p. 108.

9. As action 7 in section 6.2, p. 108.

10. As action 8 in section 6.2, p. 108.

11. As action 9 in section 6.2, p. 108.

In order to illustrate formal procedure calls we continue the example given in section 6.2, p. 108.

$F(Q, F)$ is the third procedure call and first formal procedure call:

BFS:	$\alpha+31$
MDL:	$\alpha+23$
$\Im\mathfrak{r}[0]$:	α
$\Im\mathfrak{r}[1]$:	$\alpha+7$
$\Im\mathfrak{r}[2]$:	$\alpha+23$
$\alpha+6$:	3
$\alpha+23$:	$L3$
	$\alpha+15$
	2
	$\alpha+7$
	empty
	$\alpha+31$
	$Q \oplus \alpha+15$
	$Q \oplus \alpha+7$
$\alpha+31$:	

The former value $\alpha+15$ of $\Im\mathfrak{r}[1]$ has been overwritten by the subroutine call $RELOAD\ INDEX\ REGISTERS$ after having assigned $DL := \alpha+7$.

$R(P, Q)$ is the second formal procedure call:

BFS:	$\alpha + 39$
MDL:	$\alpha + 31$
$\Im\mathfrak{r}[0]$:	α
$\Im\mathfrak{r}[1]$:	$\alpha + 15$
$\Im\mathfrak{r}[2]$:	$\alpha + 31$
$\alpha + 6$:	4
$\alpha + 31$:	$L4$
	$\alpha + 23$
	2
	$\alpha + 15$
	empty
	$\alpha + 39$
	$P \oplus \alpha$
	$Q \oplus \alpha + 7$
$\alpha + 39$:	

6.4. The instruction normal procedure exit

The task of the macro instruction

normal procedure exit

is to annul a procedure call (see section 4.10.2.4, p. 76), and to execute necessary type transfers of the values of type procedures. The actions of a call must be inverted:

1. Index registers are reloaded by the subroutine call *RELOAD INDEX REGISTERS* after the assignment $DL := \mathfrak{M}[MDL+1]$. In this case of a normal procedure exit the reloading of index registers can be further optimized if the number σ had been stored in the procedure linkage additionally. The index registers

$$\Im\mathfrak{r}[\sigma], \ldots, \Im\mathfrak{r}[0]$$

need not be reloaded. If a procedure happens to be called by a non-formal procedure call then σ is given by sP (see action 5 in section 6.2, p. 107).

2. The value of MDL is the initial address of the (old and now) new free storage and is assigned to **BFS**:

$$\textbf{BFS} := MDL.$$

3. The content of the memory cell addressed by $MDL+1$ is the (old and now) new momentary dynamic call level. It is assigned to the exceptional variable MDL:

$$MDL := \mathfrak{M}[MDL+1].$$

4. In case of a type procedure, a type transfer or compatibility check depending on the contents of **ADDR AC**, giving the type of the value delivered, and memory cell $\mathfrak{M}[\textbf{BFS}+4]$, giving the type required by the call, is executed.

5. The last action is a jump to the return address stored in the memory cell addressed by **BFS**.

In the later stages of processing the program, the formal procedure call $R(P, Q)$ is annulled by the macro instruction

normal procedure exit :

BFS:	$\alpha+31$
MDL:	$\alpha+23$
$\mathfrak{Jr}[0]$:	α
$\mathfrak{Jr}[1]$:	$\alpha+7$
$\mathfrak{Jr}[2]$:	$\alpha+23$
$\alpha+6$:	6
$\alpha+31$:	

6.5. The instruction **jump to**

The macro instruction

jump to L

executes a jump to a nonformal label L (see section 4.8, p. 59).

1. The new momentary call level is given by the value of $\mathfrak{Jr}[sPL]$:

$$MDL := \mathfrak{Jr}[sPL],$$

where sPL is the static level of that procedure P containing the declaration of L.

2. The new initial address of the free storage is given by the value of the generated variable BFS_{lL}, that is, by

$$\mathfrak{M}[MDL+5+lL],$$

and is assigned to **BFS**. lL designates the level of that array block containing the declaration of L (see section 4.9.1, p. 63).

3. The last action is a jump to the program address associated with the nonformal label L.

If one considers the fact that a jump to a nonformal label can be

a) local to a procedure body, but leading out of an array block,

b) not local to a procedure body, where the macro instruction **jump to** L may split up again[1]:

a) The effect of **array global jump to** L is action 2 before the transfer to the program address of L (action 3) is performed. MDL remains unchanged.

b) The instruction **procedure global jump to** L has to perform action 1, 2, and 3.

In our program example, **go to** $END\ P$ is a global jump out of the procedure body Q into the body P:

BFS:	$\alpha+15$
MDL:	$\alpha+7$
$\mathfrak{Ir}[0]$:	α
$\mathfrak{Ir}[1]$:	$\alpha+7$
$\mathfrak{Ir}[2]$:	$\alpha+23$ [2]
$\alpha+6$:	7
$\alpha+15$:	

6.6. The instruction formal procedure exit

The macro instruction

formal procedure exit FL

executes a jump to a formal label FL (see section 4.8):

---

[1] These cases can be already determined during the translation phase. Optimization then demands one of two different macro instructions.

[2] This value is now irrelevant.

1. The first action is the same as in section 6.3, p. 111. The actual label is given by the couple

$$L \oplus DL,$$

where L is the internal identifier of a nonformal label, and where DL is a certain dynamic call level of the smallest procedure P whose body contains the label declaration L.

1a) Here we must regard the following exceptional situation:

The discussion in section 4.10.1.1, e), p. 70, shows that the actual parameter may be the identifier $GIAP$ of a name procedure, for instance if the actual parameter is an integer label in a formal procedure call. Then $GIAP$ is to be exchanged by a corresponding label identifier as already mentioned in section 4.10.1.1, e), p. 70.

2. Index registers are reloaded by the subroutine call

$$RELOAD\ INDEX\ REGISTERS$$

(see action 7 in section 6.3, p. 111).

3. The new momentary call level is assigned

$$MDL := DL.$$

4. As action 2 in section 6.5, p. 116.

5. As action 3 in section 6.5, p. 116.

In order to illustrate such a formal procedure exit we give a suitable example of an ALGOL program:

```
begin integer N ;
   procedure P(F, FL) ;   procedure F ;   label FL ;
   begin procedure Q (R, FM) ;   procedure R ;   label FM ;
      begin N := N+1 ;
         if N = 3 then begin Q (R, END Q) ;   L3: end ;
         if N = 4 then begin F (Q, FM) ;   L4: end ;
      END Q:
      end Q ;
      N := N+1 ;
      if N = 2 then begin Q (P, END P) ;   L2: end ;
      if N = 5 then go to FL ;
   END P:
   end P ;
   N := 1 ;
   P (P, END) ;   L1:
END:
end
```

The maximum static procedure level is 2, and the fixed storage lengths of the main program and of the procedures P and Q are 7, 8, and 8, respectively.

Immediately before the execution of the go to statement **go to** FL, we have the following storage situation:

BFS:	$\alpha+39$
MDL:	$\alpha+31$
$\Im r[0]$:	α
$\Im r[1]$:	$\alpha+31$
$\Im r[2]$:	$\alpha+23$ (irrelevant)

α:	empty	$\alpha+19$:	empty
	empty		$\alpha+23$
	0		$P \oplus \alpha$
	empty		$END\ P \oplus \alpha+7$
	empty	$\alpha+23$:	$L3$
	$\alpha+7$		$\alpha+15$
	5		2
$\alpha+7$:	$L1$		$\alpha+7$
	α		empty
	1		$\alpha+31$
	α		$P \oplus \alpha$
	empty		$END\ Q \oplus \alpha+15$
	$\alpha+15$	$\alpha+31$:	$L4$
	$P \oplus \alpha$		$\alpha+23$
	$END \oplus \alpha$		1
$\alpha+15$:	$L2$		α
	$\alpha+7$		empty
	2		$\alpha+39$
$\alpha+18$:	$\alpha+7$		$Q \oplus \alpha+7$
			$END\ Q \oplus \alpha+15$
		$\alpha+39$:	

Immediately after the execution of the macro instruction

formal procedure exit FL

the following changes in the data storage have been made:

BFS:	$\alpha + 23$
MDL:	$\alpha + 15$
$\Im\mathfrak{r}[0]$:	α
$\Im\mathfrak{r}[1]$:	$\alpha + 7$
$\Im\mathfrak{r}[2]$:	$\alpha + 15$
$\alpha + 23$:	

6.7. The instructions **name address** and **name call**

If name procedures [see section 4.10.1.1, c), p. 66] are handled like function procedures, no further discussion is required. However, name procedures can never be called recursively in the following wider sense: If a name procedure $GIAP$ is called a second time before the first call has been completed, then the associated dynamic call levels of the static predecessor are also different.

Consequently, with respect to storage allocation name procedures can be handled like blocks. It is not necessary to associate index registers with them (see footnote 1 in section 4.9.1, p. 63).

For each name procedure call, the dynamic storage is to be enlarged by only three entries:

1. The return address.

2. The dynamic call level of the preceding procedure call.

3. The type of the call must be preserved as in the case of type procedure calls.

Now we enumerate the actions of the macro instruction

name address N

(see section 4.10.1.3, p. 73):

1. The first action is quite the same as in section 6.3, p. 111 and 6.6, p. 117. The actual name procedure is given by the chains

$$GIAP \oplus DL, \quad \text{or} \quad GIAP \oplus DL \oplus \text{'RI'}.$$

Here $GIAP$ is the internal identifier of a name procedure [see section 4.10.1.1, c), p. 67], DL and '**RI**' are the same as in section 6.2, procedure call.

At this stage it might happen, that the actual parameter is not a generated procedure identifier $GIAP$ but a normal type procedure identifier P [see 6.2, action 1a), p. 106]. The actual information is one of the chains

$$P \oplus DL \oplus DL1, \quad \text{or} \quad P \oplus DL \oplus DL1 \oplus \text{'\textbf{RI}'}.$$

With the help of $DL1$ which also is the same as in section 6.2 we are able to get the absolute address of the generated variable GVP (see section 4.10.2.3, p. 75):

$$DL1 + RADGVP,$$

where $RADGVP$ is the relative storage address for GVP.

$$DL1 + RADGVP$$

is coupled with the type of P and assigned to the exceptional variable **ADDR AC**. In this special case, the action of **name address** is complete, and control is transferred to the instruction following **name address** N.

2. The marker '**address**' is assigned to the memory cell $\mathfrak{M}[\textbf{BFS}+2]$.

3. As in section 6.2, p. 107.

4. As in section 6.2.

5. Index registers are reloaded by the subroutine call

$$RELOAD\ INDEX\ REGISTERS$$

(like action 7 in section 6.3, p. 111).

6. The new momentary dynamic level is given by DL:

$$MDL := DL$$

(like action 3 in section 6.6).

7. The new initial free storage address is assigned:

$$\textbf{BFS} := \textbf{BFS} + 3.$$

8. The last action is a jump to the program address marking the beginning of the translated body of $GIAP$.

The function of the macro instruction

$$\textbf{name call } N$$

does not differ very much from the scheme given above:

1. In section 4.10.1.2 it has been mentioned that exceptionally there may appear an actual parameter

$$P \oplus DL, \quad \text{or} \quad P \oplus DL \oplus \text{'\textbf{RI}'}, \quad \text{or}$$
$$P \oplus DL \oplus DL1, \quad \text{or} \quad P \oplus DL \oplus DL1 \oplus \text{'\textbf{RI}'}$$

where P is the identifier of a nonformal function procedure without any parameter. In this case the macro instruction **name call** is left and control is taken over by the macro instruction **formal procedure call**, action 3 (see section 6.3, p. 111).

2. In case of a ⟨type⟩ **name call** a type marker '**real**', '**integer**', or '**Boolean**' is assigned to $\mathfrak{M}[\mathbf{BFS}+2]$. Otherwise simply the marker '**type**' is assigned.

3.—8. as above.

6.8. The instruction **name procedure exit**

The macro instruction

name procedure exit

annuls a name procedure call and eventually executes a necessary type transfer of the value of a name call. The actions in section 6.7 have to be inverted:

1. Index registers are reloaded by the subroutine call $RELOAD$ $INDEX\ REGISTERS$ after having assigned $DL := \mathfrak{M}[\mathbf{BFS}-2]$ (compare action 1 in section 6.4, the remarks on optimization also remain valid).

2. The (old and now) new initial free storage address is assigned to **BFS** simply by the statement:

$$\mathbf{BFS} := \mathbf{BFS}-3.$$

3. The content of the memory cell addressed by **BFS**+1 is the (old and now) new momentary dynamic call level:

$$MDL := \mathfrak{M}[\mathbf{BFS}+1].$$

4. In case of annulling a ⟨type⟩ **name call** a type transfer or compatibility check depending on the contents of **ADDR AC**, giving the type of the value delivered, and memory cell $\mathfrak{M}[\mathbf{BFS}+2]$, giving the type required by the call, is executed.

5. As action 5 in section 6.4.

An example of an ALGOL program is given in order to illustrate how the macro instructions

> **name address** N,
> **integer name call** N, and
> **name procedure exit**

work:

```
begin integer J, K ;
    procedure P(N, M) ;   integer N, M ;
    begin
        N := M + K ;
    end P ;
    K := 3 ;
    P(J, K × K) ;   L1:
end
```

The target language program looks as follows (for the sake of simplicity we have only translated statements involved in name procedure handling. For the principles compare with section 4.10.1, p. 66):

```
begin integer J, K ;
    procedure P(N, M) ;   integer N, M ;
    begin
            name address N ;
        L2 : GVN := ADDR AC ;   ¹
            integer name call M ;
        L3 : AC := AC + K ;
            content of GVN := AC
    end P ;
    name procedure GIJ ;
    begin
            absolute address J ;
            AC := J
    end GIJ ;
    name procedure GIK ;
    begin
            AC := K × K ;
            ADDR AC := 'integer'
    end GIK ;
    K := 3 ;
    P(GIJ, GIK) ;   L1:
end
```

The maximum static procedure level is 1, thus 2 index registers $\Im\mathfrak{r}[0]$ and $\Im\mathfrak{r}[1]$ are to be reserved. GVN is a generated variable. The fixed storage lengths for the main program MP and for the procedure P are 8 resp. 9.

---

[1] GVN is a generated address variable.

Immediately before the execution of the macro instruction **name address** N we find the following storage situation:

ADDR AC:	irrelevant
AC:	irrelevant
BFS:	$\alpha+17$
MDL:	$\alpha+8$
$\Im r\,[0]$:	α
$\Im r\,[1]$:	$\alpha+8$

α:	empty
	empty
	0
	empty
	empty
	$\alpha+8$
	empty
$\alpha+7$:	3

$\alpha+8$:	$L1$
	α
	1
	α
	empty
	$\alpha+17$
	$GIJ\oplus\alpha$
	$GIK\oplus\alpha$
$\alpha+17$:	empty

When the name procedure GIJ is entered after the execution of **name address** N, the following changes in the storage scheme are performed:

ADDR AC:	irrelevant
AC:	irrelevant
BFS:	$\alpha+20$
MDL:	α
$\Im r\,[0]$:	α
$\Im r\,[1]$:	$\alpha+8$ (irrelevant)

$\alpha+17$:	$L2$
	$\alpha+8$
$\alpha+20$:	**'address'**

These changes are annulled when the body of GIJ is left by the macro instruction **name procedure exit**:

ADDR AC:	'integer' $\oplus \alpha + 6$
AC:	irrelevant
BFS:	$\alpha + 17$
MDL:	$\alpha + 8$
$\Im\mathfrak{r}[0]$:	α
$\Im\mathfrak{r}[1]$:	$\alpha + 8$

$\alpha + 17$:

$\alpha + 6$ is the memory address of the integer variable J, and the statement

$$GVN := \textbf{ADDR AC}$$

assigns 'integer' $\oplus \alpha + 6$ to the generated variable GVN with the memory address $\alpha + 16$. When the name procedure GIK is entered by **integer name call** M we have a storage scheme as follows:

ADDR AC:	'integer' $\oplus \alpha + 6$ (irrelevant)
AC:	irrelevant
BFS:	$\alpha + 20$
MDL:	α
$\Im\mathfrak{r}[0]$:	α
$\Im\mathfrak{r}[1]$:	$\alpha + 8$ (irrelevant)

$\alpha + 16$:	'integer' $\oplus \alpha + 6$
$\alpha + 17$:	$L3$
	$\alpha + 8$
	'integer'

$\alpha + 20$:

The body of this name procedure is left by the macro instruction **name procedure exit**:

ADDR AC:	'integer'
AC:	9
BFS:	$\alpha + 17$
MDL:	$\alpha + 8$
$\Im\mathfrak{r}[0]$:	α
$\Im\mathfrak{r}[1]$:	$\alpha + 8$

$\alpha + 17$:

AC now contains the value $K \times K = 3 \times 3 = 9$. When the procedure P is left by a normal procedure exit, the value $\mathbf{AC} + K = 9 + 3 = 12$ has been assigned to the variable J:

ADDR AC:	'integer' (irrelevant)
AC:	irrelevant
BFS:	$\alpha + 8$
MDL:	α
$\Im\mathfrak{r}[0]$:	α
$\Im\mathfrak{r}[1]$:	$\alpha + 8$ (irrelevant)

α:

empty
empty
0
empty
empty
$\alpha + 8$
12
3

$\alpha + 8$:

7. Model translator. Description

7.1. Introduction

The principles discussed in chapters 1—6 are here applied to and illustrated by a formally described translator which can serve as model for ALGOL translators. This translator system may be regarded as part of an interpreting system whose functions will include

a) Reading of ALGOL programs from an external medium into a certain machine storage area.

b) Syntax check; this means recognizing and reporting violations of the syntactical rules of ALGOL and moreover correcting them in such a way that further syntax checking is possible.

c) Translation of ALGOL programs into the machine language or a symbolic code as target language.

d) Execution of generated programs.

e) Post-mortem organization.

f) Output of generated programs for use in program libraries.

The system must contain a control mechanism which activates the execution of these activities in the sequence desired by the user. These functions require the system to be able to manipulate lists, such as the identifier list, the label list, the constant list, and the push down lists used by the passes of the translation phase.

As our aim is the description of an interpreting system which is as far as possible machine-independent, we restrict the following discussion to point c), the translation phase and point d); only a short discussion will be devoted to syntax checking.

The model translator of chapter 8 works in three passes.

In the first translator pass the translation from the internal hardware language to process language is prepared and partly performed. A list of all declared and specified identifiers is produced. Moreover, certain preparations for the implementation of recursive address calculation as listed in section 5.2 are made. The result of this pass is the modified ALGOL program from which declarators and specifiers are extracted, together with the identifier list and a list with variables not admissible for rec addr calc.

The pass which prepares the ALGOL program for the recursive address calculation according to the decomposition method follows[1]. The input ALGOL program is changed in that certain subscripted variables are replaced by the content of generated address variables and that new

---

[1] For the sake of clearness we describe the rec addr calc by a pass of its own. It could also be inserted in the decomposition pass.

statements are inserted before certain for statements. The set of process symbols is supplemented by new delimiters, and the identifier list is expanded by the declarations of generated variables.

The program and the identifier list are input for the third pass which produces the bracket-free target language program by means of the program push down and according to the rules which are extracted from the syntactical description of ALGOL as discussed in general in chapter 2. Within this pass the fixed storage allocation is done. This means that, according to the block structure and the static procedure level, a relative address is assigned to each entity. These addresses are made absolute before or during execution of the generated program by loading a suitable index register.

For programming systems used in computing centers the problem of automatic error check and detection is of great importance. Few attempts to deal with this problem theoretically have been made (cf. [123]). In practice, we rely on heuristic ad hoc methods. In order to show possibilities, a number of operations checking for syntactical errors have been included in both the translator and the run time system. In large programs one of the most serious possibilities of error which is detectable by flow analysis arises from the procedure call and actual — formal correspondence mechanism. Therefore, compatibility of procedure calls and declarations is checked in a supplement to the third pass.

The essential parts of the run time system are described in chapter 6 in an already sufficiently detailed way. Our model translator will use static type handling. We assume also that addresses are identical with integer numbers. Translators which use other type implementation methods may be constructed by replacing or omitting certain program parts. The restrictions on the ALGOL input language are exactly those already mentioned in the introduction.

7.2. Pass 1. The preparatory pass

We assume that at the start of the preparatory pass the ALGOL program is already available in the machine and consists of a suitable internal representation. The transformation of the ALGOL hardware program may be done by an input pass which will not be discussed here in great detail since it is extremely machine dependent. All transformations made by the input pass are simple codings (one to one correspondence between the ALGOL hardware program and the program stored in the machine) and are intended only to adapt the hardware symbols to machine words.

7.2.1. Input

The ALGOL input program for the preparatory pass may be regarded as a vector χ. Each component $\chi[i]$ of this vector is a single ALGOL delimiter or entity, i.e. an identifier, an integer number, a real number, a truth value, or a string. As we do not wish to take as a basis a definite machine we assume that each component $\chi[i]$ may have any length. So each entity can be stored in one component $\chi[i]$ of the vector χ^1. Comments within the ALGOL hardware program [including parameter delimiters of the form)⟨letter string⟩ :(, which are replaced by ,] have no meaning for its translation and execution. They may be easily eliminated by the input pass and have no counterpart within the vector χ.

7.2.1.1. Delimiters

In the following table all delimiters which may occur as input delimiters for the different passes are listed:

	input preparatory pass	output preparatory pass / input rec addr calc pass	input decomposition pass
arithmetic operators	$+$ $-$ $\times$ $/$ $\div$ $\uparrow$	$+$ $-$ $\times$ $/$ $\div$ $\uparrow$	$+$ $+_i$ $-$ $-_i$ $\times$ $\times_i$ $/$ $\div$ $\uparrow$ $\oplus$
relational operators	$<$ $\leqq$ $=$ $\neq$ $>$ $\geqq$	$<$ $\leqq$ $=$ $\neq$ $>$ $\geqq$	$<$ $\leqq$ $=$ $\neq$ $>$ $\geqq$
logical operators	$\neg$ $\wedge$ $\vee$ $\supset$ $\equiv$	$\neg$ $\wedge$ $\vee$ $\supset$ $\equiv$	$\neg$ $\wedge$ $\vee$ $\supset$ $\equiv$

[1] In practice it will be necessary to represent entities by internal entities of standard length.

	input preparatory pass	output preparatory pass / input rec addr calc pass	input decomposition pass
sequential operators	**go to** **if** **then** **else** **for** **do**	**go to** **if** **then** **then** Σ **else** **else** Σ **for** **do** **for end**	**go to** **if** **then** **then** Σ **else** **else** Σ **for** **for suitable** **do** **do suitable** **for end**
separators	**,** **:** **;** **:=** **step** **until** **while**	**,** **for,** **label:** **:** **;** **:=** **step** **until** **while**	**,** **for,** **label:** **:** **;** **:=** **step** **until** **while**
brackets	**(** **)** **[** **]** **begin** **end**	**(** **)** **[** **]** **begin** **block begin** **end** **block end**	**(** **)** **[** **[** **]** **begin** **block begin** **begin initial value** **begin increment** **end** **block end** **end initial value** **end increment**
declarators and specifiers	**Boolean** **integer** **real** **array** **switch** **procedure** **string** **label** **value**	**array** **switch** **procedure begin** **procedure end**	**array** **switch** **procedure begin** **procedure end**
new delimiters			**red init** **content real of** **content integer of** **content Boolean of**

The following delimiters of the original hardware ALGOL program are eliminated

comment | . | $_{10}$| ' | '

The occurrence of the following delimiter in the ALGOL program is not allowed:

own

7.2.1.2. Entities

The entities of ALGOL programs can be separated into five classes:

identifiers (e.g. $RICHARD$),
integer numbers (e.g. 15),
real numbers (e.g. $15.5_{10}3$),
truth values (e.g. **false**), and
strings (e.g. '$A-B-C$').

For our purposes it is practical to represent the entities internally by themselves enclosed in string quotes. The examples above then will look as follows:

'$RICHARD$'
'15'
'$15.5_{10}3$'
'**false**'
"$A-B-C$"

7.2.2. Output

The output of the preparatory pass will be the input for the next pass, which prepares the recursive address calculation for subscripted variables. This output consists of three lists:

1. the modified input program χ_1,
2. the identifier list (see section 7.2.2.2, p. 131),
3. the for list (see section 7.2.2.3, p. 135).

7.2.2.1. The modified input program χ_1

The modified input program χ_1 differs from the original input program χ essentially in that declarations have been extracted and stored in the identifier list. The whole set of delimiters which may appear in the output program is contained in the list on page 128. Some ALGOL delimiters which occur with more than one syntactical meaning are split up by the introduction of new delimiters. The distinction between **then** and **then** Σ resp. **else** and **else** Σ has been made in order to show explicitly that the delimiters **then** and **else** may either introduce an expression or a statement. The new delimiter **for end** serves to indicate the end of a for statement.

Each comma separating for list elements is replaced by a new delimiter **for,** . Both delimiters **for,** and **for end** are introduced in order to simplify the organization of the next pass which prepares recursive address calculation. A colon which indicates a label declaration is represented by the new delimiter **label**: . Block beginnings and ends are distinguished from compound beginnings and ends by the new delimiters **block begin** and **block end**. The list of declarators and specificators shrinks since only the program generating parts of the declarations remain within the modified input program χ_1.

7.2.2.2. The identifier list

All declared identifiers and all formal parameters are picked up and stored in the identifier list (see 3.1, p. 19) exactly in that sequence in which their declarations or specifications occur in the program. This identifier list is built up from "program structure symbols" and "identifier entries". The program structure symbols are **block begin, block end, procedure begin, procedure end**, each supplied with a marker to the corresponding begin or end symbol in this list, as in the following scheme:

Example for an identifier list

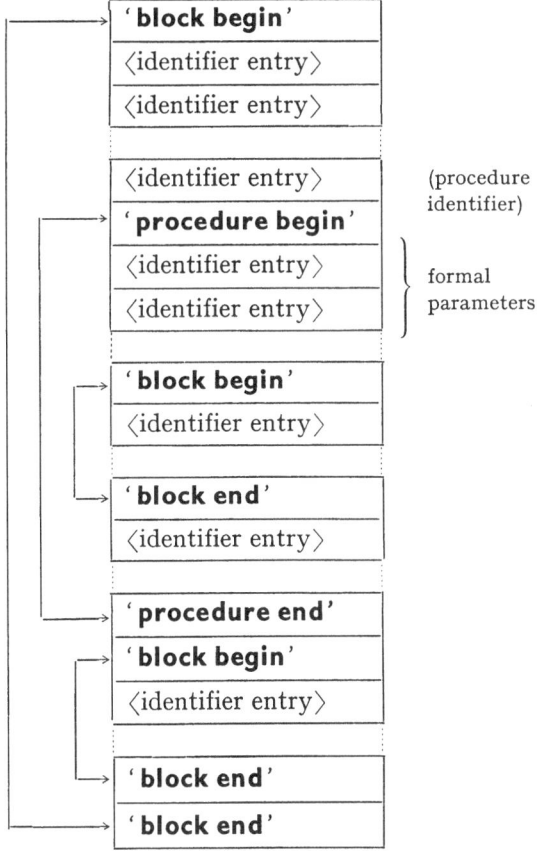

The program structure symbols consist of only three components:

(1) Internal identifier
(2) Symbol
(3) Associated internal identifier

The internal identifier has the same meaning as in the case of an identifier entry: the entry number.

The position symbol may have the following values:

Symbol:
'block begin'
'block end'
'procedure begin'
'procedure end'

The associated internal identifier is the marker mentioned above.

An identifier entry consists of the following 16 components:

(1) External identifier
(2) Internal identifier
(3) Mode
(4) Type
(5) Kind
(6) Relative address
(7) Static procedure level
(8) Array block level
(9) Program storage address
(10) Dimension
(11) Length of fixed storage or number of array identifiers with the same bound pair list
(12) Maximum array block level
(13) Associated internal identifier
(14) Output
(15) Input
(16) Error

The external identifier is the ALGOL-identifier in the internal representation, or a constant (see next section), or '**empty**'.

The internal identifier has the form of an integer number resulting from a numbering of the entries in the identifier list (in practice: the address of the identifier entry).

The mode may have the following values:

Mode:
'**normal**'
'**formal**'
'**value**'
'**standard**'
'**generated**'

It is '**standard**', if the identifier is a standard function identifier. '**generated**' does not appear in the identifier list built by the first translation pass. It is inserted by the rec addr calc pass or the decomposition pass when variables are generated.

The values of the type position may be:

Type:
'**empty**'
'**real**'
'**integer**'
'**arithmetic**'
'**Boolean**'
'**Boolean***'
'**type**'
'**string**' [1]

'**empty**' occurs if the identifier is declared or specified as **label, switch, string,** (no type) **procedure,** or if it is an unspecified formal parameter.

The fifth constituent gives the kind of the declaration or specification of the identifier

Kind:
'**variable**'
'**number**'
'**array**'
'**label**'
'**switch**'
'**string**'
'**procedure**'
'**unspecified formal parameter**'
'**array or switch**'
'**variable or label**'
'**expression procedure**' [1]
'**procedure or label**' [1]
'**expression procedure or label**' [1]

[1] This entry occurs only in connection with the transformation of actual parameters in procedure calls in pass 3.

The function of the relative address has been explained in chapter 4, p. 64. For numbers it is the fixed storage address relative to the beginning of the main program, for strings it is the address of the beginning of the string.

For variables it is the fixed storage address relative to the beginning of the main program or procedure fixed storage. Especially, for a generated variable representing the reduced initial address of an array the relative address is the address of the beginning of the information vector. For the formal parameters the relative address gives the associated cell in the "parameter-block". In most cases the relative address will be inserted later on by the decomposition pass. As the list of constants can be built up before entering the decomposition pass their relative addresses can be inserted at an earlier stage.

The program storage address for labels, switch and procedure identifiers will result from the production of target language instructions in the last pass. For a standard function this address must be delivered by the organizational system.

The dimension has importance only for procedure and array identifiers. It gives the number of parameters or the number of subscript positions, respectively. For nonformal array identifiers the number of subscript positions is simply found from the bound pair list. In the case of a formal array identifier one must search for the occurrence of subscripted variables in the body. If the number cannot be found the dimension remains empty.

The 11-th position means the length of the fixed storage of the procedure body in case of a procedure identifier or the number of array identifiers with the same bound pair list in case of an array identifier. The length of the fixed storage must include places for the local and generated variables and arrays and can only be given by the last pass.

Also the maximum array block level is only of importance for procedure identifiers and means the maximum level of array blocks local to the body (and not contained in local procedure bodies).

Certain identifier entries require a reference to another identifier entry which is given by its internal identifier. That is, for arrays the entry associated with the information vector (the generated variable representing the reduced initial address), for real (integer) numbers the number in the integer (real) representation, and for type procedure identifiers the generated variable used for the procedure value. In case of a generated name procedure identifier produced by an actual parameter which could also be a designational expression a reference to the associated generated label is given [see section 4.10.1.1, e), p. 70].

The last three positions are filled only in the case of formal parameters, and serve only to detect incorrect correspondence of actual to formal parameters: output means that the parameter appears in the left part of an assignment, input (which is of interest only if the parameter is not specified) says that the parameter appears as a primary in some expression. Error is set only if a formal parameter specified **procedure** appears with parameter lists of different length, and is used only in the final procedure call check pass.

7.2.2.3. The for list

A list, called "for list" α, is set up which will contain the following entries for each for statement occurring:

'for unsuitable'   if the loop is recognized to be unsuitable for re-cursive address calculation, that is, if the do state-ment contains a procedure call or a left part name variable,

'for unsuitable for index registers'[1]

    if the do statement contains a name or function procedure call (except standard function calls), or

'for possibly suitable'[1]   otherwise.

The entries **'for possibly suitable'** and **'for unsuitable for index registers'** are followed by a list of all variables which are not admissible (which occur as left part in an assignment statement in the do statement).

In order to have the complete structure of the nesting of the for statements we use an entry **'for end'** coupled by a two-way pointer to the matching **'for unsuitable'**, **'for unsuitable for index registers'** or **'for possibly suitable'**.

Example:

for list:

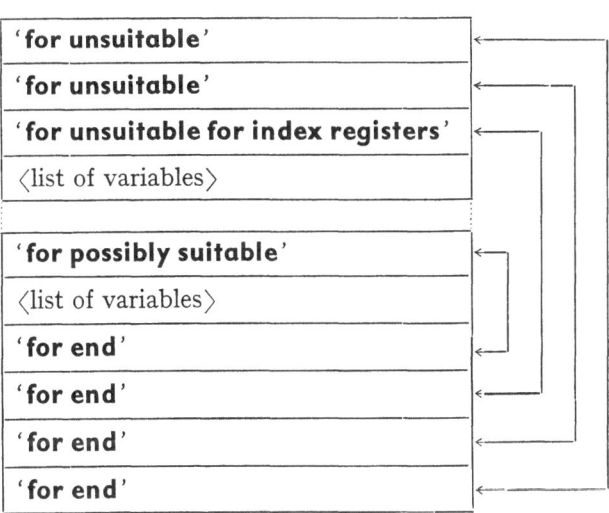

---

[1] Whether the loop is really suitable can only be recognized in the following pass.

7.3. Pass 2. The implementation of recursive address calculation

The input of the second (rec addr calc) pass is the output of the preparatory pass:

the program, called χ, consisting of the symbols listed in 7.2.1.1 and 7.2.1.2 (χ_1 in section 7.2).

the identifier list δ, described in 7.2.2.2.

the for list α, of section 7.2.2.3.

7.3.1. Output

The output of the rec addr calc pass is the modified ALGOL program and the identifier list. The for list is omitted since it will not be used further.

7.3.1.1. Program

The output program consists of two parts π and $\pi\,add$. π is the program from which the suitable for clauses and subscripted variables whose addresses are to be calculated recursively are extracted. $\pi\,add$ contains those for clauses and also the program parts "initial value" and "increment" which are produced by this pass. $\pi\,add$ may be considered to be a program vector different from π or to be inserted at the respective spots in π.

The following new delimiters are introduced (compare with the table on page 128):

'$\lceil$'	to denote index vectors ($\underset{A}{\lceil}$ is represented by the two symbols $A\lceil$).
'**red init**'	to denote the reduced initial address of the foregoing array identifier.
'**content real of**' '**content integer of**' '**content Boolean of**'	as explained in section 5.6.1,
'$\oplus$'	used for concatenation of index registers,
'**begin initial value**' '**end initial value**' '**begin increment**' '**end increment**'	used as brackets for the corresponding program parts,
'**for suitable**'	instead of '**for**', if the loop is suitable,
'**do suitable**'	instead of '**do**' if the loop is suitable.

7.3.1.2. The identifier list

The identifier list δ is extended by a list of the generated variables used for the recursive address calculation in accordance with the loop structure. This means that the additional list contains symbols for suitable loop beginnings and ends so that the allocation of fixed storage addresses within the correct block structure might easily be done in the next pass.

The new entries in the list δ are of the form: a mark '**identical**' followed by a list of variables of type **integer** which must be identified, that is, which use the same storage cell.

Example for the scheme:

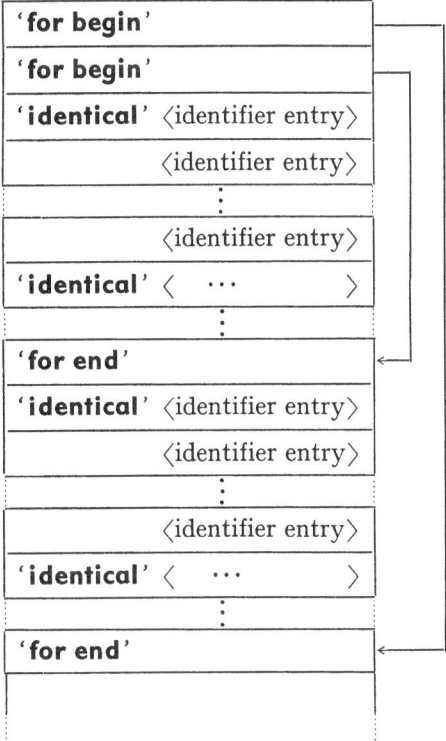

For '**for begin**' and '**for end**' the position program structure symbol is used, for '**identical**' the position mode in δ.

7.3.2. Lists used in pass 2

Since pass 2 uses a large number of different lists, a brief description of these lists is given in the following

χ source program, delivered by the preparatory pass (χ_1 in section 7.2),

δ identifier list, delivered by the preparatory pass,

α list of not admissible variables, delivered by the preparatory pass,

π output program,

$\pi\ add$ additional output program,

ι identity list, containing the generated variables which must be identified,

ω push down used for storing informations about for statements and used for the linearity test,

ξ index vector list, used for storing subscripted variables, index vectors, and expressions,

φ factor list, used for the factor of the loop variable when decomposing an index vector,

$\varkappa$ constant part list, used for the constant part when decomposing an index vector,

σ push down, used for the decomposition,

$\pi\alpha$ intermediate program storage for initial value program parts.

In the following we give short descriptions of the use and the content of the intermediate lists or push downs.

a) The push down ω:

This list contains an entry for each loop which is still relevant.

The entry for an unsuitable loop has the form:

'for unsuitable'.

The entry for a suitable loop has the form:

$\langle$entry in δ associated with the beginning of the loop$\rangle$
$\langle$beginning of the associated entries in the α-list$\rangle$
$\langle$index register counter$\rangle$
$\langle$counter x of the index vector list $\xi\rangle$
$\langle$entry number in ω of the least suitable loop containing this loop$\rangle|0$
$\langle$beginning of the loop in $\pi\rangle$
$\langle$beginning of the largest block, local to the loop, in the identifier list$\rangle$
$\langle$(generated) variable for the initial value$\rangle$
$\langle$(generated) variable for the step value$\rangle$
$\langle$loop variable$\rangle$
'for'

The free part of this list is used for the admission test for subscripted variables and expressions. Storing and deleting of delimiter and entity entries is handled like in the third pass for the decomposition of expressions. The entries for variables and expressions are '**entity**' if this variable is not the loop variable or the expression does not contain the loop variable, otherwise the entry '**loop variable**' is used.

b) The identity list ι:

ι is a two-dimensional array $\iota[1:\infty,\ 1:5]$, where each row represents an identifier entry which may have one of the forms:

'identical'|	⟨generated identifier⟩|	'integer'|	⟨index register⟩|	'empty'|
'empty'	⟨generated identifier⟩	'address'	'empty'	'local'| 'global'

When a variable is generated then a new entry marked '**identical**' is added to the list. If two identifier entries are to be identified the list is rearranged and has finally the structure:

α:	'identical'	⟨generated identifier⟩	⟨type⟩	⟨index register⟩| 'empty'	'empty'| 'local'| 'global'
	'empty'	⟨generated identifier⟩	⟨type⟩	⋮	⋮
	⋮	⋮	⋮	⋮	⋮
	'empty'	⋮	⋮	⋮	⋮
β:	'identical'	⋮	⋮	⋮	⋮
	'empty'	⋮	⋮	⋮	⋮
	⋮	⋮	⋮	⋮	⋮

list of identifiers which are identified with the entry $\iota[\alpha]$

list of identifiers which are identified with the entry $\iota[\beta]$

At first the last column is set '**empty**'. The value is changed into '**local**' if the generated variable has once occurred as left part variable in an initial value program and cannot be pushed into an outer loop. The value '**global**' is only used for numbers and declared variable entries, which are entered into the list only when they occur as increment in order to denote the associated index register.

Generated variables can not be identified with such quantities. Entries marked '**global**' are deleted from ι at the end of the respective loop. From entries marked '**local**' declarations are generated and added to δ.

c) The index vector list ξ:

A subscripted variable or an expression which is admissible for recursive address calculation is eliminated from the program π or the initial value program $\pi\alpha$ and stored in the list ξ. At the end of each loop, the corresponding part of the list is processed. The entries in ξ have in general the form:

$$v := A \text{ red init} + X[s_1, \ldots, s_n] + E \oplus \Re \mathfrak{r}[i]$$

where E is a sum of products of numbers and simple declared variables, and any of the terms on the right hand side may drop out. $\lceil$ is supplied with the first array identifier (which might be identical with A) occurring in the array declaration of $A : \dots$ **array** $X, \dots, A, \dots [\dots] ; \dots$

d) The constant part list $\varkappa$ and the factor list φ:

These auxiliary lists are used in order to store the constant part and the factor of the loop variable when an expression stored in ξ is decomposed at the end of the loop. The entries in $\varkappa$ are of the same form as those of ξ. The entries in φ have in general the form $X[u_1, \dots, u_n] + E'$ where E' is a sum of numbers or declared simple variables, and any of the terms may be empty.

e) The push down σ:

σ is an auxiliary push down used for the decomposition. The symbols contained in ξ are read one by one and stored first in both lists $\varkappa$ and φ. If certain symbols are seen to be part only of the factor or only of the constant part, then they are eliminated in $\varkappa$ or in φ respectively. For recognizing this, the push down $\sigma[1 : \infty, 1 : 3]$ is used in the well-known manner: the first component is reserved for the delimiter or for one of the entity entries: '**factor**', '**constant**', '**linear**'. In addition, each delimiter and entity entry is concatenated with the respective entry numbers in $\varkappa$ and in φ as second and third component.

f) The program storages $\pi\alpha$ and $\pi\,add$:

From the symbols stored in $\varkappa$ and in φ the instructions for the initial value program and the increment program are generated, for the subscripted variables or expressions. The initialization program is first stored in $\pi\alpha$ whereas the increment program is stored simultaneously in $\pi\,add$. When these program parts are terminated $\pi\alpha$ may be stored also in $\pi\,add$.

The instructions of $\pi\alpha$ have the form:

$$v := \langle \text{expression} \rangle ;$$
or $$\qquad v | v' := \langle \text{expression} \rangle ;$$
or $$\qquad v |\, \textbf{empty} := \langle \text{expression} \rangle \oplus \Im\mathfrak{r}[i] ;$$

where $\langle$expression$\rangle$ is an expression of the form admissible for ξ but not containing an index register. $v | v'$ means: the variable v is to be incremented by v'. $v |\, \textbf{empty}$ means: the variable v is incremented by index register modification. These additional notations are necessary for identification tests, since only two left part variables which are not incremented or two variables which have the same increment can be identified. In $\pi\ add$ instructions are stored with the same right parts as in $\pi\alpha$, but with simple variables only as left parts.

7.3.3. Program survey

The rec addr calc pass consists of the following main parts:

(1) handling of the for clause,

(2) handling of subscripted variables, index vectors, and expressions including the investigation of admissible and possible identifications,

(3) decomposition of index vectors into a constant part and a part which depends (linearly) on the controlled variable,

(4) building up of the program parts for determining initial values and for incrementing.

The following flow charts give a short picture of this pass. See pages 142 and 143.

7.4. Pass 3. Decomposition and production of target program

The task of the third pass is to decompose the program delivered by the first pass, and possibly modified by the second pass, and to produce the target program consisting of the target language sequences as described in chapter 4 and 5. The input of this pass is the output of the second pass (see the table on page 128).

7.4.1. Output

The output of this pass is the generated program together with the identifier list. The target language instructions are those explained in section 3.2.

7.4.2. Program survey

With the aid of flow charts which show program and data flow we give a picture of the decomposition phase formally described in the next chapter 8.

The reading part, labeled with $NEXT$, delivers the symbols to be processed. The identifier list is inspected if an identifier occurs in the program. The following decoding part (labeled by $PROCESS$) controls the call of different program parts in dependence of the actions on the state push down as described in section 2.5. Each entity can be pushed down immediately. See pages 144 and 145.

7.5. Editorial functions

A translator program, strictly speaking, has as its sole proper function the task of translating a correctly written program into equivalent correct machine code. In a practical system, certain editorial functions may be undertaken at the same time. One such function already discussed in this book is recursive address calculation, which modifies automatically the simple translation in the interests of target efficiency. Many editorial functions are often suggested for inclusion in a compiler. Some criterion

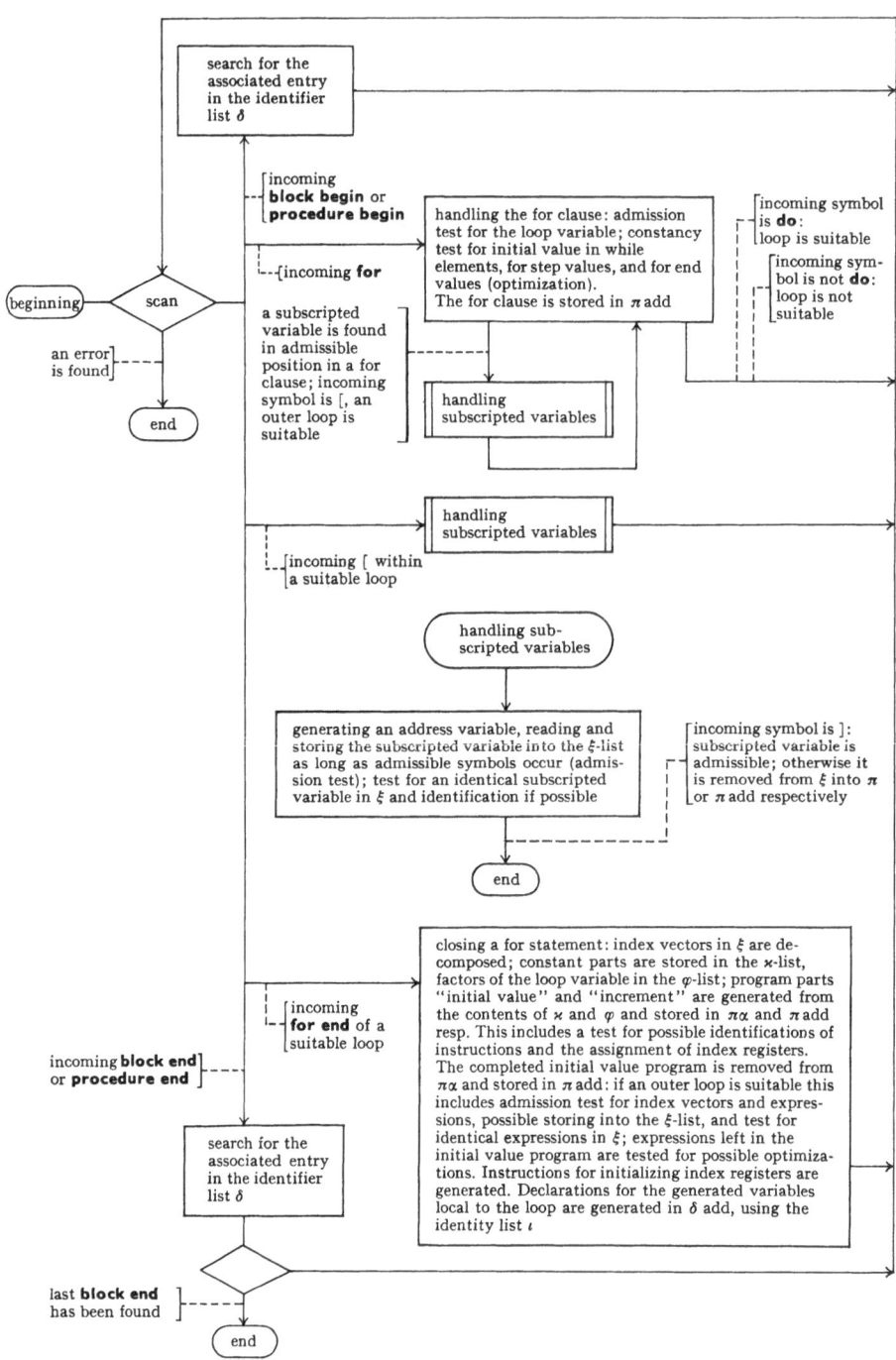

Pass 2. Program flow

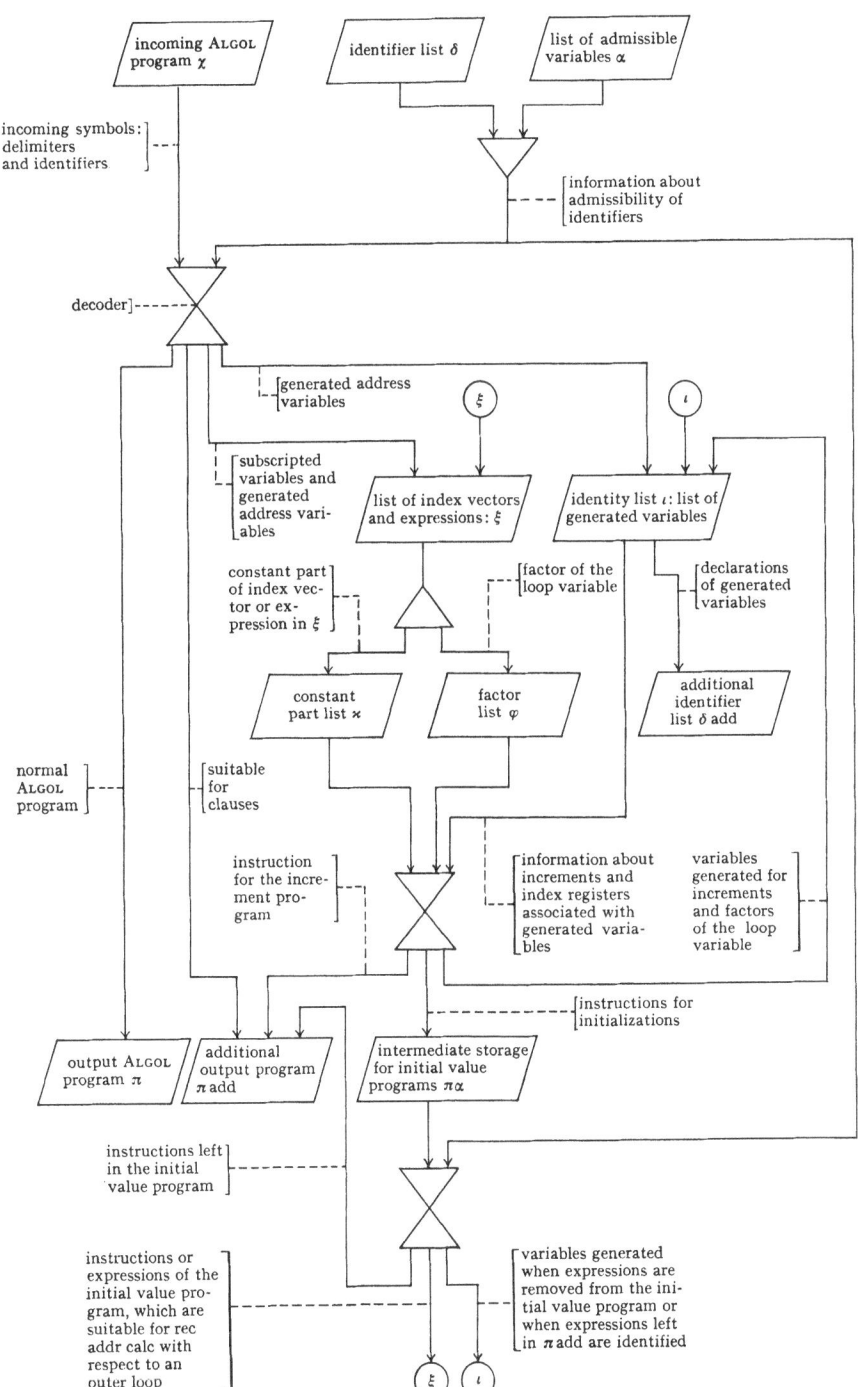

Pass 2. Information flow

is desirable to determine whether a given function should be considered
for inclusion or not. One that appears most practical is that editorial

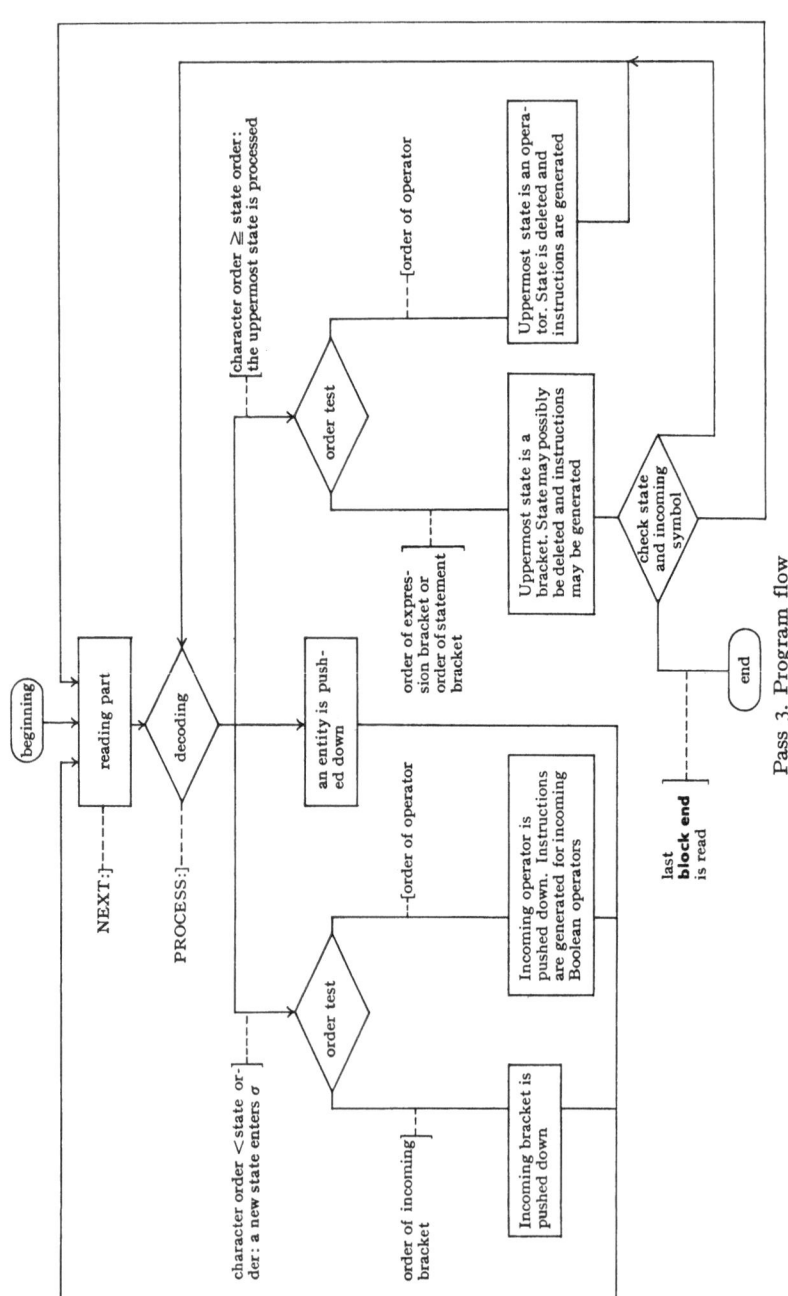

Pass 3. Program flow

functions which can be handled directly by the programmer should not be handled. Thus, recursive address calculation qualifies for inclusion, since this task can not readily be done by the programmer, while the elimination of common subexpressions, often suggested, does not.

An editorial function of a somewhat different type is the handling of incorrectly written programs. Programmers make mistakes of several

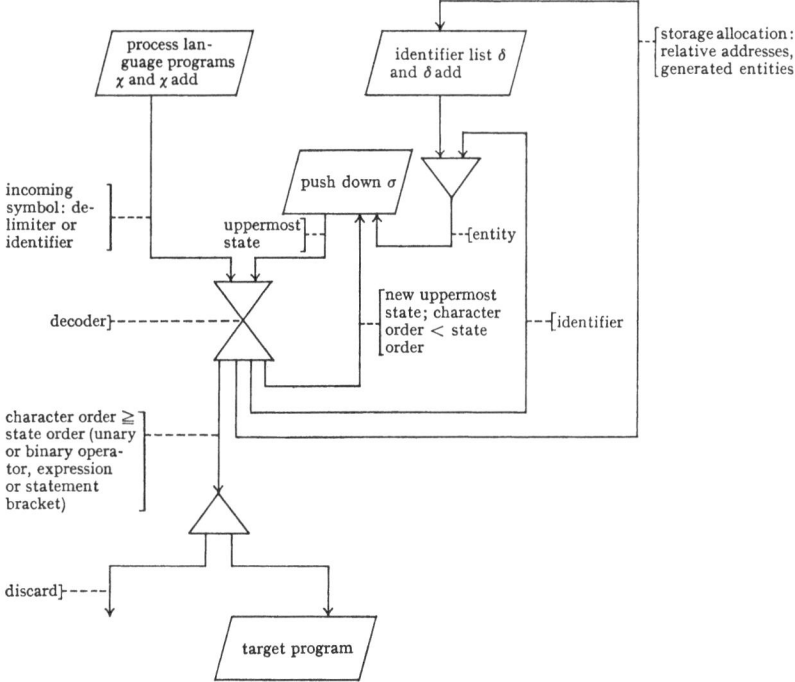

Pass 3. Information flow

types in programs which are submitted to the machine for compilation and execution. With errors that arise only because of an incorrect formulation of the computational process, the compiler system can do nothing, and they may in general not affect either the translation or execution phase directly. Errors which result in inadmissible operations at run time, such as attempted division by zero or operations that result in overflow may be detected by the run time system. These should be reported to the programmer since the ALGOL language does not have facilities to directly interrogate the appropriate toggles.

7.5.1. Syntax checking

Violations of the syntactic and semantic rules of the ALGOL Report, the conventions regarding the association of declarations and use of identifiers, and the further conventions beyond these which an actual system may have used to resolve possible ambiguities, result in an inability or inadvisability of the translation being completed. A few obvious errors, such as the misspelling of the ALGOL "word" delimiters, the presence of a semicolon before **else**, and even the omission of a declaration for a real-type global variable (though here at the cost of a drastic assumption) may be corrected; these should, however, always be brought to the attention of the programmer since his intention may have been other than that followed by the compiler.

Syntactic errors that are encountered by the machine that cannot be handled as above and which result in an interruption of the translation process should be reported to the programmer with any information which will be helpful to him in correcting the error. A general rule as to what should be given is rather difficult. For compilers used by programmers who know ALGOL only, it is useless to list machine-oriented or processor-oriented information such as absolute addresses, internal representations, and symbolic code. Information should be formulated in terms of ALGOL, the original representations of quantities and operations, and relative to the program as written by the programmer. Possible reasons for the failure to compile should be given if known. The point of error within the program at which error is detected should be pin-pointed as precisely as the recursive nature of ALGOL and ALGOL-processing permit.

Two considerations are pertinent at this point. First, the question whether only the first syntactic error should be reported or whether an effort should be made to continue processing in an effort to find additional ones is not easy to answer. The effort to continue requires the translator to recover from the first error in order to resume processing; it is difficult to avoid spurious error messages during the transition (cf. IRONS [123]).

The second consideration involves the structure of the compiler. There are advocates of a separate syntax check pass during which errors in syntax are detected and reported, after which, if there are no errors, the actual translation is undertaken. The latter then must rarely cope with an incorrectly written program. It is possible so to organize the system that only during a debugging run is the check performed while for production work the program need be submitted only to the translation. A disadvantage of this, however, is that there is considerable duplication of effort within the compiler, since the syntax check pass requires many of the actions which must be performed during the translation. Practically, also, errors often creep into completely debugged programs because of accidents; nothing is more disconcerting than to find that a previously correctly compiling and running program without explanation fails now to compile. The more efficient and preferable course is to combine the syntax check with the actual translation process and make the syntax check in all cases.

The simplest method of checking errors is the use of the full decoding matrix (see 2.5.1, p. 10). Each pair (σ_s, χ_k) which is not permitted by the syntax leads to an error message or to a program which tries to improve or correct the source program in the most plausible way. The latter requires that part of the program be changed or supplemented, or the uppermost states be changed or deleted. It is possible to detect errors using a condensed matrix, but then more involved programs are required, or error messages are necessarily less specific.

Program error detection and correction has been included in the model translator to a reasonable extent. The input pass may be made to detect orthographic errors. In connection with the identifier list or the test for suitable loops, a test for correct block structure and bracket structure within arithmetic expressions is made.

The recursive address calculation pass is called only if the input pass has not found any error. Within this second pass a check is made only within for clauses and for subscripted variables. An error immediately terminates the pass and delivers the unchanged input as output.

The decomposition pass contains a complete check for the errors which may yet occur. For errors against the syntactical rules, a matrix is used in a drastically shortened form which needs only a few storage places and allows a predecoding depending on character and state orders.

After the decomposition pass has been completed and before the target language program interpreter is entered all procedure calls are checked to determine whether call and declaration are compatible. Beyond this, all possible substitutions of formal parameters by actuals are checked for agreement with each other. Thus, the procedure *check procedure calls and substitutions of formal parameters by actuals* is able to detect at translation time a certain class of errors which ALGOL programmers normally subsume under the concept of "semantical errors".

Obviously, the check procedure can do nothing with a procedure call where the "procedure" identifier is faulty in so far as it is neither declared nor specified **procedure** nor unspecified formal. Therefore, illegal procedure calls of this type must be eliminated or properly marked in the course of the preceding syntax check operations.

7.6. Run time system. The target language program interpreter

The run time system has been patterned after the control unit of a real computer and simulates its actions: The system

1. has an instruction counter which points to a certain memory cell the content of which can be interpreted to be an instruction,

2. reads the instruction,

3. executes the instruction and, at the end,

4. assignes a new value to the instruction counter pointing to the subsequent instruction to be executed.

The body of the procedure target language program interpreter does not present code for all instructions. This is true especially for simple instructions like arithmetic and logical operations and local conditional or unconditional jumps. In present-day machines such instructions are available, and in implementations of the model translator the target language program interpreter may be considered to be an additional part of the machine control unit. More complicated instructions which deal with storage allocation and normally are not available in a computer may be realized in the form of subroutine calls.

8. Algol 60 model translator. Formal part

begin
integer
   COL δ external identifier, COL δ internal identifier,
   COL δ mode, COL δ type, COL δ kind, COL δ relative address,
   COL δ static procedure level, COL δ array block level,
   COL δ program storage address, COL δ dimension,
   COL δ length of fixed storage, COL δ number of array identifiers,
   COL δ maximum array block level, COL δ associated internal identifier,
   COL δ program structure symbol, COL δ output, COL δ input,
   COL δ error, maximum static procedure level, max ind reg,
   δ beginning, δ end, π end ;
Boolean *error ;*
string array
   $\chi[1:\infty], \chi 1[1:\infty], \chi\,add[1:\infty],$
   $\pi[1:\infty,\ 1:2], \delta[-\infty:\infty,\ 1:16], \alpha[1:\infty],$
   $\mathfrak{M}[1:\infty], \mathfrak{Ir}[0:\infty] ;$

procedure *preparatory pass* (χ)
   output: $(\chi 1, \delta, d1, \alpha, maximum\ static\ procedure\ level, error)$;
   string array $\chi, \chi 1, \delta, \alpha$;
   integer *d1, maximum static procedure level* ;
   Boolean *error* ;

comment The input for the procedure *preparatory pass* is an ALGOL program in the form of a string vector χ the declaration of which can be thought to be

$$\textbf{string array } \chi[1:\infty]\ \unicode{x0296}^{1}.$$

The possible values of the vector components $\chi[c]$ are ALGOL delimiters and entities represented in form of strings as described in sections 7.2.1.1, p. 128 and 7.2.1.2, p. 130.

The first component $\chi[1]$ of the input ALGOL program must contain the statement bracket '**begin**' i.e. the ALGOL program is assumed to be an unlabelled block or compound statement.

The output of the procedure *preparatory pass* consists of the modified ALGOL program $\chi 1$, the declaration of which may look like

$$\textbf{string array } \chi 1[1:\infty]\ \unicode{x0296},$$

the identifier list δ

$$\textbf{string array } \delta[-\infty:+\infty, 1:16]\ \unicode{x0296},$$

and the for list α

$$\textbf{string array } \alpha[1:\infty]\ \unicode{x0296}.$$

The fourth output parameter is an integer variable $d1$ denoting the lowest relevant row of the identifier list $\delta\,(d1 \leqq 0)$. The identifier list rows with row subscripts

$$d1, d1+1, \ldots, -2, -1$$

form the so called number list because here numbers, truth values, and strings of the input ALGOL program are stored.

The fifth output parameter *maximum static procedure level* denotes how many index registers $(S+1,$ if S is the maximum static procedure level) must be reserved for procedure call organization (see chapter 6). If at run time $K\,(\geqq S+1)$ index registers are available then the recursive address calculation pass may dispose of at least $K-(S+1)$ index registers for handling of admissible subscripted variables.

The sixth output parameter *error* indicates whether the preparatory pass has found a syntactical error within the input ALGOL program χ. As soon as the pass discovers an incorrect situation the logical value **true** is assigned to the formal Boolean variable *error* (compare check procedures, pp. 168–179). In this case, at the end of the preparatory pass, it is recommended to skip the recursive address calculation pass and to enter

---

[1] The symbol "$\unicode{x0296}$" means a semicolon. The proper symbol ";" may not occur within an ALGOL comment.

the translation pass immediately. More detailed descriptions of the output of the preparatory pass and complete lists of the possible values have been given in section 7.2.2, p. 130. ;

begin
  integer
      COL σ symbol pushed down,
      COL σ state,
      COL σ supplement information 1,
      COL σ supplement information 2,
      COL σ reference to the identifier list,
      COL σ number of subscripts,
      COL σ array block level,
      COL σ maximum array block level ;

comment Integer identifiers like
                    COL σ symbol pushed down
            or   *COL σ state*   etc.
have constant values during execution of the whole preparatory pass. In principle, these identifiers are superfluous, but they have been introduced in order to have more understandable denotations for columns of the push down list σ. ;

integer
      c, c max,
      c1,
      s,
      last push down column,
      d,
      beginning of the identifier list,
      last identifier list column,
      beginning of the number list,
      a,
      beginning of the for list ;

comment The integer variables
                    last push down column
                    beginning of the identifier list
                    last identifier list column
                    beginning of the number list
                    beginning of the for list
have constant values during execution of the whole preparatory pass. Like column identifiers these identifiers could be eliminated. They are used because their meaning is clearer for the reader than the meaning of certain integer numbers.

c is an integer variable used as a pointer denoting the subscript of that ALGOL program component $\chi[c]$ read by the most recent call of the procedure *read from the input program* χ. $c1$ is a pointer denoting the subscript of that modified ALGOL program component $\chi1[c1]$ which has been assigned a value by the most recent call of the procedure *store in the modified input program* $\chi1$. s and d are pointers denoting the momentarily highest relevant rows of the push down list σ and of the identifier list δ. a is an integer variable pointing to the momentarily highest relevant vector component $\alpha[a]$ of the for list α. ;

integer
  static procedure level,
  array block level,
  maximum array block level,
  location of the searched identifier,
  location of the procedure identifier,
  location of the identifier,
  location of the leading array identifier,
  dimension,
  number of parameters,
  number of array identifiers,
  $k,$
  i ;

comment The integer variables k and i have no general meaning. They are only used as controlled variables in for statements. ;

string
  mode,
  type,
  kind,
  identifier,
  input component,
  state,
  number ;
Boolean
  available formal parameter,
  inserted input component,
  behind value list ;
string array $\sigma[1:\infty, 1:7]$;

comment The string array σ is a push down list used by the preparatory pass. This list is empty when the preparatory pass is started and again when it is finished.

The array components $\sigma[s, COL\ \sigma\ symbol\ pushed\ down]$ of the first column may have the following values:

> **'empty'**
> **'begin'**
> **'switch'**
> **'procedure begin'**
> **'go to'**
> **'for'**
> **':='**
> **'step'**
> **'if'**
> **'then'**
> **'('**
> **'['** .

This list of possible values shows that the first column is used for entering by opening brackets of the Algol program χ. When the corresponding closing bracket is read these entries will be cancelled.

The array components $\sigma[s, COL\ \sigma\ state]$ of the second column may have the values

> **'empty'**
> **'declaration'**
> **'statement'**.

During execution of the preparatory pass this push down list component shows whether the pass momentarily is in an expression state, declaration state, or statement state.

The push down list components $\sigma[s, COL\ \sigma\ supplement\ information\ 1]$ of the third column may have the values

> **'empty'**
> **'block'**
> **'statement'**
> **'bound pair list'**.

Here the preparatory pass eventually deposits additional information denoting that the opening statement bracket '**begin**' introduces a new block, that the sequential operator '**if**' introduces a conditional statement, or that the opening square bracket '**[**' introduces a bound pair list.

The possible values for the push down list components $\sigma[s, COL\ \sigma\ supplement\ information\ 2]$ of the fourth column are

> **'empty'**
> **'array block'**
> **'constructed block begin'**
> **'count subscripts'**.

'**array block**' means that the corresponding block of the ALGOL program χ contains array declarations. '**constructed block begin**' means that the preparatory pass has constructed an additional block. E.g. procedure bodies and do statements will be generally made into blocks unless this has already been done by the ALGOL programmer. '**count subscripts**' means that the number of subscripts of a subscripted variable is to be determined.

The components of the last three columns eventually take up information on the following matters: A reference to a certain row of the identifier list δ, a number of subscripts, an array block level of a procedure body, or a maximum array block level of a procedure body. ;

switch
    cascade :=
      ARITHMETIC OPERATORS,
      RELATIONAL OPERATORS,
      LOGICAL OPERATORS,
      COLON,
      STEP,
      UNTIL,
      WHILE,
      GO TO,
      COLON EQUAL,
      IF,
      THEN,
      ELSE,
      FOR,
      DO,
      COMMA,
      SEMICOLON,
      OPENING ROUND BRACKET,
      CLOSING ROUND BRACKET,
      OPENING SQUARE BRACKET,
      CLOSING SQUARE BRACKET,
      BEGIN,
      END,
      BOOLEAN,
      INTEGER,
      REAL,
      ARRAY,
      SWITCH,
      PROCEDURE,
      VALUE,

> *LABEL,*
> *STRING SPECIFIER,*
> *IDENTIFIER,*
> *INTEGER NUMBER,*
> *REAL NUMBER,*
> *TRUTH VALUE,*
> *STRING ;*

>   **integer procedure** *associated integer number* (*input component*) ;
>   **string** *input component* ;
>   **code** ;

comment The integer procedure, *associated integer number*, associates a certain integer number with each possible component of the input Algol program vector χ. The associated numbers correspond in an obvious manner to the switch list elements of the switch *cascade*. Therefore no explicit procedure body is given. ;

>   **string procedure** *first constituent* (*string*) ;
>   **string** *string* ;
>   **code** ;
>   **string procedure** *second constituent* (*string*) ;
>   **string** *string* ;
>   **code** ;

comment The value of the procedure *first constituent* is the first constituent of the actual string parameter the constituents of which have eventually been concatenated by the operator $\oplus$. The procedure *second constituent* is defind analogously. Explicit procedure bodies have not been given. ;

>   **Boolean procedure** *declarator* (*symbol*) ;
>   **string** *symbol* ;
>   **begin**
>   *declarator* :=
>   *symbol* = '**real**'
> $\vee$ *symbol* = '**integer**'
> $\vee$ *symbol* = '**Boolean**'
> $\vee$ *symbol* = '**array**'
> $\vee$ *symbol* = '**procedure**'
> $\vee$ *symbol* = '**switch**'
>   **end** *declarator* ;

>   **Boolean procedure** *specifier* (*symbol*) ;
>   **string** *symbol* ;

```
begin
  specifier :=
    symbol = 'value'
    ∨ symbol = 'label'
    ∨ symbol = 'string'
  end specifier ;
Boolean procedure type declarator (symbol) ;
  string symbol ;
  begin
    type declarator :=
      symbol = 'real'
      ∨ symbol = 'integer'
      ∨ symbol = 'Boolean'
  end type declarator ;
integer procedure maximum (a, b) ;
  integer a, b ;
  begin
    maximum := if a > b then a else b
  end maximum ;
procedure read from the input program χ ;
  begin
    c max := c := c + 1 ;
    input component := χ[c] ;
    if inserted input component then
      inserted input component := false
  end read from the input program χ ;
procedure store in the modified input program χ1 (output component) ;
  string output component ;
  begin
    if ¬ inserted input component ∨ output component ≠ input
        component then
      begin
        c1 := c1 + 1 ;
        χ1[c1] := output component
      end
  end store in the modified input program χ1 ;
procedure clear push down row (s) ;
  integer s ;
  begin
    for k := 1 step 1 until last push down column do
      σ[s, k] := 'empty'
  end clear push down row ;
```

procedure *push down (symbol)* ;
  string *symbol* ;
  begin
    $s := s + 1$;
    clear push down row (s) ;
    $\sigma[s, COL \ \sigma \ symbol \ pushed \ down] := symbol$
  end *push down* ;

procedure *reset push down* ;
  begin
    $s := s - 1$
  end *reset push down* ;

procedure *push down with a reference to the identifier list δ (symbol)* ;
  string *symbol* ;
  begin
    $s := s + 1$;
    clear push down row (s) ;
    $\sigma[s, COL \ \sigma \ symbol \ pushed \ down] := symbol$;
    $\sigma[s, COL \ \sigma \ reference \ to \ the \ identifier \ list] := d$
  end *push down with a reference to the identifier list δ* ;

comment The procedure *push down with a reference to the identifier
list δ* is called after an opening statement bracket '**block begin**' or
'**procedure begin**' has been put in the form of a program structure
symbol into the identifier list (see p. 131), e.g.:

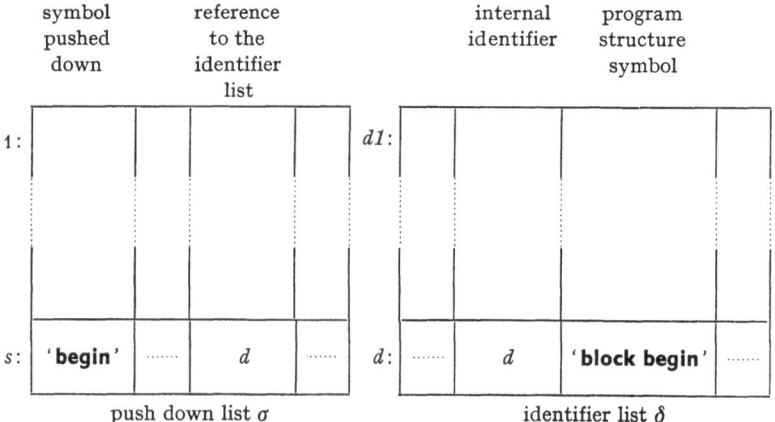

| symbol
pushed
down | reference
to the
identifier
list | | internal
identifier | program
structure
symbol |

push down list σ identifier list δ

Later, when the corresponding '**end**' from the input program χ is
read the push down list will have been cleared up to row s and the
preparatory pass is able to find the corresponding '**block begin**'
within the identifier list (d is the pointer associated with the identifier
list δ). ;

Boolean procedure *push down contains (symbol)* ;
  string *symbol* ;
  begin
    push down contains := *symbol* = σ[*s*, *COL* σ *symbol pushed down*]
  end *push down contains* ;

procedure *clear mode type and kind* ;
  begin
    mode := *type* := *kind* := 'empty'
  end *clear mode type and kind* ;

procedure *clear identifier list row (d)* ;
  integer *d* ;
  begin
    for *k* := 1 **step** 1 **until** *last identifier list column* **do**
      δ[*d, k*] := 'empty'
  end *clear identifier list row* ;

procedure *put an opening bracket in form of a program structure symbol into the identifier list* δ *(opening bracket)* ;
  string *opening bracket* ;
  begin
    d := *d*+1 ;
    clear identifier list row (d) ;
    δ[*d, COL* δ *internal identifier*] := *d* ;
    δ[*d, COL* δ *program structure symbol*] := *opening bracket*
  end *put an opening bracket in form of a program structure symbol into the identifier list* δ ;

procedure *put a closing bracket in form of a program structure symbol into the identifier list* δ *(closing bracket)* ;
  string *closing bracket* ;
  begin
    integer *d op br* ;
    d := *d*+1 ;
    clear identifier list row (d) ;
    d op br := σ[*s, COL* σ *reference to the identifier list*] ;
    δ[*d, COL* δ *internal identifier*] := *d* ;
    δ[*d, COL* δ *program structure symbol*] := *closing bracket* ;
    δ[*d, COL* δ *associated internal identifier*] := *d op br* ;
    δ[*d op br, COL* δ *associated internal identifier*] := *d*
  end *put a closing bracket in form of a program structure symbol into the identifier list* δ ;

comment The procedure *put a closing bracket in form of a program structure symbol into the identifier list* δ is called e.g. when an ' **end** ' has been read from the input program χ in connection with a push down configuration

	symbol pushed down		supple- ment infor- mation 1		reference to the identifier list	
1 :						
s .	' **begin** '	......	' **block** '	......	$d\ op\ br$	....

push down list σ

(see p. 156). After execution of the procedure the identifier list δ looks like

	internal identifier		program structure symbol		associated internal identifier	
$d1$:						
$d\ op\ br$:	......	$d\ op\ br$	' **block begin** '	....	d	
d :	......	d	' **block end** '	....	$d\ op\ br$	

identifier list δ

;

procedure *put an identifier into the identifier list* δ *(identifier)* ;
  string *identifier* ;
  begin
    $d := d + 1$;
    clear identifier list row (d) ;
    δ[d, COL δ *external identifier*] := *identifier* ;
    δ[d, COL δ *internal identifier*] := d ;
    δ[d, COL δ *mode*] := *mode* ;
    δ[d, COL δ *type*] := *type* ;
    δ[d, COL δ *kind*] := *kind* ;
    δ[d, COL δ *static procedure level*] := *static procedure level* ;
    check for the same identifier in the same block (identifier)
  end *put an identifier into the identifier list* δ ;

procedure *put a formal parameter into the identifier list* δ
  (formal parameter) ;
  string *formal parameter* ;
  begin
    $d := d + 1$;
    clear identifier list row (d) ;
    δ[d, COL δ *external identifier*] := *formal parameter* ;
    δ[d, COL δ *internal identifier*] := d ;
    δ[d, COL δ *mode*] := '**formal**' ;
    δ[d, COL δ *kind*] := '**unspecified formal parameter**' ;
    δ[d, COL δ *static procedure level*] := *static procedure level*
  end *put a formal parameter into the identifier list* δ ;

procedure *construct an equivalent real number (integer number)*
  output: *(real number)* ;
  string *integer number, real number* ;
  code ;

procedure *construct an equivalent integer number (real number)*
  output: *(integer number)* ;
  string *real number, integer number* ;
  code ;

comment The procedure *construct an equivalent real number* changes
the given input string *integer number* into an equivalent output string
real number. E.g. the integer number

> '15' becomes '15.0',
> '0'   becomes '0.0'.

The procedure *construct an equivalent integer number* changes the
given input string *real number* into an equivalent output string *integer*

number accordingly to the transfer function

$$entier\ (real\ number\ +0.5).$$

E.g. the real number

'15.0'	becomes	'15',
'15.5'	becomes	'16',
'15.5$_{10}$3'	becomes	'15500'. ;

   procedure *put a number into the number list (number, type)* ;
    value *number, type* ;
    string *number, type* ;
    begin
      for $k :=$ *beginning of the number list* **step** -1 **until** *d1* **do**
        if $\delta[k, COL\ \delta\ external\ identifier] = number$ **then**
          go to *END PROCEDURE PUT NUMBER* ;
ABOVE :
      $d1 := d1-1$;
      clear identifier list row (d1) ;
      $\delta[d1, COL\ \delta\ external\ identifier] := number$;
      $\delta[d1, COL\ \delta\ internal\ identifier] := d1$;
      $\delta[d1, COL\ \delta\ mode] := $ '**normal**' ;
      $\delta[d1, COL\ \delta\ type] := type$;
      $\delta[d1, COL\ \delta\ kind] := $ '**number**' ;
      $\delta[d1, COL\ \delta\ static\ procedure\ level] := 0$;
      if $type = $ '**real**' **then**
        begin
          construct an equivalent integer number (number, number) ;
          $type := $ '**integer**'
        end
      else
        if $type = $ '**integer**' **then**
          begin
            construct an equivalent real number (number, number) ;
            $type := $ '**real**'
          end
        else go to *END PROCEDURE PUT NUMBER* ;
      for $k := d1+1$ **step** 1 **until** *beginning of the number list* **do**
        if $\delta[k, COL\ \delta\ external\ identifier] = number$ **then**
          begin
            $\delta[d1, COL\ \delta\ associated\ internal\ identifier] := k$;
            go to *END PROCEDURE PUT NUMBER*
          end k ;
      $\delta[d1, COL\ \delta\ associated\ internal\ identifier] := d1-1$;
      go to *ABOVE* ;

END PROCEDURE PUT NUMBER:
    end *put a number into the number list* ;

comment The number list is the initial part of the identifier list δ:

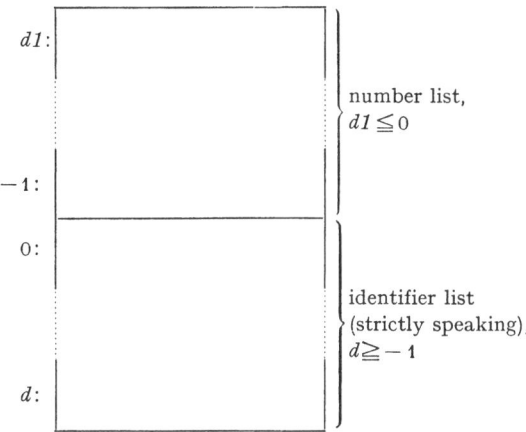

Numbers and strings occurring two or more times in the ALGOL program are stored only once in the number list. ;

    procedure *put a string into the number list* (*string*) ;
      string *string* ;
      begin
        for $k :=$ *beginning of the number list* **step** -1 **until** *d1* **do**
          if $\delta[k,$ *COL* δ *external identifier*$] =$ *string* **then**
            go to *END PROCEDURE PUT STRING* ;
        $d1 := d1 - 1$;
        clear identifier list row (*d1*) ;
        $\delta[d1,$ *COL* δ *external identifier*$] :=$ *string* ;
        $\delta[d1,$ *COL* δ *internal identifier*$] := d1$;
        $\delta[d1,$ *COL* δ *mode*$] :=$ '**normal**' ;
        $\delta[d1,$ *COL* δ *kind*$] :=$ '**string**' ;
        $\delta[d1,$ *COL* δ *static procedure level*$] := 0$;

END PROCEDURE PUT STRING:
    end *put a string into the number list* ;

    procedure *search for the identifier in the identifier list* δ
    (*IDENTIFIER NOT YET DECLARED*) ;
    label *IDENTIFIER NOT YET DECLARED* ;
    begin
      for $k := d$ **step** -1 **until** *beginning of the identifier list* **do**

begin
  if $\delta[k, COL\ \delta\ program\ structure\ symbol] = $ '**block end**'
  $\vee\ \delta[k, COL\ \delta\ program\ structure\ symbol] = $
  '**procedure end**' **then**
  begin
    $k := \delta[k, COL\ \delta\ associated\ internal\ identifier]$;
    go to $END\ LOOP\ K$
  end ;
  if $\delta[k, COL\ \delta\ external\ identifier] = identifier$ **then**
  begin
    $location\ of\ the\ searched\ identifier := k$;
    go to $END\ PROCEDURE\ SEARCH\ IDENTIFIER$
  end ;
$END\ LOOP\ K$:
  end k ;
  go to $IDENTIFIER\ NOT\ YET\ DECLARED$;
$END\ PROCEDURE\ SEARCH\ IDENTIFIER$:
  end *search for the identifier in the identifier list* δ ;

comment The procedure *search for the identifier in the identifier list* δ is called when a subscripted variable or a simple left part variable occurs. Global parameters of this procedure are the string variable *identifier* containing the input for the procedure and the integer variable *location of the searched identifier* delivering the output of the procedure. When the identifier has not yet been declared within the input program χ then the formal exit $IDENTIFIER\ NOT\ YET\ DECLARED$ is taken. This situation may occur e.g. in connection with procedure declarations. Example:

            begin
              procedure p ⅃
                begin
                  $a := 1$ ⅃
                  comment c is pointing to a ⅃
                end p ⅃
              integer a ⅃
              p
            end ;

procedure *correct the array block levels of preceding switches* ;
  begin
    $k := d+1$;

RETURN CORRECT:

    $k := k-1$;

    if $\delta[k,\ COL\ \delta\ program\ structure\ symbol] = $ ' **block begin** ' **then**

      go to *END PROCEDURE CORRECT* ;

    if $\delta[k,\ COL\ \delta\ program\ structure\ symbol] = $

      ' **procedure end** ' **then**

      begin

        $k := \delta[k,\ COL\ \delta\ associated\ internal\ identifier]$;

        go to *RETURN CORRECT*

      end ;

    if $\delta[k,\ COL\ \delta\ kind] \neq $ ' **switch** ' **then**

      go to *RETURN CORRECT* ;

    $\delta[k,\ COL\ \delta\ array\ block\ level] := array\ block\ level$;

    go to *RETURN CORRECT* ;

END PROCEDURE CORRECT:

    end *correct the array block levels of preceding switches* ;

comment If the first array declaration of an array block occurs after a switch declaration within the same block then the array block level of this switch has not been entered correctly in the identifier list (see p. 192). The array block level of such switches must be raised by 1. The procedure *correct the array block level of preceding switches* is called whenever the preparatory pass has met an array declaration within the nput program χ and has put the supplement information 2 '**array block**' into the push down list σ (see p. 191). Example:

        begin

          switch $s := L$ ¶

          array $a[1:10]$ ¶

              comment c is pointing to a ¶

        L:

        end ;

procedure *put an opening bracket in form of a program structure symbol into the for list* α *(opening bracket)* ;

string *opening bracket* ;

begin

    $a := a+1$;

    $\alpha[a] := opening\ bracket \oplus d$

end *put an opening bracket in form of a program structure symbol into the for list* α ;

comment The possible opening brackets to be used in the for list α are '**for possibly suitable**', '**for unsuitable for index registers**', or '**for unsuitable**'. When the delimiter '**do**' is read from the input program χ the procedure *put an opening bracket in form of a program structure symbol into the for list* α will be called. The opening bracket is concatenated with the value d, which points to a program structure symbol '**block begin**' within the identifier list δ (compare p. 185) ;

> **procedure** *put a closing bracket in form a program structure symbol into the for list* α *(closing bracket)* ;
> **string** *closing bracket* ;
> **begin**
>> **for** $k := a$ **step** -1 **until** *beginning of the for list* **do**
>> **begin**
>>> **if** *first constituent* $(\alpha[k]) = $ '**for unsuitable**'
>>> $\vee$ *first constituent* $(\alpha[k]) = $ '**for possibly suitable**'
>>> $\vee$ *first constituent* $(\alpha[k]) = $
>>>> '**for unsuitable for index registers**' **then**
>>> **begin**
>>>> $a := a+1$;
>>>> $\alpha[k] := $ *first constituent* $(\alpha[k]) \oplus (a-k)$;
>>>> $\alpha[a] := $ *closing bracket* $\oplus (a-k)$;
>>>> **go to** *END PROCEDURE PUT CLOSING BRACKET*
>>> **end** ;
>>> **if** *first constituent* $(\alpha[k]) = $ '**for end**' **then**
>>>> $k := k - $ *second constituent* $(\alpha[k])$
>> **end** k ;

END PROCEDURE PUT CLOSING BRACKET:
> **end** *put a closing bracket in form of a program structure symbol into the for list* α ;

comment The only possible closing bracket within the for list α is '**for end**'. It is concatenated with an integer number $(a-k)$ representing the "distance" between '**for end**' and the corresponding opening bracket '**for possibly suitable**', '**for unsuitable for index registers**', or '**for unsuitable**'. The corresponding opening bracket must be sought by running through the for list α. It is useless to mark the positions of '**for possibly suitable**' or '**for unsuitable for index registers**' in the push down by the procedure *put an opening bracket in form of a program structure symbol into the for list* α because the positions may change when identifiers are stored in the for list α (see p. 166).

for list α ;

procedure *all for statements are unsuitable (symbol)* ;
   string *symbol* ;
   begin
     for $k := a$ **step** -1 **until** *beginning of the for list* **do**
       begin
         if *first constituent* $(\alpha[k]) = $ '**for end**' **then**
           begin
             $k := k - $ *second constituent* $(\alpha[k])$;
             go to *END LOOP K*
           end ;
         if *first constituent* $(\alpha[k]) = symbol \lor$ *first constituent* $(\alpha[k]) =$
           '**for unsuitable**' **then**
           go to *END PROCEDURE UNSUITABLE* ;
         if *first constituent* $(\alpha[k]) = $ '**for possibly suitable**'
           $\lor$ *first constituent* $(\alpha[k]) =$
             '**for unsuitable for index registers**' **then**
           $\alpha[k] := symbol \oplus $ *second constituent* $(\alpha[k])$;

END LOOP K:
       end k ;
END PROCEDURE UNSUITABLE:
     end *all for statements are unsuitable* ;

comment When a procedure statement or a left part name variable occurs all surrounding for statements are unsuitable for handling by recursive address calculation. Parallel for statements are not concerned. When a function designator or a call of a name variable occurs in the do statement all surrounding loops are unsuitable for incrementing address variables by index registers. ;

> **procedure** *store the identifier in the for list* α ;
>> **begin**
>>> *search for the identifier in the identifier list*
>>>   *(IDENTIFIER NOT YET DECLARED)* ;
>>> **if** δ [*location of the searched identifier,*
>>>   *COL* δ *mode*] = '**formal**' **then**
>>>> **begin**
>>>>   *all for statements are unsuitable* ('**for unsuitable**')
>>>>   **go to** *END PROCEDURE STORE IDENTIFIER* ;
>>>> **end** ;

LOOP K:

>> **for** k := a **step** − 1 **until** *beginning of the for list* **do**
>>> **begin**
>>>> **if** *first constituent* (α[k]) = '**for end**' **then**
>>>>> **begin**
>>>>>   k := k − *second constituent* (α[k]) ;
>>>>>   **go to** *END LOOP K*
>>>>> **end** ;
>>>> **if** *first constituent* (α[k]) = '**for unsuitable**' **then**
>>>>   **go to** *END PROCEDURE STORE IDENTIFIER* ;
>>>> **if** *first constituent* (α[k]) = '**for possibly suitable**'
>>>>   ∨ *first constituent* (α[k]) =
>>>>   '**for unsuitable for index registers**' **then**
>>>>> **begin**
>>>>>   **if** *second constituent* (α[k]) < *location of the*
>>>>>   *searched identifier* **then**
>>>>>   **go to** *END PROCEDURE STORE IDENTIFIER* ;
>>>>>   **for** i := a **step** − 1 **until** k+1 **do**
>>>>>   α[i+1] := α[i] ;
>>>>>   α[k+1] := *identifier* ;
>>>>>   a := a+1 ;
>>>>>   **go to** *END LOOP K*
>>>>> **end** ;
>>>> **if** α[k] = *identifier* **then**
>>>>   **go to** *END PROCEDURE STORE IDENTIFIER* ;

END LOOP K:
>>> **end** k ;
>>> **go to** *END PROCEDURE STORE IDENTIFIER* ;

IDENTIFIER NOT YET DECLARED:
>>> *location of the searched identifier* := *beginning of the*
>>>> *identifier list* ;
>>> **go to** *LOOP K* ;

END PROCEDURE STORE IDENTIFIER:
>>> **end** *store the identifier in the for list* α ;

comment A simple nonformal or value listed identifier occurring as a left part variable is inadmissible for recursive address calculation. This identifier must be coupled with each surrounding possibly suitable for statement lying within the scope of the identifier. Input parameter for the procedure *store the identifier in the for list* α is the global string variable *identifier*. ;

>> **procedure** *eventually clear go to or colon equal in push down* ;
>> **begin**
>>> **if** *push down contains* ('**go to**') ∨ *push down contains* (':=')
>>> **then**
>>> *reset push down*
>> **end** *eventually clear go to or colon equal in push down* ;

>> **procedure** *eventually handle ends of for statements* ;
>> **begin**

END OF FOR STATEMENTS:
>>> **if** $\sigma[s, COL\ \sigma\ supplement\ information\ 2] =$
>>>> '**constructed block begin**'
>>> ∧ $\sigma[s-1, COL\ \sigma\ symbol\ pushed\ down] =$ '**for**' **then**
>>> **begin**
>>>> *put a closing bracket in form of a program structure symbol*
>>>>> *into the identifier list* δ ('**block end**') ;
>>>> *reset push down* ;
>>>> *store in the modified input program* χ1 ('**block end**') ;
>>>> *put a closing bracket in form of a program structure symbol*
>>>>> *into the for list* α ('**for end**') ;
>>>> *reset push down* ;
>>>> *store in the modified input program* χ1 ('**for end**') ;
>>>> **go to** *END OF FOR STATEMENTS*
>>> **end**
>> **end** *eventually handle ends of for statements* ;

procedure *eventually handle ends of then and for statements* ;
begin
END OF THEN AND FOR STATEMENTS:
  eventually handle ends of for statements ;
  if *push down contains* ('**then**')
   ∧ σ[*s, COL* σ *state*] = '**statement**' **then**
  begin
   reset push down ;
   go to *END OF THEN AND FOR STATEMENTS*
  end ;
 end *eventually handle ends of then and for statements* ;

comment The following pages 169—179 contain check procedures. Check procedures are unnecessary as long as the input Algol program χ is correct. In practice, check procedures are enormously important since few Algol programmers can guarantee to deliver correct programs in first attempts. In order to guarantee an efficient translation and error detecting process check procedures should fulfill the following tasks:

1. Detect syntactical errors,

2. report error situations to the Algol programmer by output procedures,

3. try to correct the configurations of lists and variables in order to produce no (or at least few) sequential errors,

4. return to the main program in a reasonable manner in order to find other syntactical errors.

See section 7.5.1. ;

procedure *announce* (*string*) ;
 string *string* ;
 code ;

comment Reports in
    check begin for or go to
    check semicolon
    check end
    check arithmetic relational or logical operator
    check colon
    check do step or while
    check if
    check then
    check else
    check opening square bracket
    check closing square bracket
    check colon equal
    check comma

are duplicated by similar reports which occur in the translation pass. They are thus unnecessary if we may suppose that this pass is always entered. ;

```
procedure insert (symbol) ;
  string symbol ;
  begin
    c := c max− 1 ;
    inserted input component := true ;
    input component := symbol ;
    go to CASCADE
  end insert ;

procedure insert matching bracket ;
  begin
    if push down contains ('switch')
      ∨ push down contains ('go to')
      ∨ push down contains (':=') then
        insert (';')
    else
    if push down contains ('for') then
      insert ('do')
    else
    if push down contains ('step') then
      insert ('until')
    else
    if push down contains ('if') then
      insert ('then')
    else
    if push down contains ('then') then
      insert ('else')
    else
    if push down contains ('(') then
      insert (')')
    else
    if push down contains ('[') then
      insert (']')
  end insert matching bracket ;

procedure check initial begin ;
  begin
    if input component = 'begin' then
      go to END CHECK INITIAL BEGIN ;
    error := true ;
```

    announce ('the ALGOL program does not start with a delimiter
      begin') ;
    go to *READ AGAIN* ;
END CHECK INITIAL BEGIN:
    end *check initial begin* ;

   procedure *check begin for or go to* ;
    begin
      if $\sigma[s, COL\ \sigma\ state] \neq$ '**empty**'
      $\wedge$ (*push down contains* ('**begin**')
      $\vee$ *push down contains* ('**then**')) **then**
        go to *END CHECK BEGIN* ;
      error := **true** ;
      announce ('a statement delimiter' $\oplus$ *input component* $\oplus$
       'does not match **begin** or **then** Σ in push down') ;
      insert matching bracket ;
END CHECK BEGIN:
    end *check begin for or go to* ;

   procedure *check semicolon* ;
    begin
      if *push down contains* ('**begin**') **then**
        go to *END CHECK SEMICOLON* ;
      error := **true** ;
      announce ('semicolon does not match **begin** in push down') ;
      insert matching bracket ;
END CHECK SEMICOLON:
    end *check semicolon* ;

   procedure *check end* ;
    begin
      if *push down contains* ('**begin**')
      $\wedge\ \sigma[s, COL\ \sigma\ supplement\ information\ 2] \neq$
        '**constructed block begin**' **then**
      go to *END CHECK END* ;
      error := **true** ;
      announce ('**end** does not match **begin** in push down') ;
      insert (';') ;
END CHECK END:
    end *check end* ;

   procedure *check arithmetic relational or logical operator* ;
    begin
      if $\sigma[s, COL\ \sigma\ state] =$ '**empty**'
      $\wedge \neg$ *push down contains* ('**go to**') **then**
        go to *END CHECK OPERATOR* ;

error := **true** ;

announce ('an operator' ⊕ *input component* ⊕ 'matches a wrong symbol' ⊕ σ[*s, COL σ symbol pushed down*] ⊕ 'in a wrong push down state' ⊕ σ[*s, COL σ state*]) ;

END CHECK OPERATOR:

    end *check arithmetic relational or logical operator* ;

  procedure *check colon* ;

    begin

      if σ[*s, COL σ supplement information 1*] =

        '**bound pair list**' **then**

        go to *END CHECK COLON* ;

      error := **true** ;

      announce ('a colon neither occurs within a bound pair list nor as a label colon') ;

END CHECK COLON:

    end *check colon* ;

  procedure *check do step or while* ;

    begin

      if *push down contains* (':=') **then**

        begin

          if σ[*s*−1, *COL σ symbol pushed down*] = '**for**' **then**

            go to *END CHECK DO* ;

          error := **true** ;

          announce (*input component* ⊕ 'does not match **for** :=

            in push down') ;

          reset push down ;

          go to *CASCADE*

        end ;

      error := **true** ;

      announce (*input component* ⊕ 'does not match :=

        in push down') ;

      if *push down contains* ('**for**') **then**

        begin

          push down (':=') ;

          go to *CASCADE*

        end ;

      if σ[*s, COL σ state*] = '**empty**' **then** *insert matching bracket* ;

      push down ('**for**') ;

      σ[*s, COL σ state*] := '**statement**' ;

      push down (':=') ;

      go to *CASCADE* ;

END CHECK DO:

  end *check do step or while* ;

  procedure *check if* ;
  begin
    if *push down contains* ('**then**') **then**
      announce ('**then** is followed by a conditional statement or
        expression')
  end *check if* ;

  procedure *check then* ;
  begin
    if *push down contains* ('**if**') **then**
      go to *END CHECK THEN* ;
    error := **true** ;
    announce ('**then** does not match **if** in push down') ;
    if $\sigma[s, COL\ \sigma\ state] \neq$ '**empty**' **then**
      begin
        push down ('**if**') ;
        $\sigma[s, COL\ \sigma\ supplement\ information\ 1] :=$ '**statement**' ;
        go to *CASCADE*
      end ;
    if $\sigma[s-1, COL\ \sigma\ symbol\ pushed\ down] =$ '**if**' **then**
      insert matching bracket ;
    push down ('**if**') ;
    go to *CASCADE* ;

END CHECK THEN:

  end *check then* ;

  procedure *check else* ;
  begin
    if *push down contains* ('**then**') **then**
      go to *END CHECK ELSE* ;
    error := **true** ;
    announce ('**else** does not match **then** in push down') ;
    if $\sigma[s, COL\ \sigma\ state] \neq$ '**empty**' **then**
      begin
        push down ('**then**') ;
        $\sigma[s, COL\ \sigma\ state] :=$ '**statement**' ;
        go to *CASCADE*
      end ;

if $\sigma[s-1, COL\ \sigma\ symbol\ pushed\ down] = $ '**then**'
  $\vee$ *push down contains* ('**if**') **then**
  insert matching bracket ;
push down ('**then**') ;
go to *CASCADE* ;

END CHECK ELSE:
  end *check else* ;

procedure *check else behind do* ;
  begin
    if $\sigma[s, COL\ \sigma\ supplement\ information\ 2] = $
      '**constructed block begin**'
      $\wedge\ \sigma[s-1, COL\ \sigma\ symbol\ pushed\ down] = $ '**for**' **then**
      announce ('**else** is terminating a for statement')
  end *check else behind do* ;

procedure *check opening square bracket* ;
  begin
    error := **true** ;
    announce ('an opening square bracket occurs without any
      preceding identifier') ;
    push down ('[') ;
    go to *STORE*
  end *check opening square bracket* ;

procedure *check closing bracket* (*opening bracket*) ;
  string *opening bracket* ;
  begin
    if *push down contains* (*opening bracket*) **then**
      go to *END CHECK CLOSING BRACKET* ;
    error := **true** ;
    announce ('a closing bracket' $\oplus$ *input component* $\oplus$
      'does not match' $\oplus$ *opening bracket* $\oplus$ 'in push down') ;
    if $\sigma[s, COL\ \sigma\ state] = $ '**empty**'
      $\wedge\ \sigma[s-1, COL\ \sigma\ symbol\ pushed\ down] = opening$
      bracket **then**
      insert matching bracket ;
    push down (*opening bracket*) ;
    go to *CASCADE* ;

END CHECK CLOSING BRACKET:
  end *check closing bracket* ;

procedure *check colon equal* ;
  begin
    if *push down contains* (':=') **then**

begin

  if $\sigma[s-1, COL\ \sigma\ symbol\ pushed\ down] \neq$ '**for**'

  $\land\ \sigma[s-1, COL\ \sigma\ symbol\ pushed\ down] \neq$ '**switch**' **then**

    go to $END\ CHECK\ COLON\ EQUAL$;

  $error :=$ **true** ;

  $announce$ ('there is more than one controlled variable in a
for statement or more than one switch identifier in a
switch declaration') ;

    go to $END\ CHECK\ COLON\ EQUAL$

  end ;

  if $\sigma[s, COL\ \sigma\ state] \neq$ '**empty**' **then**

    go to $END\ CHECK\ COLON\ EQUAL$;

  $error :=$ **true** ;

  $announce$ ('wrong push down state when colon equal comes in') ;

  $insert\ matching\ bracket$;

$END\ CHECK\ COLON\ EQUAL$:

  end $check\ colon\ equal$;

  procedure $check\ comma$;

  begin

    if $push\ down\ contains$ ('(') $\lor$ $push\ down\ contains$ ('[')

    $\lor$ ($push\ down\ contains$ (':='))

    $\land$ ($\sigma[s-1, COL\ \sigma\ symbol\ pushed\ down] =$ '**for**'

    $\lor\ \sigma[s-1, COL\ \sigma\ symbol\ pushed\ down] =$ '**switch**')) **then**

     go to $END\ CHECK\ COMMA$;

    $error :=$ **true** ;

    $announce$ ('comma matches a wrong push down configuration') ;

$END\ CHECK\ COMMA$:

  end $check\ comma$;

  procedure $check\ declaration$;

  begin

    if $\sigma[s, COL\ \sigma\ state] =$ '**declaration**' **then**

     go to $END\ CHECK\ DECLARATION$;

    $error :=$ **true** ;

    $announce$ ('the series of declarations in a block head has been
interrupted by a statement') ;

    if $\neg$ $push\ down\ contains$ ('**begin**') **then** $insert\ matching\ bracket$;

$END\ CHECK\ DECLARATION$:

  end $check\ declaration$;

procedure *check specifier* ;
  begin
    error := **true** ;
    announce ('a specifier' ⊕ input component ⊕ 'occurs outside
      a procedure heading') ;
    go to *READ*
  end *check specifier* ;

procedure *check identifier in (identifier list)* ;
  string *identifier list* ;
  begin
    if *associated integer number (input component)* = *associated*
      integer number ('Richard') **then**
      go to *END CHECK IDENTIFIER* ;
    error := **true** ;
    announce ('there is missing an identifier in a' ⊕ *identifier list*) ;
    c := *c max* − 1 ;
    input component := 'missing-identifier' ;

END CHECK IDENTIFIER:
  end *check identifier in* ;

procedure *check for the same identifier in the same block*
  (identifier) ;
  string *identifier* ;
  begin
    for k := d − 1 **step** − 1 **until** *beginning of the identifier list* **do**
      begin
        if $\delta[k, COL\ \delta\ program\ structure\ symbol]$ =
          'block begin' **then**
          go to *END PROCEDURE CHECK* ;
        if $\delta[k, COL\ \delta\ external\ identifier]$ = *identifier* **then**
          go to *ERROR PROCEDURE CHECK* ;
        if $\delta[k, COL\ \delta\ program\ structure\ symbol]$ =
          'procedure end' **then**
          k := $\delta[k, COL\ \delta\ associated\ internal\ identifier]$
      end k ;

ERROR PROCEDURE CHECK:
    error := **true** ;
    announce (*identifier* ⊕ 'has been declared twice in the same
      block') ;

END PROCEDURE CHECK:
  end *check for the same identifier in the same block* ;

procedure *check comma in (identifier list)* ;
  string *identifier list* ;
  begin
    if *input component* $=$ ',' **then**
      go to *END CHECK COMMA* ;
    error $:=$ **true** ;
    announce ('there is missing a comma in a' $\oplus$ *identifier list*) ;
    $c := c\,max - 1$;
    input component $:=$ ';' ;
    go to *END DECLARATION* ;
END CHECK COMMA:
  end *check comma in* ;

  procedure *check colon equal in switch declaration* ;
  begin
    if *input component* $=$ ':$=$' **then**
      go to *END CHECK COLON EQUAL* ;
    error $:=$ **true** ;
    announce ('a switch identifier is not followed by a colon equal') ;
    $c := c\,max - 1$;
    input component $:=$ ':$=$' ;
END CHECK COLON EQUAL:
  end *check colon equal in switch declaration* ;

  procedure *check comma in array segment* ;
  begin
    if *input component* $=$ ',' **then**
      go to *END CHECK COMMA IN ARRAY SEGMENT* ;
    error $:=$ **true** ;
    announce ('there is missing a comma in an array segment') ;
    $c := c\,max - 1$;
    input component $:=$ '[' ;
    go to *BOUND PAIR LIST* ;
END CHECK COMMA IN ARRAY SEGMENT:
  end *check comma in array segment* ;

  procedure *check opening bracket of formal parameter part* ;
  begin
    if *input component* $=$ '(' **then**
      go to *END CHECK OPENING BRACKET* ;
    error $:=$ **true** ;
    announce ('a formal parameter part does not start with an
      opening round bracket') ;
    if *input component* $=$ ')' $\lor$ *input component* $=$ ',' **then**

    go to *END CHECK OPENING BRACKET* ;
$c := c\ max - 1$;
input component := ';' ;
go to *PROCEDURE SIDE ENTRY 3* ;
END CHECK OPENING BRACKET:
    end *check opening bracket of formal parameter part* ;

  procedure *check comma in formal parameter list* ;
    begin
      if *input component* = ',' **then**
        go to *END CHECK COMMA* ;
      error := **true** ;
      announce ('there is missing a comma in a formal parameter
        list') ;
      if *input component* = ';' **then**
        go to *PROCEDURE SIDE ENTRY 3* ;
      $c := c\ max - 1$;
      input component := ';' ;
      go to *PROCEDURE SIDE ENTRY 3* ;
END CHECK COMMA:
    end *check comma in formal parameter list* ;

  procedure *check semicolon behind formal parameter part* ;
    begin
      if *input component* = ';' **then**
        go to *END CHECK SEMICOLON* ;
      error := **true** ;
      announce ('there is no semicolon behind a formal parameter
        part') ;
      if *input component* = '(' $\lor$ *input component* = ')'
      $\lor$ *input component* = ',' **then**
        go to *END CHECK SEMICOLON* ;
      $c := c\ max - 1$;
      input component := ';' ;
END CHECK SEMICOLON:
    end *check semicolon behind formal parameter part* ;

  procedure *check value listed identifier* ;
    begin
      if $\delta[k,\ COL\ \delta\ mode]$ = '**formal**' **then**
        go to *END CHECK VALUE LISTED IDENTIFIER* ;
      error := **true** ;
      announce ('a formal parameter' $\oplus$ *input component* $\oplus$ 'has
        been value listed several times') ;

END CHECK VALUE LISTED IDENTIFIER:

    end *check value listed identifier* ;

  procedure *check specified identifier* ;

    begin

      if $\delta[k, COL\ \delta\ kind] = $ '**unspecified formal parameter**' **then**

        go to *END CHECK SPECIFIED IDENTIFIER* ;

      error := **true** ;

      announce ('a formal parameter' $\oplus$ *input component* $\oplus$ 'has been specified several times') ;

END CHECK SPECIFIED IDENTIFIER:

    end *check specified identifier* ;

  procedure *check value listed or specified identifier* (*listed*) ;

    string *listed* ;

    begin

      if *available formal parameter* **then**

        go to *END CHECK IDENTIFIER* ;

      error := **true** ;

      announce (*listed* $\oplus$ 'identifier' $\oplus$ *input component* $\oplus$ 'is different from all formal parameters of a procedure declaration') ;

END CHECK IDENTIFIER:

    end *check value listed or specified identifier* ;

  procedure *check comma in value or specification list* (*list*) ;

    string *list* ;

    begin

      if *input component* = ',' **then**

        go to *END CHECK COMMA* ;

      error := **true** ;

      announce ('there is missing a comma in a' $\oplus$ *list*) ;

      if *input component* = '(' $\vee$ *input component* = ')' **then**

        go to *END CHECK COMMA* ;

      $c := c\ max - 1$;

      input component := ';' ;

      go to *PROCEDURE SIDE ENTRY 4* ;

END CHECK COMMA:

    end *check comma in value or specification list* ;

  procedure *check value listing* ;

    begin

      for $k := $ *location of the procedure identifier* $+ 2$ **step** 1 **until** d **do**

        if $\delta[k, COL\ \delta\ mode] = $ '**value**'

          $\wedge\ \delta[k, COL\ \delta\ kind] = $

            '**unspecified formal parameter**' **then**

begin
   error := **true** ;
   announce ('there is a value listed parameter' $\oplus$
      $\delta[k,\,COL\ \delta\ external\ identifier]\ \oplus$ 'which is not
      specified')
end k ;
end *check value listing* ;

procedure *check value listing 1* ;
  begin
    if *behind value list* **then**
      announce ('value list in wrong position') ;
    behind value list := **true**
  end *check value listing 1* ;

INITIALIZATIONS FOR THE PREPARATORY PASS:
  COL σ symbol pushed down := 1 ;
  COL σ state := 2 ;
  COL σ supplement information 1 := 3 ;
  COL σ supplement information 2 := 4 ;
  COL σ reference to the identifier list := 5 ;
  COL σ number of subscripts := 6 ;
  COL σ array block level := 6 ;
  COL σ maximum array block level := 7 ;

  c max := *c* := 0 ;
  c1 := 0 ;
  s := 0 ;
  last push down column := 7 ;
  d := -1 ;
  beginning of the identifier list := 0 ;
  last identifier list column := 16 ;
  d1 := 0 ;
  beginning of the number list := -1 ;
  a := 0 ;
  beginning of the for list := 1 ;
  array block level := 0 ;
  maximum array block level := 0 ;
  error := **false** ;
  static procedure level := -1 ;
  maximum static procedure level := 0 ;
  inserted input component := **false** ;

comment The following figures illustrate the appearance of liss-just after the preparatory pass has been called and the pointers ats sociated with the lists have been initialized.

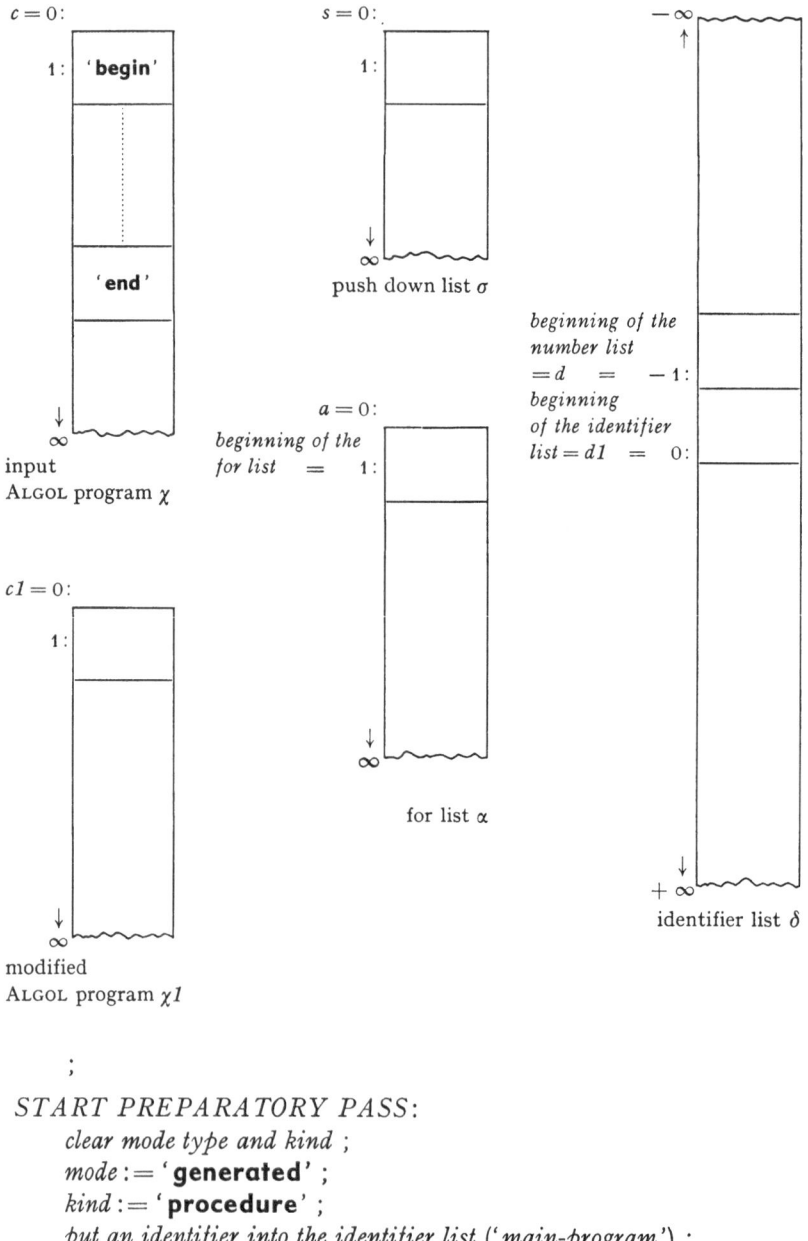

;

START PREPARATORY PASS:
    clear mode type and kind ;
    mode := '**generated**' ;
    kind := '**procedure**' ;
    put an identifier into the identifier list ('*main-program*') ;

δ [*beginning of the identifier list, COL δ dimension*] := 0 ;
push down ('**empty**') ;
put an opening bracket in form of a program structure symbol into the identifier list δ ('**block begin**') ;
push down with a reference to the identifier list δ ('**begin**') ;
σ [*s, COL σ state*] := '**statement**' ;
σ [*s, COL σ supplement information 1*] := '**block**' ;
σ [*s, COL σ supplement information 2*] :=
  '**constructed block begin**' ;
store in the modified input program $\chi 1$ ('**block begin**') ;
mode := '**standard**' ;
type := '**real**' ;
kind := '**procedure**' ;
static procedure level := 0 ;
put an identifier into the identifier list δ ('*sin*') ;
put an identifier into the identifier list δ ('*cos*') ;
put an identifier into the identifier list δ ('*abs*') ;
put an identifier into the identifier list δ ('*sqrt*') ;
put an identifier into the identifier list δ ('*arctan*') ;
put an identifier into the identifier list δ ('*ln*') ;
put an identifier into the identifier list δ ('*exp*') ;
type := '**integer**' ;
put an identifier into the identifier list δ ('*sign*') ;
put an identifier into the identifier list δ ('*entier*') ;
for k := *beginning of the identifier list* + 2 **step** 1 **until** d **do**
  δ [*k, COL δ dimension*] := 1 ;
put a number into the number list ('**true**', '**Boolean**') ;
put a number into the number list ('**false**', '**Boolean**') ;
put a number into the number list ('*0*', '**integer**') ;
put a number into the number list ('*1*', '**integer**') ;

READ AGAIN:
  read from the input program χ ;
  check initial begin ;
  put an opening bracket in form of a program structure symbol into the identifier list δ ('**block begin**') ;
  push down with a reference to the identifier list δ ('**begin**') ;
  read from the input program χ ;
  σ [*s, COL σ state*] := **if** *declarator* (*input component*) **then**
    '**declaration**' **else** '**statement**' ;
  σ [*s, COL σ supplement information 1*] := '**block**' ;
  store in the modified input program $\chi 1$ ('**block begin**') ;
  go to *CASCADE* ;

STORE:
> *store in the modified input program χ1 (input component)* ;

READ:
> *read from the input program χ* ;

CASCADE:
> **go to** *cascade [associated integer number (input component)]* ;

comment Many components of the input ALGOL program χ (e.g. arithmetic, relational, and logical operators) are handled by the preparatory pass without leaving the "main loop"

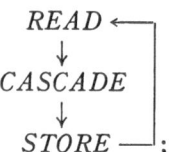

$$
\begin{array}{l}
READ \leftarrow \\
\quad \downarrow \\
CASCADE \\
\quad \downarrow \\
STORE \;\; ;
\end{array}
$$

ARITHMETIC OPERATORS:

RELATIONAL OPERATORS:

LOGICAL OPERATORS:
> *check arithmetic relational or logical operator* ;
> **go to** *STORE* ;

COLON:
> *check colon* ;
> **go to** *STORE* ;

STEP:
> *check do step or while* ;
> *push down ('***step***')* ;
> **go to** *STORE* ;

UNTIL:
> *check closing bracket ('***step***')* ;
> *reset push down* ;
> **go to** *STORE* ;

WHILE:
> *check do step or while* ;
> **go to** *STORE* ;

GO TO:
> *check begin for or go to* ;
> *push down ('***go to***')* ;
> **go to** *STORE* ;

COLON EQUAL:
    check colon equal ;
    if $\sigma[s, COL\ \sigma\ state] \neq$ '**empty**' **then**
      push down (':=') ;
    go to *STORE* ;

    comment '**go to**' and ':=' are pushed down in order to show that the preparatory pass leaves the statement state and enters the expression state (see *COL* σ *state*). If there is more than one left part variable ':=' is pushed down only once. ;

IF:
    check if ;
    state := $\sigma[s, COL\ \sigma\ state]$;
    push down ('**if**') ;
    if *state* $\neq$ '**empty**' **then**
      $\sigma[s, COL\ \sigma\ supplement\ information\ 1]$:= '**statement**' ;
    go to *STORE* ;

THEN:
    check then ;
    state := $\sigma[s, COL\ \sigma\ supplement\ information\ 1]$;
    reset push down ;
    push down ('**then**') ;
    if *state* = '**empty**' **then**
      go to *STORE* ;
    $\sigma[s, COL\ \sigma\ state]$:= '**statement**' ;
    if $\neg$ *inserted input component* **then**
      store in the modified input program $\chi 1$ ('**then** Σ') ;
    go to *READ* ;

ELSE:
    eventually clear go to or colon equal in push down ;
    check else behind do ;
    eventually handle ends of for statements ;
    check else ;
    state := $\sigma[s, COL\ \sigma\ state]$;
    reset push down ;
    if *state* = '**empty**' **then**
      go to *STORE* ;
    if $\neg$ *inserted input component* **then**
      store in the modified input program $\chi 1$ ('**else** Σ') ;
    go to *READ* ;

FOR:

    check begin for or go to ;
    push down ('**for**') ;
    $\sigma[s, COL\ \sigma\ state] := $ '**statement**' ;
    go to *STORE* ;

DO:

    check do step or while ;
    reset push down ;
    store in the modified input program $\chi 1$ ('**do**') ;
    put an opening bracket in form of a program structure symbol into the
       identifier list δ ('**block begin**') ;
    put an opening bracket in form of a program structure symbol into the
       for list α ('**for possibly suitable**') ;
    push down with a reference to the identifier list δ ('**begin**') ;
    $\sigma[s, COL\ \sigma\ state] := $ '**statement**' ;
    $\sigma[s, COL\ \sigma\ supplement\ information\ 1] := $ '**block**' ;
    $\sigma[s, COL\ \sigma\ supplement\ information\ 2] := $
      '**constructed block begin**' ;
    store in the modified input program $\chi 1$ ('**block begin**') ;
    go to *READ* ;

comment When reading '**do**' from the input program χ the preparatory pass will find the following push down configuration:

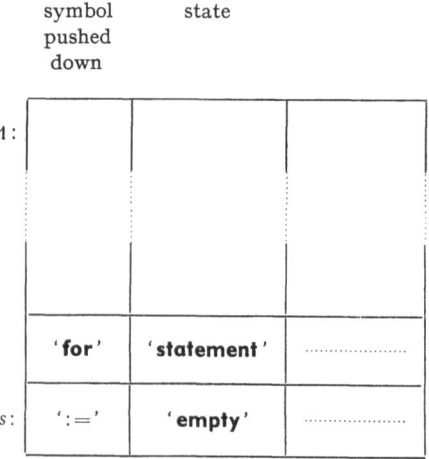

push down list σ

At the end of handling the '**do**' the push down list σ, the identifier list δ, and the for list α will look like:

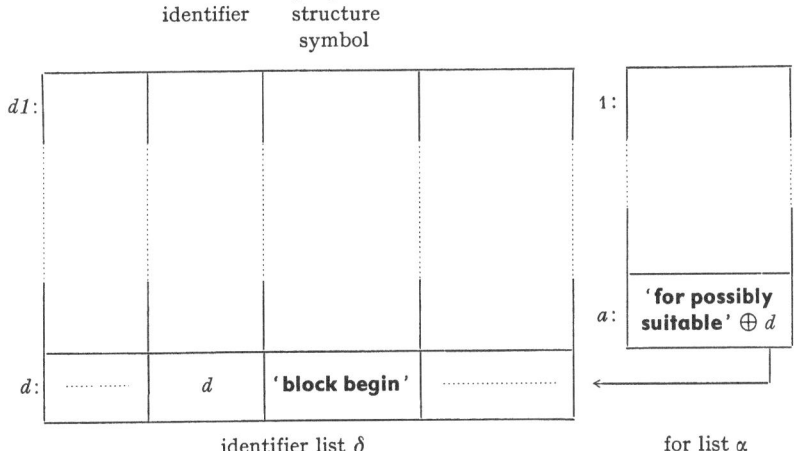

	symbol pushed down	state	supple-ment infor-mation 1	supplement information 2	reference to the identifier list	
1:						
	'**for**'	'**statement**'	'**empty**'	'**empty**'	'**empty**'	....
s:	'**begin**'	'**statement**'	'**block**'	'**constructed block begin**'	d	....

push down list σ

		internal identifier	program structure symbol				
$d1$:						1:	
						a:	'**for possibly suitable**' $\oplus\, d$
d:	....	d	'**block begin**'	....			

identifier list δ for list α ;

$COMMA$:

    check comma ;

    if $\sigma[s, COL\ \sigma\ supplement\ information\ 2] =$ '**count subscripts**' **then**
      $\sigma[s, COL\ \sigma\ number\ of\ subscripts] :=$
        $\sigma[s, COL\ \sigma\ number\ of\ subscripts] + 1$

    else

if *push down contains* (':=')
    $\wedge\ \sigma[s-1, COL\ \sigma\ symbol\ pushed\ down] = $ '**for**' **then**
    begin
        store in the modified input program $\chi 1$ ('**for,**') ;
        go to *READ*
    end ;
go to *STORE* ;

SEMICOLON:
    eventually clear go to or colon equal in push down ;
    eventually handle ends of then and for statements ;
    if $\sigma[s, COL\ \sigma\ supplement\ information\ 2] = $
        '**constructed block begin**'
    $\wedge\ \sigma[s-1, COL\ \sigma\ symbol\ pushed\ down] = $
        '**procedure begin**' **then**
    begin
        *put a closing bracket in form of a program structure symbol into
            the identifier list* δ ('**block end**') ;
        store in the modified input program $\chi 1$ ('**block end**') ;
        reset push down ;
        location of the procedure identifier :=
            $\sigma[s, COL\ \sigma\ reference\ to\ the\ identifier\ list] - 1$;
        δ[*location of the procedure identifier,
            COL* δ *maximum array block level*] := *maximum array
            block level* ;
        *put a closing bracket in form of a program structure symbol into
            the identifier list* δ ('**procedure end**') ;
        store in the modified input program $\chi 1$ ('**procedure end**') ;
        reset push down ;
        array block level := $\sigma[s, COL\ \sigma\ array\ block\ level]$;
        maximum array block level :=
            $\sigma[s, COL\ \sigma\ maximum\ array\ block\ level]$;
        static procedure level := *static procedure level* $- 1$;
        go to *BELOW*
    end procedure end ;
    if *push down contains* ('**switch**') **then**
    begin
        reset push down ;
BELOW:
        if *inserted input component* **then**
            inserted input component := **false** ;
        store in the modified input program $\chi 1$ (';') ;

       go to *END DECLARATION*
     end switch end ;
   check semicolon ;
   go to *STORE* ;
OPENING ROUND BRACKET:
   push down (' (') ;
   go to *STORE* ;
CLOSING ROUND BRACKET:
   check closing bracket (' (') ;
   reset push down ;
   go to *STORE* ;
OPENING SQUARE BRACKET:
   check opening square bracket ;
CLOSING SQUARE BRACKET:
   check closing bracket ('[') ;
   if σ[s, COL σ *supplement information 1*] =
      '**bound pair list**' **then**
     go to *END OF BOUND PAIR LIST* ;
   if σ[s, COL σ *supplement information 2*] =
      '**count subscripts**' **then**
    begin
     location of the identifier :=
       σ[s, COL σ *reference to the identifier list*] ;
     dimension := σ[s, COL σ *number of subscripts*] ;
     if δ[*location of the identifier, COL δ dimension*] = '**empty**'
       $\vee$ δ[*location of the identifier,*
        COL δ dimension] < *dimension* **then**
      begin
       δ[*location of the identifier, COL δ dimension*] := *dimension* ;
       k := *location of the identifier* ;
       identifier := δ[*location of the identifier,*
        COL δ external identifier] ;
RETURN CLOSING SQUARE BRACKET:
        k := $k-1$;
        if δ[k, COL δ *program structure symbol*] =
         '**procedure begin**' **then**
          go to *BELOW CLOSING SQUARE BRACKET* ;
        if *identifier* = δ[k, COL δ *external identifier*] **then**
         δ[k, COL δ *dimension*] := *dimension* ;
        go to *RETURN CLOSING SQUARE BRACKET*
      end dimension
    end count subscripts ;

BELOW CLOSING SQUARE BRACKET:
   reset push down ;
   go to *STORE* ;

  comment The rather sophisticated situation

$$\delta\,[location\ of\ the\ identifier,\ COL\ \delta\ dimension] < dimension$$

may come up, e.g., when the following first piece of a correct Algol program χ has been read:

begin
  procedure $p\,(a,\,a)$ ⌐
    array a ⌐
    begin
      begin
        procedure q ⌐
          begin
            $a[1]:=2$
          end q ⌐
        array $a[1:10]$ ⌐
      end ⌐
      $a[1,\,1]\;\cdots$ ⌐
      comment c is pointing to] ⌐

This program contains different arrays denoted by the same identifier a and an array declaration a following a procedure declaration p in the same block.

It is worthwhile to discuss the situation where the array declaration is

$$\textbf{array } a[1:10,\ 1:10,\ 1:10] ⌐$$

instead of

$$\textbf{array } a[1:10] ⌐$$

and the assignment statement

$$a[1]:=2$$

is replaced by

$$a[1,\,1,\,1]:=2\,.$$

The preparatory pass delivers an incorrect number of subscripts for the formal array a. The number delivered is 3 which is larger than the correct number 2. The error is not serious since the translation pass uses the number of subscripts only in order to reserve enough storage places for the information vectors of formal arrays (see section 4.9.2, p. 65). In order to enable the run time instructions

        procedure call   and
        formal procedure call

(see sections 6.2 and 6.3, p. 104) to check the correspondence of actual and formal parameters the translation pass will deliver the exact numbers of subscripts. The translation pass can find the exact numbers as the complete identifier list is at its disposal.

The piece of program after *RETURN OPENING SQUARE BRACKET* is necessary as there may be ALGOL procedure declarations (see e.g. $p(a, a)$ in the above example) with several identical formal parameters. ;

BEGIN:

> *check begin for or go to* ;
> *read from the input program* χ ;
> **if** *declarator (input component)* **then**
>> **begin**
>>> *put an opening bracket in form of a program structure symbol into the identifier list* δ ('**block begin**') ;
>>> *push down with a reference to the identifier list* δ ('**begin**') ;
>>> $\sigma[s, COL\ \sigma\ state] :=$ '**declaration**' ;
>>> $\sigma[s, COL\ \sigma\ supplement\ information\ 1] :=$ '**block**' ;
>>> *store in the modified input program* $\chi 1$ ('**block begin**')
>> **end**
>> **else**
>>> **begin**
>>>> *push down* ('**begin**') ;
>>>> $\sigma[s, COL\ \sigma\ state] :=$ '**statement**' ;
>>>> *store in the modified input program* $\chi 1$ ('**begin**')
>>> **end** ;
>> **go to** *CASCADE* ;

END:

> *eventually clear go to or colon equal in push down* ;
> *eventually handle ends of then and for statements* ;
> *check end* ;
> **if** $\sigma[s, COL\ \sigma\ supplement\ information\ 1] =$ '**block**' **then**
>> **begin**
>>> *put a closing bracket in form of a program structure symbol into the identifier list* δ ('**block end**') ;
>>> **if** $\sigma[s, COL\ \sigma\ supplement\ information\ 2] =$ '**array block**' **then**
>>> *array block level* := *array block level* $- 1$;
>>> *store in the modified input program* $\chi 1$ ('**block end**')
>> **end**
>> **else** *store in the modified input program* $\chi 1$ ('**end**') ;
> *reset push down* ;

if $s > 2$ **then**
  go to *READ* ;
put a closing bracket in form of a program structure symbol into the
  identifier list δ ('**block end**') ;
store in the modified input program $\chi 1$ ('**block end**') ;
δ [*beginning of the identifier list,*
  COL δ *maximum array block level*] := *maximum array block level* ;
for $k :=$ *beginning of the for list* **step** 1 **until** a **do**
  $\alpha[k] :=$ *first constituent* $(\alpha[k])$;
go to *FINISH PREPARATORY PASS* ;

BOOLEAN:

INTEGER:

REAL:
  check declaration ;
  clear mode type and kind ;
  type := *input component* ;
  read from the input program χ ;
  if *input component* = '**array**' **then**
    go to *ARRAY SIDE ENTRY 1* ;
  if *input component* = '**procedure**' **then**
    go to *PROCEDURE SIDE ENTRY 1* ;
  check identifier in ('type list') ;
  mode := '**normal**' ;
  kind := '**variable**' ;

TYPE RETURN:
  put an identifier into the identifier list δ (*input component*) ;
  read from the input program χ ;
  if *input component* $\neq$ ';' **then**
    begin
      check comma in ('type list') ;
      read from the input program χ ;
      check identifier in ('type list') ;
      go to *TYPE RETURN*
    end ;

END DECLARATION:
  read from the input program χ ;
  if $\neg$ *declarator* (*input component*) **then**
    $\sigma[s, COL \ \sigma \ state] :=$ '**statement**' ;
  go to *CASCADE* ;

ARRAY:

   check declaration ;
   clear mode type and kind ;
   type := '**real**' ;

ARRAY SIDE ENTRY 1:

   if $\sigma[s, COL\ \sigma\ supplement\ information\ 2]$ = '**empty**' **then**
     begin
       $\sigma[s, COL\ \sigma\ supplement\ information\ 2]$:= '**array block**' ;
       array block level := *array block level* + 1 ;
       maximum array block level :=
         maximum (*array block level, maximum array block level*) ;
       correct the array block levels of preceding switches
     end ;
   mode := '**normal**' ;
   kind := '**array**' ;

ARRAY SIDE ENTRY 2:

   number of array identifiers := 0 ;
   location of the leading array identifier := $d + 1$;
   store in the modified input program $\chi1$ ('**array**') ;
   read from the input program χ ;
   check identifier in ('array segment') ;
   store in the modified input program $\chi1$ (*input component*) ;

ARRAY RETURN:

   number of array identifiers := *number of array identifiers* + 1 ;
   put an identifier into the identifier list δ (*input component*) ;
   read from the input program χ ;
   if *input component* ≠ '[' **then**
     begin
       check comma in array segment ;
       read from the input program χ ;
       check identifier in ('array segment') ;
       go to *ARRAY RETURN*
     end ;

BOUND PAIR LIST:

   δ[*location of the leading array identifier*,
     COL δ *number of array identifiers*] := *number of array identifiers* ;
   δ[*location of the leading array identifier*,
     COL δ *array block level*] := *array block level* ;
   push down ('[') ;
   $\sigma[s, COL\ \sigma\ supplement\ information\ 1]$:= '**bound pair list**' ;

$\sigma[s,\ COL\ \sigma\ supplement\ information\ 2] := \text{'count subscripts'}$;
$\sigma[s,\ COL\ \sigma\ number\ of\ subscripts] := 1$;
go to $STORE$;

END OF BOUND PAIR LIST:

    for $k := location\ of\ the\ leading\ array\ identifier$ **step** 1 **until**
      location of the leading array identifier + *number of array*
        identifiers −1 **do**
      $\delta[k,\ COL\ \delta\ dimension] := \sigma[s,\ COL\ \sigma\ number\ of\ subscripts]$;
    reset push down ;
    store in the modified input program $\chi 1$ (']') ;
    store in the modified input program $\chi 1$ (';') ;
    read from the input program χ ;
    if *input component* $=$ ';' **then**
      go to *END DECLARATION* ;
    check comma in ('array list') ;
    go to *ARRAY SIDE ENTRY 2* ;

SWITCH:

    check declaration ;
    clear mode type and kind ;
    $mode := \text{'normal'}$;
    $kind := \text{'switch'}$;
    push down ('**switch**') ;
    store in the modified input program $\chi 1$ ('**switch**') ;
    read from the input program χ ;
    check identifier in ('switch declaration') ;
    put an identifier into the identifier list δ (*input component*) ;
    $\delta[d,\ COL\ \delta\ array\ block\ level] := array\ block\ level$;
    store in the modified input program $\chi 1$ (*input component*) ;
    read from the input program χ ;
    push down (':=') ;
    check colon equal in switch declaration ;
    go to $STORE$;

PROCEDURE:

    check declaration ;
    clear mode type and kind ;

PROCEDURE SIDE ENTRY 1:

    $mode := \text{'normal'}$;
    $kind := \text{'procedure'}$;
    store in the modified input program $\chi 1$ ('**procedure begin**') ;
    read from the input program χ ;
    check identifier in ('procedure declaration') ;

store in the modified input program χ1 *(input component)* ;
store in the modified input program χ1 *(';')* ;
σ[s, COL σ *array block level*] := *array block level* ;
σ[s, COL σ *maximum array block level*] :=
   maximum array block level ;
array block level := 0 ;
maximum array block level := 0 ;
put an identifier into the identifier list δ *(input component)* ;
static procedure level := *static procedure level* +1 ;
maximum static procedure level :=
   maximum (static procedure level, maximum static procedure level) ;
location of the procedure identifier := d ;
put an opening bracket in form of a program structure symbol into the
   identifier list δ *('* **procedure begin** *')* ;
push down with a reference to the identifier list δ *('* **procedure begin** *')*;
number of parameters := 0 ;
read from the input program χ ;
if *input component* = ';' **then**
   go to *PROCEDURE SIDE ENTRY 3* ;
check opening bracket of formal parameter part ;

PROCEDURE SIDE ENTRY 2:
   read from the input program χ ;
   check identifier in (' **formal parameter list** *')* ;
   put a formal parameter into the identifier list δ *(input component)* ;
   number of parameters := *number of parameters* +1 ;
   read from the input program χ ;
   if *input component* ≠ ')' **then**
      begin
         check comma in formal parameter list ;
         go to *PROCEDURE SIDE ENTRY 2*
      end ;
   read from the input program χ ;
   check semicolon behind formal parameter part ;

PROCEDURE SIDE ENTRY 3:
   δ[*location of the procedure identifier, COL* δ *dimension*] :=
      number of parameters ;
   behind value list := **false** ;

PROCEDURE SIDE ENTRY 4:
   read from the input program χ ;
   if *declarator (input component)* ∨ *specifier (input component)* **then**

begin
  if *input component =* ' **value**' **then**
    begin
      check value listing 1 ;

PROCEDURE VALUE RETURN:
      read from the input program χ ;
      check identifier in ('value list') ;
      available formal parameter := **false** ;
      for $k :=$ *location of the procedure identifier* $+ 2$ **step**
        1 **until** d **do**
        if $\delta[k,$ *COL* δ *external identifier*$] =$ *input component* **then**
          begin
            check value listed identifier ;
            available formal parameter := **true** ;
            $\delta[k,$ *COL* δ *mode*$] :=$ ' **value**'
          end k ;
      check value listed or specified identifier ('value listed') ;
      read from the input program χ ;
      if *input component =* ';' **then**
        go to *PROCEDURE SIDE ENTRY 4* ;
      check comma in value or specification list ('value list') ;
      go to *PROCEDURE VALUE RETURN*
    end value ;
   behind value list := **true** ;
   type := ' **empty**' ;
   if *type declarator* (*input component*) **then**
    begin
      type := *input component* ;
      read from the input program χ ;
      if *input component =* ' **array**'
        $\lor$ *input component =* ' **procedure**' **then**
        go to *SPECIFICATION 1* ;
      check identifier in ('specification list') ;
      kind := ' **variable**' ;
      go to *SPECIFICATION 2*
    end *type declarator* ;
   if *input component =* ' **array**' **then**
    type := ' **real**' ;

SPECIFICATION 1:
   kind := *input component* ;

PROCEDURE SPECIFICATION RETURN:
> read from the input program χ ;
> check identifier in ('specification list') ;

'*PECIFICATION 2*:
> available formal parameter := **false** ;
> **for** k := location of the procedure identifier $+2$ **step**
>   1 **until** d **do**
>   **if** $\delta[k,\ COL\ \delta\ external\ identifier] =$ input component **then**
>   **begin**
>   check specified identifier ;
>   available formal parameter := **true** ;
> $\delta[k,\ COL\ \delta\ type] :=$ type ;
> $\delta[k,\ COL\ \delta\ kind] :=$ kind
>   **end** k ;
> check value listed or specified identifier ('specified') ;
> read from the input program χ ;
> **if** input component $=$ ';' **then**
>   **go to** *PROCEDURE SIDE ENTRY 4* ;
> check comma in value or specification list ('specification list') ;
> **go to** *PROCEDURE SPECIFICATION RETURN*
> **end** declarator or specifier ;
> check value listing ;
> put an opening bracket in form of a program structure symbol into the
>   identifier list δ ('**block begin**') ;
> push down with a reference to the identifier list δ ('**begin**') ;
> $\sigma[s,\ COL\ \sigma\ state] :=$ '**statement**' ;
> $\sigma[s,\ COL\ \sigma\ supplement\ information\ 1] :=$ '**block**' ;
> $\sigma[s,\ COL\ \sigma\ supplement\ information\ 2] :=$
>   '**constructed block begin**' ;
> store in the modified input program $\chi 1$ ('**block begin**') ;
> **go to** *CASCADE* ;

VALUE:

LABEL:

STRING SPECIFIER:
> check specifier ;

IDENTIFIER:
> store in the modified input program $\chi 1$ (input component) ;
> identifier := input component ;
> read from the input program χ ;
> **if** input component $=$ ':=' **then**

begin
   store the identifier in the for list α ;
   go to *COLON EQUAL*
end ;
if *input component* = ':' $\wedge$ $\sigma[s, COL\ \sigma\ state] \neq$ '**empty**' **then**
  begin
   clear mode type and kind ;
   mode := '**normal**' ;
   kind := '**label**' ;
   put an identifier into the identifier list δ *(identifier)* ;
   $\delta[d, COL\ \delta\ array\ block\ level] := array\ block\ level$;
   store in the modified input program $\chi 1$ ('**label** :') ;
   go to *READ*
  end ;
if *input component* = '[' **then**
  begin
   push down ('[') ;
   search for the identifier in the identifier list δ *(STORE)* ;
   if ($\delta[location\ of\ the\ searched\ identifier, COL\ \delta\ mode]$ = '**formal**'
     $\vee$ $\delta[location\ of\ the\ searched\ identifier, COL\ \delta\ mode]$ = '**value**')
    $\wedge$ ($\delta[location\ of\ the\ searched\ identifier, COL\ \delta\ kind]$ = '**array**'
     $\vee$ $\delta[location\ of\ the\ searched\ identifier, COL\ \delta\ kind]$ =
      '**unspecified formal parameter**') **then**
    begin
     $\sigma[s, COL\ \sigma\ supplement\ information\ 2]$:=
      '**count subscripts**' ;
     $\sigma[s, COL\ \sigma\ reference\ to\ the\ identifier\ list]$:=
      location of the searched identifier ;
     $\sigma[s, COL\ \sigma\ number\ of\ subscripts]$:= 1
    end ;
   go to *STORE*
  end ;
if (*input component* = '(' $\vee$ *input component* = ';'
  $\vee$ *input component* = '**end**' $\vee$ *input component* = '**else**')
  $\wedge$ $\sigma[s, COL\ \sigma\ state] \neq$ '**empty**' **then**
  begin
   all for statements are unsuitable ('**for unsuitable**') ;
   go to *CASCADE*
  end ;
if $\sigma[s, COL\ \sigma\ state]$ = '**empty**' **then**
  begin
   search for the identifier in the identifier list δ
    (NOT YET DECLARED) ;

if δ [*location of the searched identifier, COL δ kind*] =
    'procedure'
  $\vee$ δ [*location of the searched identifier, COL δ kind*] =
    'unspecified formal parameter'
  $\vee$ δ [*location of the searched identifier, COL δ mode*] =
    'formal'
  $\wedge$ δ [*location of the searched identifier, COL δ kind*] =
    'variable' then

NOT YET DECLARED:
    all for statements are unsuitable
    (**'for unsuitable for index registers'**)
  end ;
 go to *CASCADE* ;

comment Simple left part variables are inadmissible for recursive address calculation and therefore are stored in the for list α.

Do statements containing procedure statements are unsuitable for recursive address calculation. Function designators and calls of name variables do not produce such difficulties because side effects are not allowed, but they cause the surrounding loops to be unsuitable for index register use in rec addr calc.

When the preparatory pass meets a formal subscripted variable then the number of subscripts must be determined. Having done so, the translation pass can handle formal subscripted variables in the same way as nonformal subscripted variables. ;

INTEGER NUMBER:
    store in the modified input program $\chi 1$ (input component) ;
    number := *input component* ;
    put a number into the number list (number, **'integer'**) ;
    read from the input program χ ;
    if *input component* = ':' $\wedge$ σ [*s, COL σ state*] $\neq$ **'empty' then**
      begin
        mode := **'normal'** ;
        type := **'integer'** ;
        kind := **'label'** ;
        put an identifier into the identifier list δ (number) ;
        δ [*d, COL δ array block level*] := *array block level* ;
        *store in the modified input program $\chi 1$ (***'label* :'**) ;
      go to *READ*
      end ;
  go to *CASCADE* ;

REAL NUMBER:
  *put a number into the number list (input component, '***real***')* ;
  go to *STORE* ;
TRUTH VALUE:
  go to *STORE* ;
STRING:
  put a string into the number list (input component) ;
  go to *STORE* ;
FINISH PREPARATORY PASS:
  end *preparatory pass* ;

procedure *recursive address calculation pass* (χ, δ,
            begin identifier list, α, *maximum static procedure level*)
            output: (π, π *add*, δ, *begin identifier list*,
            end identifier list, max ind reg, error) ;
  value *maximum static procedure level* ;
  integer *begin identifier list, end identifier list,*
          maximum static procedure level, max ind reg ;
  Boolean *error* ;
  string array χ, π, π *add*, δ, α ;
  begin
    integer
        chi, k, p, p add, int id, delta, delta add, a, x, w, f, s, id list,
        last suitable for in ω, *begin for block in* δ, *v, index register, pα,*
        COL ι identical, COL ι external identifier, COL ι type,
        COL ι index register, COL ι local ;

    Boolean *suitable loop* ;
    string *incoming symbol, variable* ;
    string array
        $\xi[1:\infty]$, $\omega[1:\infty]$, $\iota[1:\infty, 1:5]$,
        $\varkappa[1:\infty]$, $\varphi[1:\infty]$, $\sigma[1:\infty, 1:3]$, $\pi\alpha[1:\infty]$;
    Boolean procedure *delimiter* (x) ;
      string x ;
      code ;
    Boolean procedure *generated variable* (x) ;
      string x ;
      code ;

Boolean procedure *index register name* (x) ;
  string x ;
  code ;

integer procedure *character order* (x) ;
  string x ;
  code ;

integer procedure *state order* (x) ;
  string x ;
  code ;

comment The values of the functions *character order* and *state order* are the same as those of the procedures used in pass 3 and are listed on pages 246—248 ;

procedure *read* χ ;
  begin $chi := chi + 1$; *incoming symbol* $:= \chi[chi]$ **end** ;

procedure *read* ξ ;
  begin
L: $x := x + 1$; *incoming symbol* $:= \xi[x]$;
    if *incoming symbol* $=$ '**empty**' **then** **go to** L
  end ;

procedure *read* $\pi\alpha$;
  begin $p\alpha := p\alpha + 1$; *incoming symbol* $:= \pi\alpha[p\alpha]$ **end** ;

comment Several of the following procedure declarations are given only by abbreviated notations: The letter n in a procedure identifier gives the number of parameters and denotes that a proper declaration is needed for n$=$1, and/or n$=$2, and/or other values of n. Practically the call of such a procedure with a specific number n of parameters can be replaced by n calls of the appropriate procedure with only one parameter. The "meta-identifier" xi must be replaced by $x1, x2, \ldots$ in dependence of the value of the loop variable i. ;

procedure *store* n *in* $\pi(x1, x2, \ldots, xn)$;
  string $x1, x2, \ldots, xn$;
  begin
    integer i ;
    for $i := 1$ **step** 1 **until** n **do**
      begin $p := p + 1$; $\pi[p] := xi$ **end**
  end ;

```
procedure store n in π add (x1, x2, ..., xn) ;
  string x1, x2, ..., xn ;
  begin
    integer i ;
    for i := 1 step 1 until n do
        begin p add := p add+1 ;  π add [p add] := xi end
  end ;
procedure store n in πα (x1, x2, ..., xn) ;
  string x1, x2, ..., xn ;
  begin
    integer i ;
    for i := 1 step 1 until n do
        begin pα := pα+1 ; πα [pα] := xi end
  end ;
procedure push down n in ω (x1, x2, ..., xn) ;
  string x1, x2, ..., xn ;
  begin
    integer i ;
    for i := 1 step 1 until n do
        begin w := w+1 ; ω [w] := xi end
  end ;
procedure store in ϰ (x) ;
  string x ;
  begin k := k+1 ; ϰ [k] := x end ;
procedure store in φ (x) ;
  string x ;
  begin f := f+1 ; φ [f] := x end ;
procedure push down n in ξ (x1, x2, ..., xn) ;
  string x1, x2, ..., xn ;
  begin
    integer i ;
    for i := 1 step 1 until n do
        begin x := x+1 ; ξ [x] := xi end
  end ;
procedure push down in σ (x1, x2, x3) ;
  string x1, x2, x3 ;
  begin
    σ [s+1, 1] := x1 ;  σ [s+1, 2] := x2 ;  σ [s+1, 3] := x3 ;
    s := s+1 ;
  end ;
```

```
      procedure push down in ι(a, b, c, d, e) ;
         string a, b, c, d, e ;
         begin
            id list := id list + 1 ;
            ι[id list, COL ι identical] := a ;
            ι[id list, COL ι external identifier] := b ;
            ι[id list, COL ι type] := c ;
            ι[id list, COL ι index register] := d ;
            ι[id list, COL ι local] := e ;
         end ;

      procedure generate variable of type (type) result: (variable) ;
         string type, variable ;
         begin
         v := v + 1 ;  variable := 'generated-' ⊕ v ;
         push down in ι ('identical', variable, type, 'empty', 'empty')
         end ;

      Boolean procedure identical (v1, v2) ;
         string v1, v2 ;
         begin
            integer i ;  string v3 ;
            if v1 = v2 then
L3:            begin
                  identical := true ;
                  go to END IDENTICAL
               end ;
            for i := id list step − 1 until 1 do
               begin
                  if ι[i, COL ι external identifier] = v1 then go to L1 ;
                  if ι[i, COL ι external identifier] = v2 then go to L2
               end ;

L1:      v3 := v2 ;  v2 := v1 ;  v1 := v3 ;

L2:      for i := i step − 1 until 1 do
               begin
                  if ι[i, COL ι external identifier] = v1 then go to L3 ;
                  if ι[i, COL ι identical] = 'identical' then
                     begin identical := false ;
                        go to END IDENTICAL end
               end ;

END IDENTICAL:
         end ;
```

```
        procedure identify (v1, v2) ;
          string v1, v2 ;
          begin
            integer i, end list 1, end list 2, end id, l, k ;
            string ι4, ι5 ;
            end list 1 := id list ;
            for i := end list 1 step −1 until 1 do
              begin
                if ι[i, COL ι external identifier] = v1 ∨
                   ι[i, COL ι external identifier] = v2 then
                   go to L1
                else
                   if ι[i, COL ι identical] = 'identical' then
                      end list 1 := i−1
              end ;
L1:         end list 2 := end list 1 ;
            for i := i−1 step −1 until 1 do
              begin
                if ι[i, COL ι external identifier] = v1 ∨
                   ι[i, COL ι external identifier] = v2 then
                   go to L2
                else
                   if ι[i, COL ι identical] = 'identical' then
                      end list 2 := i−1
              end ;
L2:         if end list 1 = end list 2 then go to END IDENTIFY ;
            ι4 := if ι[end list 1, COL ι index register] ≠ 'empty' then
                     ι[end list 1, COL ι index register]
                  else ι[end list 2, COL ι index register] ;
            ι5 := if ι[end list 1, COL ι local] ≠ 'empty' then
                     ι[end list 1, COL ι local]
                  else ι[end list 2, COL ι local] ;
            end id := id list ;
            for i := end list 1 step −1 until 1 do
              begin
                push down in ι ('empty', ι[i, COL ι external identifier],
                   ι[i, COL ι type], ι4, ι5) ;
                if ι[i, COL ι identical] = 'identical' then go to L3
              end ;
L3:         l := end list 1+1−i ;
            for i := i−1 step −1 until end list 2+1 do
              for k := 1 step 1 until 5 do ι[i+l, k] := ι[i, k] ;
            l := end list 2 − end id ;
```

```
    for i := end id + 1 step 1 until id list do
       for k := 1 step 1 until 5 do ι[i+l, k] := ι[i, k] ;
    id list := end id ;
    if ι4 ≠ ι[end list 2, COL ι index register] V
       ι5 ≠ ι[end list 2, COL ι local] then
       begin
          for i := end list 2 step − 1 until 1 do
             begin
                ι[i, COL ι index register] := ι4 ;
                ι[i, COL ι local] := ι5 ;
                if ι[i, COL ι external identifier] = 'identical' then
                   go to END IDENTIFY
             end
       end ;
END IDENTIFY :
    end ;

    procedure mark (n, x, v) ;
       value n ; integer n ; string x, v ;
       begin
          integer i, end list ;
          end list := id list ;
          for i := id list step − 1 until 1 do
             if ι[i, COL ι external identifier] = v then
                go to L
             else
                if ι[i, COL ι identical] = 'identical' then
                   end list := i − 1 ;
L :        for i := end list step − 1 until 1 do
             begin
                ι[i, n] := x ;
                if ι[i, COL ι identical] = 'identical' then go to END
             end ;
END :
       end mark ;

    procedure generate index register (v1, v2) ;
       integer v2 ; string v1 ;
```

comment For the sake of simplicity we make here the assumption that sufficiently many index registers are available in the machine. The counter for the index registers starts with the input parameter *maximum static procedure level*. In practical implementations a test must be inserted to determine whether the maximum number K of index registers is reached. In this case address variables for storing addresses of further subscripted variables must be handled like other generated variables without index registers. It might be useful to keep a special list of all address variables and, later on, to assign index registers only to those variables which occur most frequently ;

> **begin**
> $v2 := index\ register := index\ register + 1$;
> $mark\ (COL\ \iota\ index\ register,\ index\ register,\ v1)$;
>   **if** $max\ ind\ reg < index\ register$ **then**
> $max\ ind\ reg := index\ register$
> **end** ;

> **procedure** *search next for list entry* αa ;
>   **begin**
> $L:$ $a := a + 1$;
>   **if** $\alpha[a] \neq$ '**for unsuitable**' $\wedge$ $\alpha[a] \neq$
>   '**for possibly suitable**' $\wedge$ $\alpha[a] \neq$
>   '**for unsuitable for index registers**' **then**
>   **go to** L
>   **end** ;

> **Boolean procedure** *admissible variable* $(int\ id,\ for\ clause)$;
>   **value** *int id, for clause* ;
>   **integer** *int id* ; **Boolean** *for clause* ;
>   **begin**
>   **integer** *begin a* ; **string** *id* ;
>   *admissible variable* := **false** ;
>   **if** $\neg$ ((**if** *for clause* **then** **false** **else** *local* $(int\ id)$) $\vee$
> $\delta[int\ id,\ COL\ \delta\ mode] =$ '**formal**' $\vee$
> $\delta[int\ id,\ COL\ \delta\ kind] \neq$ '**variable**') **then**
>   **begin**
> $id := \delta[int\ id,\ COL\ \delta\ external\ identifier]$;
> $begin\ a :=$ **if** *for clause* **then**
> a
>   **else** $\omega[last\ suitable\ for\ in\ \omega - 9]$;
> $L:$ $begin\ a := begin\ a + 1$;
>   **if** $\alpha[begin\ a] = id$ **then**
>   **go to** $L1$
>   **else**
>   **if** $\alpha[begin\ a] \neq$ '**for end**' $\wedge$
> $\alpha[begin\ a] \neq$ '**for possibly suitable**' $\wedge$

$\alpha\,[begin\ a] \neq$ **'for unsuitable for index registers'**$\wedge$
$\alpha\,[begin\ a] \neq$ **'for unsuitable'** **then**
    go to L ;
$admissible\ variable := $ **true** ;
L1:   **end**
    end admissible variable ;

  procedure *handling block begin* ;
    begin
$L:$ $delta := delta + 1$;
      if $\delta\,[delta, COL\ \delta\ program\ structure\ symbol] \neq$
        'block begin'$\wedge$
        $\delta\,[delta, COL\ \delta\ program\ structure\ symbol] \neq$
          'procedure begin' **then**
        go to L
    end ;

  procedure *handling block end* ;
    begin
$L:$ $delta := delta + 1$;
      if $\delta\,[delta, COL\ \delta\ program\ structure\ symbol] \neq$ **'block end'**$\wedge$
        $\delta\,[delta, COL\ \delta\ program\ structure\ symbol] \neq$
          'procedure end' **then**
        go to L ;
      if $\delta\,[delta, COL\ \delta\ associated\ internal\ identifier] = 1$ **then**
        go to $END\ PASS$
    end ;

  procedure *deliver the first array identifier (int id A)*
              result: $(int\ id\ B, B)$;
    value *int id* A ; **integer** *int id* A, *int id* B ; **string** B ;
    begin
      integer i ;
      for $i := int\ id\ A$ **step** -1 **until** 1 **do**
        if $\delta\,[i, COL\ \delta\ number\ of\ array\ identifiers] \neq$ **'empty'** **then**
          go to L ;
$L:$ $B := \delta\,[i, COL\ \delta\ external\ identifier]$;
      $int\ id\ B := i$
    end ;

  Boolean procedure *local (int id)* ;
    value *int id* ; **integer** *int id* ;
    $local := int\ id > \omega\,[last\ suitable\ for\ in\ \omega - 4] \wedge$
          $int\ id < \delta\,[\omega\,[last\ suitable\ for\ in\ \omega - 4],$
          $COL\ \delta\ associated\ internal\ identifier]$;

procedure *search in identifier list for* (*identifier*)
result: (*int id*) ;
  integer *int id* ; **string** *identifier* ;

comment The action of this procedure is essentially the same as that of the procedure *search for an identifier in the identifier list* (...) of pass 3, where the necessary explanations are given (see p. 258) ;

  begin
    integer *i*, *begin search*, *end search* ;
    begin search :=
      δ[*delta, COL δ associated internal identifier*] $-$ 1 ;
L: *end search* := *begin identifier list* ;
SEARCH:
    for *i* := *begin search* **step** $-$1 **until** *end search* **do**
      begin
        if δ[*i, COL δ program structure symbol*] = '**block end**' $\vee$
          δ[*i, COL δ program structure symbol*] =
            '**procedure end**' **then**
        begin
          i := δ[*i, COL δ associated internal identifier*] ;
          go to *END SEARCH LOOP*
        end ;
        if δ[*i, COL δ program structure symbol*] =
          '**block begin**' $\vee$
          δ[*i, COL δ program structure symbol*] =
          '**procedure begin**' **then**
        begin
          end search := *begin search* $+$ 2 ;
          begin search :=
            δ[*i, COL δ associated internal identifier*] $-$ 1 ;
          go to *SEARCH*
        end ;
        if δ[*i, COL δ external identifier*] = *identifier* $\wedge$
          $\neg$ (δ[*i, COL δ type*] = '**integer**' $\wedge$
          δ[*i, COL δ kind*] = '**label**') **then**
          go to *END SEARCH FOR AN IDENTIFIER* ;
END SEARCH LOOP:
      end ;
    if *end search* = *begin identifier list* **then**
      go to *ERROR FOUND* ;
    begin search :=
      δ[*begin search* $+$ 1, *COL δ associated internal identifier*] $-$ 1 ;
    go to *L* ;

END SEARCH FOR AN IDENTIFIER:
        int id := i
     end ;

    procedure *generate declarations* ;
     begin
       integer i, j, k ;
       for i := 1 **step** 1 **until** *id list* **do**
         begin
           if $\iota[i, COL \ \iota \ local]$ = '**local**' **then**
             begin
               delta add := *delta add* + 1 ;
               $\delta[delta \ add, COL \ \delta \ external \ identifier]$:=
                 $\iota[i, COL \ \iota \ external \ identifier]$;
               $\delta[delta \ add, COL \ \delta \ internal \ identifier]$:= *delta add* ;
               $\delta[delta \ add, COL \ \delta \ mode]$:= $\iota[i, COL \ \iota \ identical]$;
               $\delta[delta \ add, COL \ \delta \ type]$:= '**integer**' ;
               $\delta[delta \ add, COL \ \delta \ kind]$:= '**variable**' ;
               for i := 6 **step** 1 **until** 16 **do** $\delta[delta \ add, j]$:=
                 '**empty**'
             end
         end ;
       delta add := *delta add* + 1 ;
       $\delta[delta \ add, COL \ \delta \ internal \ identifier]$:= *delta add* ;
       $\delta[delta \ add, COL \ \delta \ program \ structure \ symbol]$:= '**for end**' ;
       $\delta[delta \ add, COL \ \delta \ associated \ internal \ identifier]$:= $\omega[w - 10]$;
       $\delta[\omega[w - 10], COL \ \delta \ associated \ internal \ identifier]$:= *delta add* ;

    comment Elimination of the irrelevant entries in ι ;

       j := 0 ;
       for i := 1 **step** 1 **until** *id list* **do**
         if $\iota[i, COL \ \iota \ local]$ $\neq$ '**empty**' **then**
           begin
             j := j + 1 ;
             if $j \neq i$ **then**
               begin
                 for k := 1 **step** 1 **until** 4 **do** $\iota[j, k]$:= $\iota[i, k]$;
                 $\iota[j, 5]$:= '**empty**'
               end
           end ;
       id list := j
     end generate declarations ;

```
procedure delete from (l1) to: (l2) in: (list, l) ;
  value l1, l2 ;  integer l1, l2, l ;  string array list ;
  begin
    integer i ;
    if l = l2 then
      l := l1 − 1
    else
      begin
        l := l + l1 − l2 − 1 ;
        for i := l1 step 1 until l do
          list [i] := list [i − l1 + l2 + 1]
      end
  end ;
```

comment This pass does not contain a complete error check. Syntactical errors which are easy to detect or which might lead to undesired actions of the next pass are checked. They cause the breaking-off of the recursive address calculation pass without a special message. Checks are made in for list elements and in subscript expressions, by means of the following three procedures ;

```
procedure check delimiter in expression ;
  begin
    Boolean procedure admissible delimiter (x) ;
      string x ;
      code ;
    if ¬ admissible delimiter (incoming symbol) then
      go to ERROR FOUND
  end ;
```

comment The Boolean procedure *admissible delimiter* has the following values:

$$
admissible\ delimiter\ (x) = \begin{cases}
\textbf{true, if } x = `+' \mid `-' \mid `\times' \mid `/' \mid `\div' \mid `\uparrow' \mid \\
\qquad `<' \mid `\leq' \mid `=' \mid `\neq' \mid `\geq' \mid \\
\qquad `>' \mid `\neg' \mid `\wedge' \mid `\vee' \mid `\supset' \mid `\equiv' \mid \\
\qquad `(' \mid `)' \mid `[' \mid `]' \mid `,' \mid \\
\qquad `\textbf{if}' \mid `\textbf{then}' \mid `\textbf{else}' \\
\textbf{false, } otherwise ;
\end{cases}
$$

```
procedure check delimiter in subscript expression
  (NON LINEAR, ERROR) ;
  label NON LINEAR, ERROR ;
  begin
    if incoming symbol = `+' ∨ incoming symbol = `−' then
```

begin
  if $\omega[w] = `+` \lor \omega[w] = `-` \lor \omega[w] = `\times`$ **then**
    go to $ERROR$
end
else
  if *incoming symbol* $= `\times` \lor$ *incoming symbol* $= `\oplus` \lor$
    incoming symbol $= `\bar{[}` \lor$ *incoming symbol* $= `$ **red init**$`\lor$
    incoming symbol $= `)` \lor$ *incoming symbol* $= `]` \lor$
    incoming symbol $= `,` \lor$ *incoming symbol* $= `;`$ **then**
    begin
      if *delimiter* $(\omega[w])$ **then go to** $ERROR$
    end
  else
    if *incoming symbol* $= `[`$ **then**
      begin
        if *delimiter* $(\omega[w])$ **then**
          go to $ERROR$
          else go to $NON\ LINEAR$
      end
    else
      if *incoming symbol* $= `(`$ **then**
        begin
          if $\neg$ *delimiter* $(\omega[w]) \land$
            $\delta[int\ id,\ COL\ \delta\ mode] \neq `$ **standard**$`$ **then**
            go to $NON\ LINEAR$
        end
      else
        if $\omega[w] = `(` \lor \neg$ *delimiter* $(\omega[w])$ **then**
          go to $NON\ LINEAR$
          else go to $ERROR$;
$CORRECT$:
    end ;

  procedure *check subscript number* (*int id, index number*) ;
    value *int id, index number* ;
    integer *int id, index number* ;

  comment At the first occurrence of a component of a specific formal
array no test is made, since the number of indices given in the identifier
list δ may not be correct (compare with pass one, p. 188). The index
number found is inserted in δ with a mark (here the unary minus sign)
which indicates to the next pass that a correct index number has already

been found. If the entry for the formal array in δ contains the minus sign, the test must be made ;

> **begin**
>   **if** $\delta[int\ id,\ COL\ \delta\ mode] \neq$ '**normal**' $\wedge$
> $\delta[int\ id,\ COL\ \delta\ dimension] > 0$ **then**
> $\delta[int\ id,\ COL\ \delta\ dimension] := -\ index\ number$
>   **else**
>   **if** $index\ number \neq abs(\delta[int\ id,\ COL\ \delta\ dimension])$ **then**
>   **go to** $ERROR\ FOUND$
> **end** ;

> **procedure** *handling the for clause* ;

comment The for clause stored in χ will be read and, if suitable, stored in the additional program storage $\pi\ add$, otherwise stored in π.

The following program includes the test whether the loop variable is an admissible variable of type **integer**, whether the entities within the step elements and the expressions before **while** are admissible variables, standard function designators, or constants only, whether the end value remains constant, and whether the for clause contains more than one for list element.

The global variable *suitable loop* is set **true** if the loop is recognized to be suitable for recursive address calculation.

Subscripted variables within initial and end elements may be checked for recursive calculation with respect to an outer suitable loop ;

> **begin**
>   **integer** *begin for, list elements, begin initial value, begin until, i* ;
>   **Boolean** *initial value constant, step found* ;
>   **string** *loop variable, initial variable, step variable* ;
>   *search next for list entry* αa ;
>   **if** $\alpha[a] =$ '**for unsuitable**' **then**
>   **go to** $START\ READING\ PROGRAM$;
> $POSSIBLY\ SUITABLE$:
>   *store 1 in* $\pi\ add$ ('**for suitable**') ;
>   *read* χ ; *store 1 in* $\pi\ add$ (*incoming symbol*) ;
>   *begin for* $:= p\ add - 1$;
>   *search in identifier list for* (*incoming symbol*) *result*: (*int id*) ;
>   **if** $\neg$ (*admissible variable* (*int id*, **true**) $\wedge$
> $\delta[int\ id,\ COL\ \delta\ kind] =$ '**integer**') **then**
>   **go to** $UNSUITABLE$;
>   *loop variable* $:=$ *incoming symbol* ;
>   *read* χ ; *store 1 in* $\pi\ add$ (*incoming symbol*) ;

INITIALIZATION:
    step found := **false** ;
    initial variable := *step variable* := '**empty**' ;
    list elements := 1 ;

INITIAL VALUE:
    initial value constant := **true** ;
    begin initial value := *p add* ;

L1:   *read* χ ; *store 1 in* π *add (incoming symbol)* ;

    if *delimiter (incoming symbol)* **then**

      begin

        if *incoming symbol* = '[' $\wedge$ *suitable loop* **then**

          begin

            handling subscripted variables (π *add, p add,*
              store 1 in π *add*) ;

            go to *L1*

          end ;

        if *incoming symbol* = '**while**' **then**

          go to *WHILE ELEMENT* ;

        if *incoming symbol* = '**do**' **then go to** *DO* ;

        if *incoming symbol* = '**for,**' **then go to** *FOR COMMA* ;

        if *incoming symbol* = '**step**' **then**

          begin

            if *begin initial value* $+ 2 = p$ *add* $\wedge$
              (δ [*int id*, *COL* δ *kind*] = '**number**' $\vee$
              δ [*int id*, *COL* δ *kind*] = '**variable**' $\wedge$
              δ [*int id*, *COL* δ *mode*] $\neq$ '**formal**') $\wedge$
              δ [*int id*, *COL* δ *type*] = '**integer**' **then**
              initial variable := π *add* [*p add* $-$ 1] ;

comment The initial variable *GVa* can be identified with the initial value, if this exists of only a number or a simple declared or value listed variable of type **integer** ;

            step found := **true** ;

            go to *STEP VALUE*

          end ;

        check delimiter in expression ;

        go to *L1*

      end handling delimiters within the initial value expression ;

    search in identifier list for (incoming symbol) result: (*int id*) ;

14*

        if ¬ (δ [*int id, COL* δ *kind*] = '**number**' ∨
              δ [*int id, COL* δ *mode*] = '**standard**' ∨
              admissible variable (*int id*, **true**) ∧ *incoming symbol* $\neq$
              loop variable) **then**
          initial value constant := **false** ;
        go to *L1* ;

STEP VALUE :
        read χ ; *store 1 in* π *add* (*incoming symbol*) ;
        if *delimiter* (*incoming symbol*) **then**
          begin
            if *incoming symbol* $\neq$ '**until**' **then**
              begin
                check delimiter in expression ;
                go to *STEP VALUE*
              end ;
            if π *add* [*p add* − 2] $\neq$ '**step**' **then go to** *END VALUE* ;
            if δ [*int id, COL* δ *type*] = '**integer**' **then**
              step variable := π *add* [*p add* − 1] ;
            go to *END VALUE*
          end delimiter ;
        search in identifier list for (*incoming symbol*) *result* : (*int id*) ;
        if δ [*int id, COL* δ *kind*] = '**number**' ∨
        admissible variable (*int id*, **true**) ∧ *incoming symbol* $\neq$
        loop variable
        ∨ δ [*int id, COL* δ *mode*] = '**standard**' **then**
          go to *STEP VALUE* ;
        go to *UNSUITABLE* ;

END VALUE :
        begin until := *p add* ;

L2 :   *read* χ ; *store 1 in* π *add* (*incoming symbol*) ;
        if *delimiter* (*incoming symbol*) **then**
          begin
            if *incoming symbol* $\neq$ '**for,**' ∧ *incoming symbol* $\neq$ '**do**' **then**
              begin
                check delimiter in expression ;
                go to *L2*
              end ;
            π *add* [*begin until*] := '**until constant**' ;
            if *incoming symbol* = '**for,**' **then go to** *FOR COMMA* ;
            go to *DO*
          end delimiter in the end value ;
        search in identifier list for (*incoming symbol*) *result* : (*int id*) ;

if $\delta[int\ id,\ COL\ \delta\ kind] = $ '**number**' $\lor$
$\delta[int\ id,\ COL\ \delta\ mode] = $ '**standard**' $\lor$
admissible variable (*int id*, **true**) $\land$
incoming symbol $\neq$ *loop variable* **then**
   go to $L2$;
go to *READ REST 2* ;

WHILE ELEMENT:
   if $\neg$ *initial value constant* **then go to** *UNSUITABLE* ;
   go to *READ REST OF EXPRESSION* ;

FOR COMMA:
   list elements := *list elements* + 1 ;
   go to *INITIAL VALUE* ;

READ REST OF EXPRESSION:
   read χ ; *store 1 in* π *add* (*incoming symbol*) ;

READ REST 1:
   if $\neg$ *delimiter* (*incoming symbol*) **then**
     begin
      search in identifier list for (*incoming symbol*) *result*: (*int id*) ;

READ REST 2:
     if $\delta[int\ id,\ COL\ \delta\ kind] = $ '**procedure**' $\land$
      $\delta[int\ id,\ COL\ \delta\ mode] \neq $ '**standard**' $\lor$
      $\delta[int\ id,\ COL\ \delta\ kind] = $
       '**unspecified formal parameter**' $\lor$
      $\delta[int\ id,\ COL\ \delta\ mode] = $ '**formal**' $\land$
      $\delta[int\ id,\ COL\ \delta\ kind] = $ '**variable**' **then**
      $\alpha[a] := $ '**for unsuitable for index registers**' ;

comment Name and function calls in end values and while elements must be handled like those in the do statement (cf. 5.6.3, p. 100, 7.2.2.3, p. 135, and 8, pass 1, p. 197) ;

      go to *READ REST OF EXPRESSION*
     end entity ;
   if *incoming symbol* = '[' $\land$ *suitable loop* **then**
     begin
      handling subscripted variables (π *add*, p *add*, *store 1 in* π *add*) ;
      go to *READ REST OF EXPRESSION*
     end ;
   if *incoming symbol* = '**for,**' **then go to** *FOR COMMA* ;
   if *incoming symbol* = '**do**' **then go to** *DO* ;
   check delimiter in expression ;
   go to *READ REST OF EXPRESSION* ;

UNSUITABLE:

    for $i :=$ *begin for* $+1$ **step** 1 **until** p *add* **do**
      store 1 in $\pi(\pi \, add\,[i])$;
    $p \, add :=$ *begin for* -1 ;
    push down 1 in ω ('**for unsuitable**') ;
    go to *END FOR CLAUSE* ;

DO: **if** ¬ *step found* **then go to** *UNSUITABLE* ;
    if *list elements* > 1 ∨ *initial variable* = '**empty**' **then**
      initial variable := *loop variable* ;
    if *list elements* > 1 ∨ *step variable* = '**empty**' **then**
      begin
        generate variable of type ('**integer**') *result*: (*step variable*) ;
        ι [*id list, COL* ι *local*] := '**local**'
      end ;
    for *begin for block in* $\delta :=$ *delta* $+1$ **step** 1 **until**
        end identifier list **do**
      if δ [*begin for block in* δ, *COL* δ *program structure symbol*] =
        '**block begin**' **then**
        go to L ;
L: *delta add* := *delta add* $+1$;
    δ [*delta add, COL* δ *internal identifier*] := *delta add* ;
    δ [*delta add, COL* δ *program structure symbol*] := '**for begin**' ;
    index register := *maximum static procedure level* ;
    push down 11 in ω (*delta add, a, index register, x,*
      last suitable for in ω, $p+2$, *begin for block in* δ,
      initial variable, step variable, loop variable, '**for**') ;
    last suitable for in $\omega := w$;
    suitable loop := **true** ;
    $\pi[p]$:= '**do suitable**' ;
    store 6 in π (*begin for, 0, 0, list elements,*
      initial variable, step variable) ;

END FOR CLAUSE:
    end handling the for clause ;

    procedure *handling subscripted variables* (π, p, *store*) ;
    integer p ; **array** π ; **procedure** *store* ;

comment If a subscripted variable occurs within a suitable loop
the program *handling subscripted variables* is called. If the array identifier
is global to the respective suitable loop the procedure *admission test* is
called which performs the necessary checks. If all subscript expressions
turn out to be zero the subscripted variable can immediately be replaced

by '*A*' '**red init**' in the program storage π. A suitable subscripted variable is extracted from π, stored in ξ and replaced by '**content** ⟨type⟩ **of**' '*GV*' in π, where *GV* is a generated address variable. In ξ a test for identifications is made. Left part variables with identical right parts are identified ;

> **begin**
>> **integer** *begin in* ξ, *end expression, int id A, int id B, i* ;
>> **Boolean** *admissible, zero* ;
>> **string** *A, B, GV* ;
>> $A := \pi[p-1]$;
>> **if** *delimiter* (*A*) **then go to** *ERROR FOUND* ;
>> *search in identifier list for* (*A*) *result*: (*int id A*) ;
>> **if** *local* (*int id A*) $\lor$ $\delta[int\ id\ A,\ COL\ \delta\ kind] \neq$ '**array**' **then**
>>> **go to** *END HANDLING* ;
>> *deliver the first array identifier* (*int id A*) *result*: (*int id B, B*) ;
>> *generate variable of type* ('**address**') *result*: (*GV*) ;
>> *begin in* $\xi := x$;
>> *push down* 7 *in* ξ (*GV*, ':=', *A*, '**red init**', '+', *B*, '$\bar{[}$') ;
>> *admission test* (*w*, **true**, *read* χ)
>>> *result*: (*admissible, end expression, zero*) ;
>> **if** $\neg$ *admissible* **then**
>>> **begin**
>>>> **for** $i :=$ *begin in* $\xi + 8$ **step** 1 **until** *x* **do** *store* ($\xi[i]$) ;
>>>> $x :=$ *begin in* ξ ;
>>>> **go to** *END HANDLING*
>>> **end** ;
>> **if** *zero* **then**
>>> **begin**
>>>> $\pi[p-1] :=$ '**content**' $\oplus$ $\delta[int\ id\ A,\ COL\ \delta\ type] \oplus$ '**of**' ;
>>>> $\pi[p] := A$; *store* ('**red init**') ;
>>>> $x :=$ *begin in* ξ ;
>>>> **go to** *END HANDLING*
>>> **end** ;
>> *test for possible identifications* (*begin in* $\xi + 2$) ;
>> $\pi[p-1] :=$ '**content**' $\oplus$ $\delta[int\ id\ A,\ COL\ \delta\ type] \oplus$ '**of**' ;
>> $\pi[p] := GV$;

END HANDLING:
> **end** handling subscripted variables ;

> **procedure** *handling expressions* (*increment*) ;
>> **string** *increment* ;

comment When at the end of a loop the initial value program is transferred from $\pi\alpha$ to π *add* the right part expressions are tested whether they are suitable for recursive address calculation with respect to an outer loop. We require that the entire expression, or at least an index vector contained in it, is suitable. Other subexpressions are not treated here even if they are otherwise suitable. Suitable expressions are extracted from $\pi\alpha$ or π *add* resp., and stored in ξ like subscripted variables. An entire instruction (including the left part variable) may be pushed into the ξ-list, if the left part variable remains unchanged during the inner loop, or is only incremented by index register modification. Otherwise an error would be produced since the value of a variable which is used for recursive calculation in an outer loop must remain unchanged in the inner loop ;

> **begin**
>> **integer** i, *begin in* ξ, *end expression* ;
>> **Boolean** *admissible, zero* ;
>> **string** GV ;
>> *begin in* $\xi := x$;
>> *push down 2 in* ξ ('**empty**', ':=') ;
>> *admission test* (w, **false**, *read* $\pi\alpha$)
>>> *result*: (*admissible, end expression, zero*) ;
>> **if** $\neg$ *admissible* **then**
>>> **begin**
>>>> **if** *end expression* $=$ *begin in* $\xi + 2$ **then**
>>>>> **begin**
>>>>>> **for** $i :=$ *begin in* $\xi + 3$ **step** 1 **until** x **do**
>>>>>>> *store 1 in* π *add* ($\xi[i]$) ;
>>>>>> $x :=$ *begin in* ξ ;
>>>>>> **go to** *END HANDLING*
>>>>> **end** ;

PARTLY ADMISSIBLE:
>>>>> *generate variable of type* ('**integer**') *result*: (GV) ;
>>>>> $\xi[$*begin in* $\xi + 1] := GV$;
>>>>> *store 1 in* π *add* (GV) ;
>>>>> **for** $i :=$ *end expression* $+1$ **step** 1 **until** x **do**
>>>>>> *store 1 in* π *add* ($\xi[i]$) ;
>>>>> $x :=$ *end expression* $+1$;
>>>>> $\xi[x] :=$ ';'
>>>> **end**
>>> **else**
>>>> **begin**

ADMISSIBLE:

     if *increment* $=$ '**empty**' **then**
      begin
       $\xi\,[begin\ in\ \xi+1] := \pi\ add\,[p\ add-1]$;
       $p\ add := p\ add-2$
      end
     else
      begin
       generate variable of type ('**integer**')
        result: $(\xi\,[begin\ in\ \xi+1])$;
       store 2 in $\pi\ add$ $(\xi\,[begin\ in\ \xi+1],\ ';')$
      end
    end ;
     test for possible identifications $(begin\ in\ \xi+2)$;

END HANDLING:

    end handling expressions ;

   procedure *admission test* (*begin in* ω, *chi, read*)
          result: (*admissible, end expression, all positions zero*) ;
    value *begin in* ω, *chi* ; **integer** *begin in* ω, *end expression* ;
    Boolean *chi, admissible, all positions zero* ; **procedure** *read* ;

   comment *admission test* is one of the main parts of the recursive address calculation pass, and is called by the procedures *handling subscripted variables* and *handling expressions*.

   If *chi* has the value **true** then a subscripted variable in χ must be treated, otherwise an expression in $\pi\alpha$. The symbols are read one by one and stored in ξ.

   An (index) expression is admissible if it is linear in the loop variable, and if it contains only integer constants, index registers, admissible variables of type **integer** and integer standard function designators, and delimiters

$$\text{'(', ')', '+', '-', '×', '⊕', \textbf{red init}, '[', ']', ','},$$

or any other delimiter admissible in expressions within actual parameter expressions.

   If a not admissible symbol occurs the content of ξ is removed, the result variable *admissible* is set **false**, and the test is stopped.

   If only a subexpression is suitable then *admissible* is also set **false** and *end expression* delivers the end of the subexpression in ξ. Otherwise *admissible* is **true** and the value of *end expression* is irrelevant. The last result parameter is set **true** if all index expressions of a subscripted variable are zero, otherwise **false**. For the test the push down ω is used in the way described in 7.3.2. Irrelevant embracing round brackets are eliminated ;

```
        begin
            integer begin in ξ, i, e, index number, number of parentheses,
                    int id A ;
            Boolean index vector, zero ;
            string loop variable ;
            loop variable := ω [last suitable for in ω − 1] ;
            index vector := false ; admissible := true ;
            e := x ;
            if chi then go to START INDEX VECTOR ;
            push down 1 in ω (':=') ;
READ:   read ;
            push down 1 in ξ (incoming symbol) ;
HANDLING A DELIMITER:
            if delimiter (incoming symbol) then
                begin
PROCESS:    check delimiter in subscript expression
                    (NOT ADMISSIBLE, ERROR FOUND) ;
                if character order (incoming symbol) < (if delimiter (ω [w])
                then state order (ω [w]) else state order (ω [w − 1])) then
                begin
                    if incoming symbol = 'red init' then go to READ;
                    if incoming symbol = '[' then
                        begin
                            if index vector then
                                go to NOT ADMISSIBLE ;
START INDEX VECTOR:
                            push down 1 in ω (incoming symbol) ;
                            index number := 1 ; int id A := int id ;
                            zero := index vector := true ;
L1:                         begin in ξ := x ;
L2:                         read ;
                            push down 1 in ξ (incoming symbol) ;
                            if incoming symbol = '(' then
                                begin
                                    push down 1 in ω ('(poss irrelevant') ;
                                    go to L2
                                end ;
                            go to HANDLING A DELIMITER
                        end beginning of an index vector ;
                    push down 1 in ω (incoming symbol) ;
```

if ¬ *index vector* ∧ *incoming symbol* = '+' **then**
  $e := x - 1$;
go to *READ*
end order (incoming symbol) < state order ;
if $\omega[w-1] = $ '+' ∨ $\omega[w-1] = $ '−' ∨
$\omega[w-1] = $ '⊕' **then**
begin
  if *delimiter* $(\omega[w-2])$ **then**

UNARY OPERATOR:
      begin
        $\omega[w-1] := \omega[w]$; $w := w - 1$;
        go to *PROCESS*
      end ;

BINARY OPERATOR:
      if $\omega[w] = $ '**loop variable**' **then**
        $\omega[w-2] := $ '**loop variable**' ;
      $w := w - 2$;
      go to *PROCESS*
    end ;

MULTIPLICATION:
      if $\omega[w-1] = $ '×' **then**
      begin
        if $\omega[w] = $ '**loop variable**' ∧
          $\omega[w-2] = $ '**loop variable**' **then**
          go to *NOT ADMISSIBLE* ;
        go to *BINARY OPERATOR*
      end ;

ROUND BRACKET:
    if *incoming symbol* = ')' **then**
      begin
        if $\omega[w-1] = $ '(' **then**
          begin $\omega[w-1] := \omega[w]$; $w := w - 1$;
          go to *READ* **end** ;

POSSIBLY IRRELEVANT:
        $\omega[w-1] := \omega[w]$; $w := w - 1$;
        number of parentheses := 1 ;
L5:   *read* ; *push down 1 in* ξ (*incoming symbol*) ;
        if *incoming symbol* = ')' **then**

begin
  number of parentheses :=
    number of parentheses + 1 ;
  $\omega[w-1] := \omega[w]$; $w := w-1$;
  go to *L5*
end ;
if *incoming symbol* $\neq$ ']' $\wedge$
  incoming symbol $\neq$ ',' **then**
  go to *HANDLING A DELIMITER* ;

ELIMINATE IRRELEVANT ROUND BRACKETS:
  for $i := 1$ **step** 1 **until** *number of parentheses* **do**
    $\xi[begin\ in\ \xi+i] := $ 'empty' ;
  $x := x - number\ of\ parentheses + 1$;
  $\xi[x] := incoming\ symbol$;
  end round bracket ;

SQUARE BRACKET:
  if *incoming symbol* = ']' **then**
    begin
    $w := w-2$;
    if *chi* **then**

HANDLE SUBSCRIPTED VARIABLES:
      begin
      if $\neg\ (\xi[x] = $ '0' $\wedge\ (\xi[x-1] = $ ',' $\vee$
        $\xi[x-1] = $ '[')) **then**
        all positions zero := **false**
      else *all positions zero* := *zero* ;
      push down 1 in ξ (';') ;
      check subscript number (int id A, index number) ;
      go to *END ADMISSION TEST*
      end handle subscripted variables
    else
      begin
      $e := x$;
      index vector := **false** ;
      go to *READ*
      end handle expressions
    end
  else
    if *incoming symbol* = ',' **then**

begin
  if ¬ ($\xi[x] = $ '0' ∧ ($\xi[x-1] = $ ',' ∨
  $\xi[x-1] = $ '[')) **then**
  zero := **false** ;
  $w := w-1$; *index number* := *index number* $+1$;
  go to *L1*
end ;

COLON EQUAL:
  $w := w-2$;
  go to *END ADMISSION TEST* ;

NOT ADMISSIBLE:
  if ¬ *index vector* **then** *end expression* := *e* ;
  admissible := **false** ;
  $w := begin\ in\ \omega$;
  go to *END ADMISSION TEST*
end handling a delimiter ;

HANDLING AN ENTITY:
  if *generated variable (incoming symbol)* **then**
    go to *NOT ADMISSIBLE* ;
  if ¬ *delimiter* ($\omega[w]$) **then go to** *ERROR FOUND* ;
  if *index register name (incoming symbol)* **then go to** *L3* ;
  search in identifier list for (incoming symbol)
    result: *(int id)* ;
  if $\delta[int\ id,\ COL\ \delta\ kind] = $ '**array**' **then**
    begin
      if *local (int id)* ∨ *chi* **then**
        go to *NOT ADMISSIBLE*
      else
        begin
          push down 1 in ω ('**entity**') ;
          go to *READ*
        end
    end ;
  if ($\delta[int\ id,\ COL\ \delta\ kind] = $ '**number**' ∨
  admissible variable (int id, **false**$)$) ∧
  $\delta[int\ id,\ COL\ \delta\ type] = $ '**integer**' **then**
  begin
    if *incoming symbol* $\neq$ *loop variable* **then**
L3:   *push down 1 in* ω ('**entity**')
    else *push down 1 in* ω ('**loop variable**') ;
    go to *READ*
  end number or admissible variable ;

STANDARD FUNCTION:

> **if** $\delta\,[int\ id,\ COL\ \delta\ mode] =$ '**standard**' $\wedge$
> $\delta\,[int\ id,\ COL\ \delta\ type] =$ '**integer**' **then**
> **begin**
>   *push down 1 in* ω ('**entity**') ;

L6:   *read ; push down 1 in* ξ *(incoming symbol)* ;

>   **if** *incoming symbol* $=$ '(' **then**
>   **begin**
>   *push down 1 in* ω ('(') ;
>   **go to** *L6*
>   **end** ;
>   **if** *incoming symbol* $=$ ')' **then**
>   **begin**
> $w := w - 1$;
>   **if** $\omega\,[w] =$ '**entity**' **then go to** *READ* ;
>   **go to** *L6*
>   **end** ;
>   **if** *delimiter (incoming symbol)* **then**
>   **begin** *check delimiter in expression* ; **go to** *L6* **end** ;
>   *search in identifier list for (incoming symbol)*
>   *result*: *(int id)* ;
>   **if** $\delta\,[int\ id,\ COL\ \delta\ kind] =$ '**number**' $\vee$
> $\delta\,[int\ id,\ COL\ \delta\ mode] =$ '**standard**' $\vee$
>   *admissible variable (int id,* **false**$) \wedge$
>   *incoming symbol* $\neq$ *loop variable* **then**
>   **go to** *L6*
>   **end** standard function ;
>   **go to** *NOT ADMISSIBLE* ;

END ADMISSION TEST:

>   **end** ;

> **procedure** *test for possible identifications (begin in* ξ) ;
>   **value** *begin in* ξ ; **integer** *begin in* ξ ;

comment If the ξ-list contains an assignment statement the right part of which is identical to the right part of the assignment statement stored as the last one then the respective left part variables are identified if these generated variables are declared to be of the same type. The procedure is called within *handling subscripted variables* and within *handling expressions* ;

>   **begin**
>   **integer** *begin expression, i, j, k* ;
>   **for** $j := $ *begin in* $\xi - 2$ **step** -1 **until**
> $\omega\,[$*last suitable for in* $\omega - 7] + 1$ **do**

begin
   for $i := x$ **step** -1 **until** *begin in* ξ **do**
     if $\xi[i] = \xi[j]$ **then**
      $j := j - 1$
     else go to *L1* ;

EXPRESSION FOUND:
     for $i := id \ list$ **step** -1 **until** 1 **do**
      if $\iota[i, COL \ \iota \ external \ identifier] = \xi[j] \ \vee$
       $\iota[i, COL \ \iota \ external \ identifier] = \xi[begin \ in \ \xi - 1]$ **then**
       go to *L2* ;
L2:   **for** $k := i - 1$ **step** -1 **until** 1 **do**
      if $\iota[k, COL \ \iota \ external \ identifier] = \xi[j] \ \vee$
       $\iota[k, COL \ \iota \ external \ identifier] = \xi[begin \ in \ \xi - 1]$ **then**
       go to *L3* ;
L3:   **if** $\iota[i, COL \ \iota \ type] \neq \iota[k, COL \ \iota \ type]$ **then go to** *L1* ;
     identify $(\xi[j], \xi[begin \ in \ \xi - 1])$;
     $x := begin \ in \ \xi - 2$;
     go to *END TEST* ;
L1:   **end** ;
EXPRESSION NOT FOUND:
END TEST:
   end test for possible identifications ;

  procedure *closing a for statement* ;

  comment If a '**for end**' occurs in χ the procedure *closing a for statement* is called.

At the end of a suitable loop the appropriate instructions in ξ are read. Each one is decomposed and then instructions for the initial value program and the increment program are generated. The completed initial value program in $\pi\alpha$ then is read again, stored in $\pi \ add$ and checked for expressions which may be pushed into an outer loop. For the instructions remaining in $\pi \ add$ possible optimizations are performed: equal subexpressions are extracted and replaced by generated variables. The initial value program is completed by the instructions for loading the index registers used.

The program could be still optimized insofar as the initial value program and the increment program could be eliminated including the beginning and end delimiters if no instruction is stored, but this is not done here ;

  begin
    integer *end list, begin* π *add, i* ;
    string *increment* ;
    if $\omega[w] = $ '**for unsuitable**' **then**

begin
   $w := w - 1$;
   go to *END CLOSING A FOR STATEMENT*
end ;
if $x \neq \omega[w-7]$ **then**
begin

SUBSCRIPTED VARIABLES HAVE BEEN FOUND:
   store 1 in π *add* ('**begin increment**') ;
   $\pi[\omega[w-5]+1] := p$ *add* ;
   end list $:= x$; $x := \omega[w-7]$;
   $p\alpha := 0$; *store 1 in* $\pi\alpha$ (';') ;

L0: $k := f := 0$;

READ: *decomposition* ;

L1:   *generate program* ;
   if $x \neq$ *end list* **then go to** *L0* ;
   store 2 in π *add* ('**end increment**',
    '**begin initial value**') ;
   $\pi[\omega[w-5]] := p$ *add* ;
   last suitable for in $\omega := \omega[w-6]$;
   suitable loop $:=$ *last suitable for in* $\omega \neq 0$;
   $x := \omega[w-7]$;
   end list $:= p\alpha$; $p\alpha := 1$; *begin* π *add* $:= p$ *add* $+1$;

L20:   *increment* $:=$ '**empty**' ;

L2:   *read* $\pi\alpha$;
   if *incoming symbol* $=$ '|' **then**
     begin
       read $\pi\alpha$;
       increment $:=$ *incoming symbol* ;
       go to *L2*
     end ;
   store 1 in π *add* (*incoming symbol*) ;
   if *incoming symbol* $=$ ':=' $\wedge$ *suitable loop* **then**
    handling expressions (*increment*) ;
   if *incoming symbol* $=$ ';' **then**
     begin
       if p *add* $>$ *begin* π *add* **then**
        begin
         mark (*COL* ι *local*, '**local**', π *add* [*begin* π *add*]) ;
         perform optimizations in π *add* (*begin* π *add*) ;
         begin π *add* $:= p$ *add* $+1$
        end ;

if $p\alpha \neq end$ *list* **then go to** $L20$

end

else go to $L2$;

for $i := \omega[w-8]+1$ **step** 1 **until** *index register* **do**
store 4 in π add ('$\Re[$' $\oplus i \oplus$ ']', ':=', '0', ';') ;

if *suitable loop* **then**

begin

if ω [*last suitable for in* $\omega-8$] $<$ *index register* **then**
ω [*last suitable for in* $\omega-8$] $:=$ *index register*

else *index register* $:= \omega$ [*last suitable for in* $\omega-8$]

end ;

store 1 in π *add* ('**end initial value**')

end subscripted variables have been found

else

begin

store 4 in π *add* ('**begin increment**', '**end increment**',
'**begin initial value**', '**end initial value**') ;

$\pi[\omega[w-5]] := p\ add - 1$;

$\pi[\omega[w-5]+1] := p\ add - 3$

end ;

generate declarations ;

$w := w - 11$;

END CLOSING A FOR STATEMENT:

end ;

procedure *decomposition* ;

comment An index vector or an expression stored in ξ is decomposed into a constant part and the factor of the loop variable. The constant part is stored in $\varkappa$, the factor stored in φ (compare with 7.3.2).

When this procedure is called no syntactical error may occur since a complete test is made before in *admission test*.

The push down σ is used for finding out which symbols are only constituents of the constant part or only of the factor and which must be deleted in φ or in $\varkappa$ respectively.

Parentheses turning out to be irrelevant in the decomposed parts, factors one and unary '$+$'-operators are eliminated, and index vectors with empty subscript expressions are replaced by zero ;

begin

integer *number of parentheses, i, j* ;

string *loop variable* ;

Boolean *constant part zero, factor zero* ;
constant part zero := *factor zero* := **true** ;
$s := 0$;
read ξ ; *store in* $\varkappa$ *(incoming symbol)* ;
read ξ ;
push down in σ *(':=', k, f)* ;
loop variable := $\omega[w-1]$;
go to *READ 1* ;

STORE:
  store in φ *(incoming symbol)* ;

READ 1:
  store in $\varkappa$ *(incoming symbol)* ;

READ: *read* ξ ;
  if $\neg$ *delimiter (incoming symbol)* **then**
   begin
    if *loop variable* = *incoming symbol* **then**
     begin
     push down in σ *('**factor**', k+1, f+1)* ;
     incoming symbol := '1'
     end
    else *push down in* σ *('**constant**', k+1, f+1)* ;
    go to *STORE*
   end ;

STANDARD FUNCTION:
  if *incoming symbol* = '(' $\wedge$ $\neg$ *delimiter* $(\sigma[s, 1])$ **then**
   begin
   number of parentheses := 1 ;
L1:   *store in* φ *(incoming symbol)* ;
   store in $\varkappa$ *(incoming symbol)* ;
   read ξ ;
   if *incoming symbol* = '(' **then**
    number of parentheses := *number of parentheses* + 1
   else
    if *incoming symbol* = ')' **then**
     begin
      number of parentheses := *number of parentheses* − 1 ;
      if *number of parentheses* = 0 **then go to** *STORE*
     end ;
    go to *L1*
   end read standard function ;

REPETITION:

    if if *delimiter* $(\sigma[s, 1])$ **then true else**

        state order $(\sigma[s-1, 1]) > $ *character order* $(incoming\ symbol)$

        then

      begin

        if *incoming symbol* $= \ '+' \wedge$ *delimiter* $(\sigma[s, 1])$ **then**

          go to *READ* ;

        if *incoming symbol* $=$ **'red init'** **then go to** *STORE* ;

        push down in σ $(incoming\ symbol,\ k+1,\ f+1)$;

        go to *STORE*

      end ;

ROUND BRACKET:

    if $\sigma[s-1, 1] = \ '('$ **then**

      begin

        $i := \sigma[s-1, 2]$; $j := \sigma[s-1, 3]$;

        store in φ $(incoming\ symbol)$;

        store in $\varkappa$ $(incoming\ symbol)$;

        if $i = k-2$ **then**

          begin $\varkappa[k-2] := \varkappa[k-1]$; $k := k-2$ **end**

        else

          if $(\varkappa[i-1] = \ '[' \vee \varkappa[i-1] = \ ',' \vee \varkappa[i-1] = \ '+') \wedge$

           character order $(\xi[x+1]) \geq$ *state order* $('+')$ **then**

           begin

             delete from **(if** $\varkappa[i-1] = \ '+' \wedge$

             $\varkappa[i+1] = \ '-'$ **then** $i-1$ **else** i) *to*: (i)

             in: $(\varkappa, k)$;

             $k := k-1$

           end ;

        if $j = f-2$ **then**

          begin $\varphi[f-2] := \varphi[f-1]$; $f := f-2$ **end**

        else

          if $(\varphi[j-1] = \ '[' \vee \varphi[j-1] = \ ',' \vee \varphi[j-1] = \ '+') \wedge$

           character order $(\xi[x+1]) \geq$ *state order* $('+')$ **then**

           begin

             delete from **(if** $\varphi[j-1] = \ '+' \wedge$

             $\varphi[j+1] = \ '-'$ **then** $j-1$ **else** j)

             to: (j) *in*: (φ, f) ;

             $f := f-1$

           end ;

DELETE: $\sigma[s-1, 1] := \sigma[s, 1]$;

        $s := s-1$;

        go to *READ*

      end round bracket ;

ARITHMETIC OPERATORS:

    if $\sigma[s-1, 1] = \text{`}-\text{'}\wedge$ *delimiter* $(\sigma[s-2, 1])$ **then**

      begin $\sigma[s-1, 1] := \sigma[s, 1]$; $s := s-1$;

      go to *REPETITION* **end** ;

    if $\sigma[s-1, 1] = \text{`}+\text{'} \vee \sigma[s-1, 1] = \text{`}-\text{'} \vee$

      $\sigma[s-1, 1] = \text{`}\oplus\text{'}$ **then**

    begin

      if $\sigma[s-2, 1] = \text{`constant'} \wedge$

      $\sigma[s, 1] \neq \text{`constant'}$ **then**

      begin

        delete from $(\sigma[s-2, 3])$ *to*: (**if** $\sigma[s-1, 1] = \text{`}+\text{'}$ **then**

          $\sigma[s-1, 3]$ **else** $\sigma[s-1, 3]-1$) *in*: (φ, f) ;

        $\sigma[s-2, 1] := \text{`linear'}$

      end

      else

        if $\sigma[s, 1] = \text{`constant'} \wedge$

        $\sigma[s-2, 1] \neq \text{`constant'}$ **then**

        delete from $(\sigma[s-1, 3])$ *to*: (f) *in*: (φ, f) ;

      if $\sigma[s-2, 1] = \text{`factor'} \wedge \sigma[s, 1] \neq \text{`factor'}$ **then**

      begin

        delete from $(\sigma[s-2, 2])$ *to*: (**if** $\sigma[s-1, 1] = \text{`}+\text{'}$ **then**

          $\sigma[s-1, 2]$ **else if** $\sigma[s-1, 1] = \text{`}-\text{'}$ **then**

          $\sigma[s-1, 2]-1$ **else** $\sigma[s-1, 2]-2$) *in*: $(\varkappa, k)$;

        if $\sigma[s-1, 1] = \text{`}\oplus\text{'}$ **then** $\varkappa[\sigma[s-2, 2]] := \text{`}0\text{'}$;

        $\sigma[s-2, 1] := \text{`linear'}$

      end

      else

        if $\sigma[s, 1] = \text{`factor'} \wedge \sigma[s-2, 1] \neq \text{`factor'}$ **then**

        delete from $(\sigma[s-1, 2])$ *to*: (k) *in*: $(\varkappa, k)$;

      $s := s-2$;

      go to *REPETITION*

    end operators $+$, $-$, or $\oplus$;

    if $\sigma[s-1, 1] = \text{`}\times\text{'}$ **then**

    begin

      if $\varkappa[\sigma[s-1, 2]-1] = \text{`}1\text{'}$ **then**

      delete from $(\sigma[s-1, 2]-1)$ *to*: $(\sigma[s-1, 2])$ *in*: $(\varkappa, k)$

      else

        if $\varkappa[k] = \text{`}1\text{'}$ **then** $k := k-2$;

      if $\varphi[\sigma[s-1, 3]-1] = \text{`}1\text{'}$ **then**

      delete from $(\sigma[s-1, 3]-1)$ *to*: $(\sigma[s-1, 3])$ *in*: (φ, f)

      else

        if $\varphi[f] = \text{`}1\text{'}$ **then** $f := f-2$;

if $\sigma[s-2, 1] =$ '**constant**' **then** $\sigma[s-2, 1] := \sigma[s, 1]$;
$s := s-2$;
 go to *REPETITION*
end operator $\times$;
if $\sigma[s-1, 1] =$ '$[$' **then**
 begin
  if $\sigma[s, 1] =$ '**factor**' **then**
   begin
    delete from $(\sigma[s, 2])$ *to*: (k) *in*: $(\varkappa, k)$;
    store in $\varkappa$ ('0')
   end
  else
   if $\sigma[s, 1] =$ '**constant**' **then**
    begin
     delete from $(\sigma[s, 3])$ *to*: (f) *in*: (φ, f) ;
     store in φ('0')
    end ;
  if $\neg\ (\varkappa[k] =$ '0' $\wedge\ (\varkappa[k-1] =$ '$[$' $\vee\ \varkappa[k-1] =$ '$,$')) **then**
   constant part zero := **false** ;
  if $\neg\ (\varphi[f] =$ '0' $\wedge\ (\varphi[f-1] =$ '$[$' $\vee\ \varphi[f-1] =$ '$,$')) **then**
   factor zero := **false** ;
  if *incoming symbol* $=$ '$,$' **then**
   begin $s := s-1$; **go to** *STORE* **end** ;

END OF AN INDEX VECTOR:
  if *factor zero* **then**
   $\sigma[s-2, 1] :=$ '**constant**'
  else
   if *constant part zero* **then**
    $\sigma[s-2, 1] :=$ '**factor**'
   else $\sigma[s-2, 1] :=$ '**linear**' ;
  $s := s-2$;
  go to *STORE*
  end incoming '$,$' or '$]$' ;

COLON EQUAL:
 if $\sigma[s, 1] =$ '**factor**' **then**
  begin
   delete from $(\sigma[s, 2])$ *to*: (k) *in*: $(\varkappa, k)$;
   store in $\varkappa$ ('0')
  end
 else
  if $\sigma[s, 1] =$ '**constant**' **then**

begin
    delete from $(\sigma[s, 3])$ *to:* (f) *in:* (φ, f) ;
    store in φ ('0')
end ;

END DECOMPOSITION:

    end ;

    procedure *generate program* ;

    comment This procedure is called in *closing a for statement* after an expression of ξ has been decomposed. From the contents of the $\varkappa$- and the φ-list the instructions for the initial value program and the increment program are generated and stored in $\pi\alpha$ or π *add* respectively. A test is made whether $\pi\alpha$ contains already an instruction with identical right part. Then in certain case the two left part variables may be identified and the new instruction can be deleted again ;

    begin
        integer *begin* $\pi\alpha$, *begin expr, end expr, i, ind reg incr* ;
        string *initial element, step element, factor, increment*
        begin $\pi\alpha := p\alpha$;
        factor $:= $ *increment* $:= $ '**empty**' ;
        step element $:= \omega[w-2]$;
        initial element $:= \omega[w-3]$;

GENERATE VALUE OF FACTOR AND INCREMENT:
        if $\varphi[1] = $ '0' $\wedge f=1$ **then**
          begin
           factor $:= $ *increment* $:= $ '0' ;
           go to *GENERATE INITIAL VALUE*
          end ;
        if $\varphi[1] = $ '1' $\wedge f=1$ **then**
          begin
           factor $:= $ '1' ; *increment* $:= $ *step element* ;
           go to *GENERATE INITIAL VALUE*
          end ;
        if $f=1$ **then**
          begin

FACTOR IS NUMBER OR VARIABLE:
           factor $:= \varphi[1]$; **go to** *GENERATE INCREMENT*
          end ;

GENERAL CASE OF FACTOR:

    store 2 in $\pi\alpha$ (`'empty'`, `':='`) ;

    for $i := 1$ **step** 1 **until** f **do** *store 1 in* $\pi\alpha(\varphi[f])$;

    store 1 in $\pi\alpha$ (`';'`) ;

    search in $\pi\alpha$ *for* (*begin* $\pi\alpha + 2$, $p\alpha$, *begin* $\pi\alpha$)

      result: (*begin expr, end expr*) *exit*: (*NOT FOUND*) ;

    factor := $\pi\alpha[\textit{begin expr} - 1]$;

    $p\alpha$:= *begin* $\pi\alpha$;

    go to *GENERATE INCREMENT* ;

NOT FOUND:

    generate variable of type (`'integer'`) *result*: (*factor*) ;

    $\pi\alpha[\textit{begin } \pi\alpha + 1]$:= *factor* ;

GENERATE INCREMENT:

    if *step element* = `'1'` **then**

      begin

        increment := *factor* ;

        go to *GENERATE INITIAL VALUE*

      end ;

    store 6 in $\pi\alpha$ (`'empty'`, `':='`, *step element*, `'×'`, *factor*, `';'`) ;

    search in $\pi\alpha$ *for* ($p\alpha - 4$, $p\alpha$, *begin* $\pi\alpha$)

      result: (*begin expr, end expr*) *exit*: (*INCR NOT FOUND*) ;

    increment := $\pi\alpha$ [*begin expr* -1] ;

    $p\alpha := p\alpha - 6$;

    go to *GENERATE INITIAL VALUE* ;

INCR NOT FOUND:

    generate variable of type (`'integer'`) *result*: (*increment*) ;

    $\pi\alpha[p\alpha - 5]$:= *increment* ;

GENERATE INITIAL VALUE:

    begin $\pi\alpha := p\alpha$;

    if *increment* = `'0'` **then**

      store 1 in $\pi\alpha(\varkappa[1])$

    else *store 3 in* $\pi\alpha(\varkappa[1]$, `'|'`, *increment*) ;

    for $i := 2$ **step** 1 **until if** $\varkappa[k-1] =$ `'⊕'` **then**

    $k-2$ **else** k **do**

      store 1 in $\pi\alpha(\varkappa[i])$;

    if $\neg$ (*initial element* = `'0'` $\vee$ *factor* = `'0'`) **then**

      begin

        if $\pi\alpha[p\alpha] =$ `'0'` $\wedge$ $\pi\alpha[p\alpha - 1] =$ `':='` **then**

          $p\alpha := p\alpha - 1$

        else *store 1 in* $\pi\alpha$(`'+'`) ;

if *initial element* = '1' **then**
   store 1 in $\pi\alpha$ *(factor)*
else
   if *factor* = '1' **then**
     store 1 in $\pi\alpha$ *(initial element)*
   else *store 3 in* $\pi\alpha$ *(initial element,* '×'*, factor)*
end ;
for $i :=$ *id list* **step** -1 **until** 1 **do**
  if $\iota[i, COL \iota$ *external identifier*$] = \varkappa[1] \wedge$
  $\iota[i, COL \iota type] \neq$ '**address**' **then**
   go to *USE NO INDEX REGISTER* ;
if *increment* = '0' $\vee$ $\alpha[\omega[w-9]] =$
   '**for unsuitable for index registers**' **then**
  go to *USE NO INDEX REGISTER* ;
if $\varkappa[k-1] =$ '$\oplus$' **then**
  begin
   store 2 in $\pi\alpha(\varkappa[k-1], \varkappa[k])$;
   go to *USE NO INDEX REGISTER*
  end ;

GENERATE INDEX REGISTER AND INCREMENT:
  for $i :=$ *id list* **step** -1 **until** 1 **do**
   if $\iota[i, COL \iota$ *external identifier*$] =$ *increment* **then**
    begin
     ind reg incr $:= \iota[i, COL \iota$ *index register*$]$;
     go to *L1*
    end ;
  push down in ι ('**identical**'*, increment,* '**integer**'*,* '**empty**'*,*
  '**global**') ;
  go to *L2* ;

L1:   **if** *ind reg incr* = '**empty**' **then**
L2:   **begin**
     generate index register (increment) result: *(ind reg incr)* ;
     store 6 in π *add* ('$\Re\mathfrak{r}[$' $\oplus$ *ind reg incr* $\oplus$ ']', ':=',
       '$\Re\mathfrak{r}[$' $\oplus$ *ind reg incr* $\oplus$ ']', '+', *increment*, ';')
    end ;
  store 2 in $\pi\alpha($'$\oplus$'*,* '$\Re\mathfrak{r}[$' $\oplus$ *ind reg incr* $\oplus$ ']') ;
  $\pi\alpha[begin \, \pi\alpha + 3] :=$ '**empty**' ;

USE NO INDEX REGISTER:
  store 1 in $\pi\alpha$ (';') ;
  search in $\pi\alpha$ *for (begin* $\pi\alpha + 2$*, pα, begin* $\pi\alpha$*)*
   result: *(begin expr, end expr)*
   exit: *(INITIAL VALUE NOT FOUND)* ;

identify $(\varkappa[1], \pi\alpha\ [begin\ expr - 1])$;
$p\alpha := begin\ \pi\alpha$;
go to *END GENERATE PROGRAM* ;

INITIAL VALUE NOT FOUND:
    if $\pi\alpha[begin\ \pi\alpha+2] = '|' \wedge$
    $\pi\alpha[begin\ \pi\alpha+3] \neq$ '**empty**' **then**
    store 6 in π *add* $(\varkappa[1], ':=', \varkappa[1], '+', increment, ';')$;

END GENERATE PROGRAM:
    end generate program ;

    procedure *search in* $\pi\alpha$ *for* (*begin expr, end expr, end search*)
                    result: (*begin expr found, end expr found*)
                    exit: (*NOT FOUND*) ;
    value *begin expr, end expr, end search* ;
    integer *begin expr, end expr, end search, begin expr found,*
           end expr found ;
    label *NOT FOUND* ;

comment If a new instruction has been added to the auxiliary program storage $\pi\alpha$ then a check follows whether $\pi\alpha$ already contains an instruction with the same right part. In this case the result variables are supplied with the respective beginning and end in $\pi\alpha$, otherwise a jump to the formal label *NOT FOUND* is executed ;

    begin
      integer i, j ;
      for $i := 3$ **step** 1 **until** *end search* $- 2$ **do**
        begin
          if $\pi\alpha[begin\ expr] = \pi\alpha[i] \wedge \pi\alpha[i-2] \neq '|'$ **then**
            begin
              for $j := 0$ **step** 1 **until** *end expr* $-$ *begin expr* **do**
                if $\pi\alpha[i+j] \neq \pi\alpha[begin\ expr+j]$ **then go to** L ;
              begin expr found $:= i$;
              end expr found $:= i + end\ expr - begin\ expr$;
              go to *END SEARCH* ;
L: $i := j+i$
            end
        end ;
      go to *NOT FOUND* ;
END SEARCH:
    end search in $\pi\alpha$ for expression ;

    procedure *perform optimizations in* π *add* (*begin in* π *add*) ;
    value *begin in* π *add* ; **integer** *begin in* π *add* ;

comment Instructions of the initial value programs which can not be pushed into an outer loop are optimized in such a way that identical subexpressions of the right parts are extracted, assigned to a generated variable, and replaced by this variable. Two auxiliary procedures *search in π add* and *replace* are used ;

> **begin**
>   **integer** *start expr, beg expr, end expr, beg expr found,*
>   *end expr found, i, j* ;
>   **string** *variable found* ;
>   *start expr* := *beg expr* := *begin in π add* + 2 ;

L1:   *search in π add for (start expr, begin in π add* − 1)
>   *result*: (*end expr, beg expr found, end expr found*) ;
>   **if** *end expr* < *start expr* **then go to** *NOT FOUND* ;
>   **if** π *add* [*beg expr found* − 1] = ':=' ∧
> π *add* [*end expr found* + 1] = ';' **then**

L2:   **begin**
>   **if** π *add* [*start expr* − 1] = ':=' ∧
> π *add* [*end expr* + 1] = ';' **then**

IDENTICAL RIGHT PARTS:

>   **begin**
>   *variable found* := π *add* [*beg expr* − 2] ;
>   *p add* := *beg expr* − 1 ;
>   **for** *i* := *p add* **step** − 1 **until** *beg expr found* **do**
> π *add* [*i*] := π *add* [*i* − 2] ;
> π *add* [*beg expr found* − 2] := *variable found* ;
> π *add* [*beg expr found* − 1] := ':=' ;
>   **go to** *END PERFORM OPTIMIZATION*
>   **end** identical right parts ;
>   *variable found* := π *add* [*beg expr found* − 2] ;
>   **if** *end expr* − *start expr* < 2 **then go to** *NOT FOUND* ;

VARIABLE FOUND CAN BE USED:

> π *add* [*start expr*] := *variable found* ;
>   *i* := *start expr* ; *j* := *end expr* ;
>   **for** *j* := *j* + 1 **while** π *add* [*j* − 1] ≠ ';' **do**
>   **begin** *i* := *i* + 1 ; π *add* [*i*] := π *add* [*j*] **end** ;
>   *p add* := *p add* + *start expr* − *end expr* ;
>   *start expr* := *beg expr* ;
>   **if** π *add* [*start expr* + 1] = ';' **then**
>   **go to** *END PERFORM OPTIMIZATION* ;
>   **go to** *L1*
>   **end** variable found can be used ;

if π *add* [*start expr* -1] $= \, ':=' \, \wedge$
   π *add* [*end expr* $+1$] $= \, ';'$ **then**
  begin
    replace (*beg expr found, end expr found,*
      π *add* [*begin in* π *add*]) ;
    p add := *begin in* π *add* $+3$;
    go to *END PERFORM OPTIMIZATION*
  end ;
generate variable of type ('**integer**') *result*: (*variable found*) ;
replace (*beg expr found, end expr found, variable found*) ;
start expr := *start expr* $+4$;
end expr := *end expr* $+4$;
beg expr := *start expr* $+2$;
go to *VARIABLE FOUND CAN BE USED* ;

NOT FOUND:
  start expr := *start expr* $+1$;
  if π *add* [*start expr*] $= \, '['$ **then**
    begin
L3:   *start expr* := *start expr* $+1$;
     if π *add* [*start expr*] $\neq \, ']'$ **then go to** *L3* ;
     go to *NOT FOUND*
    end ;
  if π *add* [*start expr*] $= \, '+'$ **then**
    begin
     if *start expr* $-$ *beg expr* >3 **then** *beg expr* := *start expr* $+1$;
     start expr := *start expr* $+1$;
     go to *L1*
    end ;
  if π *add* [*start expr*] $\neq \, ';'$ **then go to** *NOT FOUND* ;

END PERFORM OPTIMIZATION:
  end ;

  procedure *search in* π *add for* (*start expr, start search*)
        result: (*end expr, beg expr found, end expr found*) ;
  value *start expr, start search* ;
  integer *start expr, start search, end expr, beg expr found,*
        end expr found ;

comment The procedure delivers the beginning and the end in π *add* of an expression which is found to be identical with the sub-expression stored from *start expr* to *end expr* ;

  begin
    integer *i, j, end, found from* ;

```
          Boolean index vector, search ;
          end := start expr − 1 ;
          search := index vector := false ;
          if π add [start expr +1] = ';' then go to END SEARCH ;
          for i := π [ω [w− 5]] +2 step 1 until start search do
             begin
               if ¬ search then
                  begin
L1:                  if (π add [i] = ':=' ∨ π add [i] = '+' ∧
                        ¬ index vector) ∧ π add [i+2] ≠ ';' then
                        begin
                          found from := i+1 ;
                          j := start expr ;
                          search := true ;
                          go to L3
                        end ;
                     if π add [i] = '[' then
                        index vector := true
                     else
                        if π add [i] = ']' then
                          index vector := false ;
                     go to L3
                  end do not search ;
TEST:
               if (π add [i] = ';' ∧ (π add [j] = ';' ∨ π add [j] = '+') ∨
                     π add [i] = '+' ∧ π add [j] = ';') ∧
                     j > start expr +2 then
                  end := j− 1 ;
               if π add [i] ≠ π add [j] ∨ π add [i] = ';' then
                  begin
                     if end = start expr − 1 then
                        begin search := false ; go to L1 end ;
                     go to END SEARCH
                  end ;
               if π add [i] = '+' ∧ j > start expr +2 ∧ ¬ index vector then
                  begin end := j− 1 ; go to L2 end ;
               if π add [i] = '[' then
                  index vector := true
               else
                  if π add [i] = ']' then index vector := false ;
L2:            j := j+1 ;
L3:         end ;
```

END SEARCH:
     end expr := *end* ;
     beg expr found := *found from* ;
     end expr found := *found from* + *end* − *start expr*
     end search in π *add* for expressions ;

     procedure *replace* (*begin expr*, *end expr*, *v*) ;
       value *begin expr*, *end expr* ;
       integer *begin expr*, *end expr* ;
       string *v* ;

comment A subexpression beginning at *begin expr* and ending at *end expr* is extracted from an assignment statement and stored before this as assignment statement with a generated variable *v* as left part variable. The subexpression then is replaced by *v*, e.g.

$$\ldots v1 := \ldots \langle \text{subexpression} \rangle \ldots \text{ʕ} \ldots$$

is replaced by

$$\ldots v := \langle \text{subexpression} \rangle \text{ ʕ } v1 := \ldots v \ldots \text{ ʕ} \ldots ;$$

     begin
       integer *i*, *m* ;
       for *i* := *p add* **step** − 1 **until** *end expr* + 1 **do**
         π *add* [*i*+4] := π *add* [*i*] ;
       padd := *padd* + 4 ;
       for *i* := *begin expr* − 1 **step** − 1 **until** 1 **do**
         if π *add* [*i*] = ';' ∨ π *add* [*i*] = '**begin initial value**' **then**
           go to *L*
         else *store 1 in* π *add* (π *add* [*i*]) ;
L: π *add* [*i*+1] := *v* ;
       π *add* [*i*+2] := ':=' ;
       m := *begin expr* − *i* − 3 ;
       if *m* ≠ 0 **then**
         for *i* := *begin expr* **step** 1 **until** *end expr* **do**
           π *add* [*i*−*m*] := π *add* [*i*] ;
       i := *end expr* − *m* + 1 ; π *add* [*i*] := ';' ;
       for *i* := *i* + 1 **step** 1 **until** *end expr* + 3 **do**
         begin
           π *add* [*i*] := π *add* [*p add*] ;   *p add* := *p add* − 1
         end ;
       π *add* [*end expr* + 4] := *v*
     end replace ;

PROGRAM OF THE RECURSIVE ADDRESS CALCULATION
PASS:

INITIALIZATIONS:

    COL ι identical := 1 ;
    COL ι external identifier := 2 ;
    COL ι type := 3 ;
    COL ι index register := 4 ;
    COL ι local := 5 ;
    chi := *p* := *p add* := *delta* := *id list* := *a* := *x* := *w* :=
       max ind reg := *v* := *last suitable for in* ω := 0 ;
    delta add := δ[1, *COL* δ *associated internal identifier*] ;
    suitable loop := **false** ;

START READING PROGRAM:
    read χ ; *store 1 in* π (*incoming symbol*) ;
    if *incoming symbol* = '**block begin**' ∨
    incoming symbol = '**procedure begin**' **then**
    handling block begin
    else
      if *incoming symbol* = '**block end**' ∨
      incoming symbol = '**procedure end**' **then**
      handling block end
      else
        if *incoming symbol* = '**for**' **then**
          handling the for clause
        else
          if *incoming symbol* = '[' ∧ *suitable loop* ∧
          π[*p*−2] ≠ '**array**' **then**
          handling subscripted variables (π, *p*, *store 1 in* π)
          else
            if *incoming symbol* = '**for end**' **then**
              closing a for statement ;
    go to *START READING PROGRAM* ;

ERROR FOUND:
    error := **true** ;
    end identifier list := δ[1, *COL* δ *associated internal identifier*] ;
    go to *L* ;

END OF THE PASS:
    end identifier list := *delta add* ;
L:
  end recursive address calculation pass ;

procedure *decomposition and generation pass* (χ, χ *add*, δ,
              begin identifier list, end identifier list, error)
              result: (π, π *end, begin identifier list,*
              end identifier list, δ, *error*) ;
  integer *begin identifier list, end identifier list*, π *end* ;

Boolean *error* ;
string array χ, χ *add*, δ, π ;
begin
  integer
    del, mark, internal identifier, symbol, instruction, mode,
    information state, type, information label, kind,
    information identifier, information type, information 0,
    information 1, information 2, information 3, information 4,
    supplement information, information error,
    internal identifier of true, internal identifier of false,
    int id of number 0, int id of number 1,
    order opening bracket, order expression bracket,
    order statement bracket,
    s, k, k add, k1, p, p1, delta, end delta, delta add,
    i, int id, accumulator loaded, relative address counter,
    array block level counter, static procedure level counter,
    max address, entry number of free storage cell,
    entry number of procedure identifier, int id of parameter,
    number of parameter, undefined left part variable,
    begin switch list, begin subscript bounds, begin of loop var,
    begin step, name loop var, list elements, initial variable,
    step variable, end variable, loop variable,
    label jump over, label entry number, new label entry number,
    jump and, jump or, label return, label test exhaustion,
    next list element, label do statement, begin initial value,
    begin increment, do statement ;

  Boolean
    normal program, array block, content of, subroutine 1,
    subroutine 2, not bool, simple list element, negative, fixed step,
    incoming symbol correct ;
  string
    incoming symbol, state, specification, operation,
    conditional order, relational operator, type variable, type wanted,
    type of left part, truth value, type of AC, kind of loop var,
    type of loop var ;
  string array $\sigma[1:\infty, 1:7]$;

comment Two variables *state* and *specification* are used which, in addition to the push down, determine the specific manner of translation. They may take the following values:

$$state = \begin{cases} \text{'expression'}, & \text{if an arithmetic or Boolean expression} \\ & \text{is expected} \\ \text{'statement'}, & \text{otherwise} \end{cases}$$

$$specification = \begin{cases} \text{'empty'}, & \text{outside of actual parameters,} \\ \text{'expression'}, & \text{if an arithmetic or Boolean actual} \\ & \text{expression is expected,} \\ \text{'designational'}, & \text{if an actual designational expression} \\ & \text{is expected,} \\ \text{'otherwise'}, & \text{if an actual procedure, array,} \\ & \text{or switch identifier or an actual string} \\ & \text{is expected,} \\ \text{'undefined'}, & \text{otherwise} \end{cases}$$

state serves first, to distinguish integer labels from integer numbers, and second, to supplement specifications and to detect errors against given declarations and specifications in a rather simple way. The use of *specification* has a similar aim especially for actual parameters. The most critical value is '**undefined**'. If the actual parameter gives no further information two programs must be generated (see 4.10.1.1), otherwise '**undefined**' may be replaced by '**expression**' or '**designational**' (see also the procedures *change situation to specification expression (label)*). *state* and *specification* could be incorporated in σ. We do this only in certain cases ('**if**' | '**procedure call**' etc.) and sometimes only for the purpose of syntax check ('$:=$' | '$[$' etc.) ;

comment The arrays used in this pass have the following meaning.

χ	source program, delivered by the foregoing pass,
χ *add*	additional source program containing suitable for clauses, initial value, and increment programs, delivered by the second pass,
δ	identifier list, delivered by the first and second pass,
π	target program,
σ	push down list, combining state push down and auxiliary push down (see section 2.5).

χ, χ *add*, and δ do not require further explanations, as they have been used also in the former passes.

a) The target program π:

π is a two-dimensional array $\pi[1:\infty, 1:2]$. The instructions generated are stored row-wise: the first component is used for labels, the second for the instruction, in general concatenated from instruction part and internal identifier. The two integer variables *mark* and *instruction* are used to denote the first and second column respectively.

π: *mark*:	*instruction*:
⟨list of labels⟩	⟨instruction⟩
⋮	⋮

The position "*mark*" is '**empty**' or contains the labels marking the following target language instruction (which may also be a simple identifier) on the position "*instruction*".

b) The push down σ is a two-dimensional array $\sigma[1:\infty, 1:7]$. Each column corresponds to one entry of the push down: $\sigma[s, 1], \ldots, \sigma[s, 7]$ is σ_s. In order to facilitate the readability of the pass we use identifiers instead of the index numbers $1, \ldots, 7$ with the following meaning:

1	2	3	4
del	*symbol* | *internal identifier* | *information 4*	*mode* | *information identifier* | *information label* | *information 0*	*type* | *information state* | *information 1*

5	6	7
kind | *information type* | *information 2*	*supplement information* | *information 3*	*information error*

An entry σ_s may be a state or an entity. The column names are used partly for state entries and partly for entity entries.

b1) state entries:

A state entry has the form:

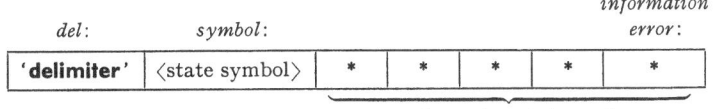

							information error:
del:	*symbol*:						
'**delimiter**'	⟨state symbol⟩	*	*	*	*	*	

supplementary information
connected with the state entry

The following types of supplementary information are used:

information label   is used for storing labels of foreward jumps, and is stored together with the state symbols **then**, **then** Σ, **else**, **else** Σ, **switch**:=, **do**, **do n**, **do suitable**, **do n suitable**, **procedure call**,

information state	is used to push down the value of the variable *state* together with the state symbols [, [̄, **if**, :=, **procedure call**,
information identifier	is used for storing the internal identifiers of the information vector together with the state symbols [and [̄,
information type	is used for storing the type of the expression after **then** together with **else**,
information 1	is used for storing a label together with the state symbols **do, do n, do suitable, do n suitable**,
information 2	is used for storing the value of the variable *specification* with the state symbols [, [̄, **if**, and **procedure call** and for the kind of the for list element together with **do** and **do suitable**,
information 3	is used for storing the relative address counter together with **block begin, array block begin, do suitable, do n suitable**,
information error	is set '**error announced**' if the state is used in an error message, otherwise '**empty**'.

In connection with procedure calls the following additional entries occur:

information 3	is used for storing the output program address counter p,
information 2	is used for storing the input program address counter k,
information 1	is used for counting the number of parameters,
information 0	is used for storing the internal identifier of a formal parameter.

In connection with the procedure identifier of a procedure declaration the following entries occur:

information label	is used for the jump leading over the declaration,
information 1	is used for saving the "beginning of free storage" cell,
information 2	is used for saving the counter for the maximum relative address number,
information 3	is used for saving the relative address counter,
information 4	is used for storing the array block level counter.

b 2) entity entries:

An entity entry has the form:

del	internal identifier	mode	type	kind	supplement information	information error
'entity'	'empty' | ⟨internal identifier⟩	'normal' | 'standard' | 'value' | 'formal' | 'generated'	'empty' | 'real' | 'integer' | 'Boolean' | 'Boolean *' | 'type' | 'arithmetic'	'number' | 'variable' | 'array' | 'procedure' | 'label' | 'switch' | 'string' | 'number cellar' | 'AC' | 'relation' | 'open jump' | 'open jump relation' | 'designational' | 'switch designator' | 'unspecified formal parameter' | 'array or switch' | 'variable or label'	'empty' | 'result true' | 'result false' | 'content of'	'empty' | 'inserted'

The entries *internal identifier, mode, type,* and *kind* can simply be taken from the identifier list in the case the entity is an incoming identifier. Otherwise they are generated. The component *supplement information* is '**content of**' for (generated) address variables, '**result true**' or '**result false**' for an '**open jump**' entry, and '**empty**' otherwise. If the entity is in any form found to be incorrect, or if the entity is inserted by the check procedures then the column *information error* is set '**inserted**', normally '**empty**'.

b3) In special cases (procedure calls) a whole entry in σ is used for supplementary information. See the comment of the program part

ACTUAL PARAMETERS IN PROCEDURE CALLS, p. 320 ;

comment The following switches are remainders of the decoding matrix. They are used within the reading and decoding part labelled by *READ NEXT SYMBOL.* With the incoming symbols or states suitable numbers: *opening bracket nr, expression bracket nr,* and *statement bracket nr* are associated by the following code procedures whose function values are given by tables. *incoming Boolean nr* and *Boolean operator nr* are used for switches within the handling of Boolean operators. ;

switch *INCOMING OPENING BRACKET* :=
   INCOMING ROUND BRACKET,
   INCOMING SQUARE BRACKET,
   INCOMING IF,
   INCOMING COLON EQUAL,
   INCOMING LABEL COLON,
   INCOMING GO TO,
   INCOMING REDUCED INITIAL ADDRESS,
   INCOMING CONTENT REAL OF,
   INCOMING CONTENT INTEGER OF,
   INCOMING CONTENT BOOLEAN OF,
   INCOMING FOR,
   INCOMING FOR SUITABLE,
   INCOMING DO SUITABLE,
   INCOMING BEGIN INITIAL VALUE,
   INCOMING BEGIN INCREMENT,
   INCOMING BEGIN,
   INCOMING BLOCK BEGIN,
   INCOMING ARRAY,
   INCOMING SWITCH,
   INCOMING PROCEDURE BEGIN ;

switch *EXPRESSION BRACKET PAIR* :=
  ROUND BRACKET,
  ROUND BRACKET NOT,
  SQUARE BRACKET,
  IF,
  THEN EXPRESSION,
  ELSE EXPRESSION,
  COLON EQUAL,
  GO TO,
  FOR,
  FOR N,
  FOR COLON EQUAL,
  STEP,
  STEP INSERTED,
  UNTIL,
  WHILE,
  FOR COLON EQUAL SUITABLE,
  STEP SUITABLE,
  UNTIL SUITABLE,
  UNTIL CONSTANT,
  ACTUAL PARAMETERS IN PROCEDURE CALLS,
  ARRAY SQUARE BRACKET,
  ARRAY COLON,
  SWITCH COLON EQUAL,
  PROCEDURE BEGIN,
  ENTITY ON STATEMENT LEVEL ;

switch *STATEMENT BRACKET PAIR* :=
  THEN STATEMENT,
  ELSE STATEMENT,
  DO,
  DO N,
  DO SUITABLE,
  DO N SUITABLE,
  BEGIN,
  BLOCK BEGIN,
  ARRAY BLOCK BEGIN,
  BEGIN INITIAL VALUE,
  BEGIN INCREMENT,
  PROCEDURE BODY BEGIN ;

integer procedure *character order* (x) ;
  string x ;
  code ;

integer procedure *state order* (x) ;
  string x ;
  code ;

integer procedure *opening bracket nr* (x) ;
  string x ;
  code ;

integer procedure *incoming Boolean nr* (x) ;
  string x ;
  code ;

integer procedure *expression bracket nr* (x) ;
  string x ;
  code ;

integer procedure *statement bracket nr* (x) ;
  string x ;
  code ;

integer procedure *Boolean operator nr* (x) ;
  string x ;
  code ;

comment The type procedures listed above associate an integer value with each incoming delimiter, or state symbol respectively. In actual implementations, it seems advisable to incorporate these values as separate components into the internal coded representations of the different symbols. In this case, evaluation of the function designators simply means extraction of the respective component of the code representation of the argument symbol.

The following table contains the values of the code procedures *character order, opening bracket nr, incoming Boolean nr* depending on the incoming delimiter as argument.

incoming character	*character order*	*opening bracket nr*	*incoming Boolean nr*
‘ (’	1	1	
‘ [’ | ‘ [’	1	2	
‘ if ’	1	3	
‘ := ’	1	4	
‘ label : ’	1	5	
‘ go to ’	1	6	
‘ red init ’	1	7	
‘ content real of ’	1	8	
‘ content integer of ’	1	9	
‘ content Boolean of ’	1	10	
‘ for ’	1	11	
‘ for suitable ’	1	12	

incoming character	character order	opening bracket nr	incoming Boolean nr
'do suitable'	1	13	
'array'	1	18	
'switch'	1	19	
'procedure begin'	1	20	
'begin initial value'	2	14	
'begin increment'	2	15	
'begin'	2	16	
'block begin'	2	17	
'↑'	3		
'×' | '×$_i$' | '/' | '÷'	4		
'+' | '+$_i$' | '−' | '−$_i$' | '⊕'	5		
'<' | '≦' | '=' | '≠' | '≧' | '>'	6		
'¬'	7		
'∧'	8		1
'∨'	9		2
'⊃'	10		3
'≡'	11		
')'	12		
']'	12		
','	12		
':'	12		
'then'	12		
'else'	12		
'step'	12		
'until'	12		
'while'	12		
'for,'	12		
'do'	12		
'for end'	13		
'then Σ**'**	13		
'else Σ**'**	13		
'?'	13		
'end'	13		
'block end'	13		
'procedure end'	13		

The following table contains the values of the code procedures *state order, expression bracket nr, statement bracket nr, Boolean operator nr* depending on the uppermost state as parameter.

state	state order	expression bracket nr	statement bracket nr	Boolean operator nr
'↑' | '↑$_{ii}$' | '↑$_{ri}$'	3			
'×' | '×$_i$' | '/' | '÷'	4			
'+' | '+$_i$' | '−' | '−$_i$' | '−$_u$' | '⊕'	5			
'<' | '≦' | '=' | '≠' | '≧' | '>'	6			
'¬'	7			

state	state order	expression bracket nr	statement bracket nr	Boolean operator nr
'∧' | '∧ not'	8			1
'∨' | '∨ not'	9			2
'⊃' | '⊃ not'	10			3
'≡'	11			4
'('	12	1		
'(not'	12	2		
'[' | '['	12	3		
'if'	12	4		
'then'	12	5		
'else'	12	6		
':='	12	7		
'go to'	12	8		
'for' | 'for suitable'	12	9		
'for n'	12	10		
'for :='	12	11		
'for inserted 1'	12			
'for inserted 2'	12			
'for n inserted'	12	10		
'step'	12	12		
'step inserted'	12	13		
'until'	12	14		
'while'	12	15		
'for := suitable'	12	16		
'step suitable'	12	17		
'until suitable'	12	18		
'until constant'	12	19		
'procedure call'	12	20		
'array ['	12	21		
'array :'	12	22		
'array'	12			
'switch'	12			
'switch :='	12	23		
'procedure begin'	12	24		
'begin'	13	25	7	
'block begin'	13	25	8	
'array block begin'	13	25	9	
'then Σ'	13	25	1	
'else Σ'	13	25	2	
'do'	13	25	3	
'do n'	13	25	4	
'do suitable'	13	25	5	
'do n suitable'	13	25	6	
'begin initial value'	13	25	10	
'begin increment'	13	25	11	
'procedure body begin'	13	25	12	

;

comment The values of the following code procedures can be given by function tables.

delimiter (x) has the value **true** if the actual string x is an ALGOL or a generated delimiter, otherwise **false**. *index register* (x) resp. *generated identifier* (x) have the value **true** if the actual entity x represents an index register, or is an identifier generated by the recursive address calculation pass resp., otherwise **false**. The procedure *value is zero* delivers the value **true** if the actual parameter is a number representing in any form the value 0, otherwise **false**. ;

  Boolean procedure *delimiter* (x) ;
    string x ;
    code ;

  Boolean procedure *index register* (x) ;
    string x ;
    code ;

  Boolean procedure *generated identifier* (x) ;
    string x ;
    code ;

  Boolean procedure *value is zero* (x) ;
    string x ;
    code ;

  Boolean procedure *Boolean operator* (x) ;
    string x ;
    code ;

  Boolean procedure *relational op* (x) ;
    string x ;
    code ;

  Boolean procedure *nonsymmetric operator* (x) ;
    string x ;
    code ;

  Boolean procedure *integer operator* (x) ;
    string x ;
    code ;

comment The values of the type procedures *Boolean operator*, *relational op*, *nonsymmetric operator*, and *integer operator* are given by the following function table, depending on the parameter x. The evaluation can be handled in a way similar to that described in the case of the procedures *character order* (x) etc. mentioned before.

x	Boolean operator (x)	relational op (x)	nonsymmetric operator (x)	integer operator (x)
'↑'	false	false	true	false
'↑$_{ii}$'	./.	./.	true	./.
'↑$_{ri}$'	./.	./.	true	./.
'×'	false	false	false	false
'×$_i$'	false	false	false	true
'/'	false	false	true	false
'÷'	false	false	true	false
'+'	false	false	false	false
'+$_i$'	false	false	false	true
'−'	false	false	true	false
'−$_i$'	false	false	true	true
'<'	false	true	./.	./.
'≦'	false	true	./.	./.
'='	false	true	./.	./.
'≠'	false	true	./.	./.
'≧'	false	true	./.	./.
'>'	false	true	./.	./.
'∧'	true	./.	./.	./.
'∧ not'	true	./.	./.	./.
'∨'	true	./.	./.	./.
'∨ not'	true	./.	./.	./.
'⊃'	true	./.	./.	./.
'⊃ not'	true	./.	./.	./.
'≡'	true	./.	./.	./.

;

```
string procedure symmetric operator (x) ;
   string x ;
   code ;

string procedure int op (x) ;
   string x ;
   code ;

string procedure not rel op (x) ;
   string x ;
   code ;

string procedure type 1 (x) ;
   string x ;
   code ;

string procedure type 2 (x) ;
   string x ;
   code ;
```

string procedure *type result* (x) ;
   string x ;
   code ;

comment The values of the string procedures *symmetric operator*, *int op*, *not rel op*, *type 1*, *type 2*, and *type result* are given by the following function table, depending on the parameter x.

x	*symmetric operator* (x)	*int op* (x)	*not rel op* (x)	*type 1* (x) *type result* (x)	*type 2* (x)
$'\uparrow_{ii}'$	./.	./.	./.	'integer'	'integer'
$'\uparrow_{ri}'$	./.	./.	./.	'real'	'integer'
$'\uparrow'$	./.	./.	./.	'real'	'real'
$'\times'$	$'\times'$	$'\times_i'$	./.	'real'	'real'
$'\times_i'$	$'\times_i'$	./.	./.	'integer'	'integer'
$'/'$	./.	./.	./.	'real'	'real'
$'\div'$	./.	./.	./.	'integer'	'integer'
$'+'$	$'+'$	$'+_i'$	./.	'real'	'real'
$'+_i'$	$'+_i'$	./.	./.	'integer'	'integer'
$'-'$	$'+'$	$'-_i'$	./.	'real'	'real'
$'-_i'$	$'+_i'$	./.	./.	'integer'	'integer'
$'<'$	$'>'$	./.	$'\geq'$	./.	./.
$'\leq'$	$'\geq'$	./.	$'>'$	./.	./.
$'='$	$'='$	./.	$'\neq'$	./.	./.
$'\neq'$	$'\neq'$	./.	$'='$	./.	./.
$'\geq'$	$'\leq'$	./.	$'<'$	./.	./.
$'>'$	$'<'$	./.	$'\leq'$	./.	./.
$'\equiv'$	$'\equiv'$	./.	./.	./.	./.

;

   string procedure *instruction function* (x, y) ;
      string x, y ;
      code ;

   string procedure *not bool operator* (x) ;
      string x ;
      code ;

   string procedure *result of Boolean operation* (x) ;
      string x ;
      code ;

comment The following function tables give the values of the string procedure *instruction function* depending on the two parameters x and y, and the values of the procedures *not bool operator* and *result of Boolean operation* depending on the parameter x.

instruction function (x, y)		
x ⟍ y	'∧' | 'V not' | '⊃ not'	'∧ not' | 'V' | '⊃'
'<'	'if AC ≧ 0 then local jump to'	'if AC < 0 then local jump to'
'≦'	'if AC > 0 then local jump to'	'if AC ≦ 0 then local jump to'
'='	'if AC + 0 then local jump to'	'if AC = 0 then local jump to'
'+'	'if AC = 0 then local jump to'	'if AC + 0 then local jump to'
'≧'	'if AC < 0 then local jump to'	'if AC ≧ 0 then local jump to'
'>'	'if AC ≦ 0 then local jump to'	'if AC > 0 then local jump to'
'empty'	'if ⌐ AC then local jump to'	'if AC then local jump to'

x	result of Boolean operation (x)	not bool operator (x)
'∧'	'result false'	'∧ not'
'∧ not'	'result true'	./.
'V'	'result true'	'V not'
'V not'	'result false'	./.
'⊃'	'result true'	'⊃ not'
'⊃ not'	'result false'	./.

;

```
    procedure read ;
      begin
        if normal program then
          begin
            k := k+1 ;
            incoming symbol := χ [k]
          end read normal program
        else
          begin
            k add := k add +1 ;
            incoming symbol := χ add [k add]
          end read program generated by the recursive address
                calculation pass ;
        incoming symbol correct := true
      end read ;

    procedure compile (x) ;
      string x ;
      begin
        p := p+1 ;
        π [p, instruction] := x ⊕ ';' ;
        π [p+1, mark] := 'empty'
      end compile ;

    procedure compile label (label entry number) ;
      value label entry number ; integer label entry number ;
```

comment For practical implementations it is sufficient either to store the label in the program (to produce a program with symbolic addresses) or to insert the program storage address in the identifier list (to produce a relative addressed machine language program). We have provided both ;

if ¬ *error* **then**
  begin
    $\pi[p+1, mark] := \pi[p+1, mark] \oplus$ *label entry number* $\oplus$
    '**label** :' ;
    $\delta[$*label entry number, COL* δ *program storage address*$] := p+1$
  end compile label ;

comment At the end of the pass the standard function programs must be added to the generated program. For this purpose the following two code procedures are needed. The first one delivers the number of storage places needed by the standard function with entry number i in the identifier list δ. The second procedure serves for storing this standard procedure in the program storage π. The explicit actions depend on the interpretive system used. ;

integer procedure *length of standard function* (i) ;
  value i ; **integer** i ;
  code ;

procedure *store standard function in* (i) ;
  value i ; **integer** i ;
  code ;

procedure *push down* (x) ;
  string x ;
  begin
    clear sigma $(s+1)$;
    $\sigma[s+1, del] := $ '**delimiter**' ;
    $\sigma[s+1, symbol] := x$;
    $s := s+1$
  end push down ;

procedure *push down entity* $(x1, x2, x3, x4, x5)$;
  string $x1, x2, x3, x4, x5$;
  begin
    clear sigma $(s+1)$;
    $\sigma[s+1, del] := $ '**entity**' ;
    $\sigma[s+1, internal\ identifier] := x1$;
    $\sigma[s+1, mode] := x2$;

$\sigma[s+1, type] := x3$;
$\sigma[s+1, kind] := x4$;
$\sigma[s+1, supplement\ information] := x5$;
$s := s+1$
end push down entity ;

procedure *clear sigma* (s) ;
  value s ; **integer** s ;
  begin
    integer i ;
    for $i := 1$ **step** 1 **until** 7 **do** $\sigma[s, i] := \,$'**empty**'
  end clear sigma s ;

procedure *delete* (i) ;
  value i ; **integer** i ;
  $s := s-i$;

procedure *clear end delta* ;
  begin
    integer i ;
    for $i := 1$ **step** 1 **until** 16 **do** $\delta[end\ delta, i] := \,$'**empty**'
  end clear end delta ;

integer procedure *max* (x, y) ;
  value x, y ; **integer** x, y ;
  $max := $ **if** $x \geq y$ **then** x **else** y ;

comment The following procedures are concerned with the identifier list δ: with storage allocation, generation of new variables or labels, and the search for identifier entries. The identifier list has the following form:

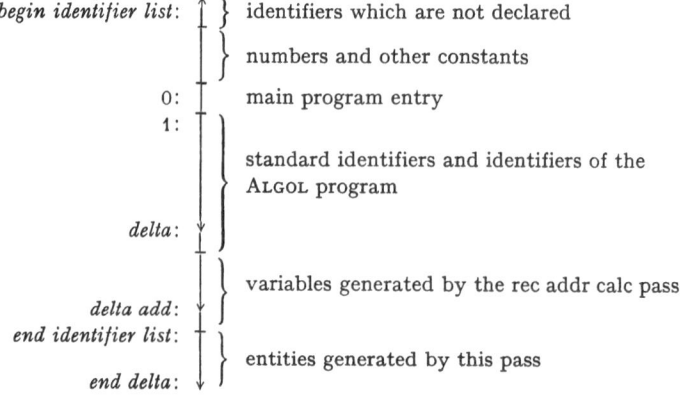

begin identifier list: } identifiers which are not declared
} numbers and other constants
0: main program entry
1:
} standard identifiers and identifiers of the Algol program
delta:
} variables generated by the rec addr calc pass
delta add:
end identifier list:
} entities generated by this pass
end delta:

The initial values of *begin identifier list* and *end identifier list* are delivered by the first or second pass resp., the value of the variable *end delta* gives always the relevant end of the identifier list during decomposition. The values of the variable *delta* gives that entry number of the list which corresponds to the ALGOL program part just handled, *delta add* is the counter for the list of generated variables ;

```
procedure storage allocation for blocks (array block) ;
   Boolean array block ;
   begin
      integer i, j, end of block ;
      for delta := delta + 1 step 1 until end identifier list do
         if δ[delta, COL δ program structure symbol] =
            'block begin' then
            go to NEW BLOCK FOUND ;

NEW BLOCK FOUND:
      array block := false ;
      end of block := δ[delta, COL δ associated internal identifier] ;
      for i := delta + 1 step 1 until end of block − 1 do
         begin
            if δ[i, COL δ program structure symbol] =
               'block begin' then
               go to END STORAGE ALLOCATION ;
            if δ[i, COL δ program structure symbol] =
               'procedure begin' then
               begin
                  i := δ[i, COL δ associated internal identifier] ;
                  go to END HANDLING A NEW IDENTIFIER
               end ;
            if δ[i, COL δ kind] = 'variable' then
               begin
                  relative address counter := relative address counter + 1 ;
                  δ[i, COL δ relative address] := relative address counter ;
                  go to END HANDLING A NEW IDENTIFIER
               end ;
            if δ[i, COL δ kind] = 'array' then
               begin
                  array block := true ;
                  δ[i, COL δ associated internal identifier] :=
                     end delta + 1 ;
                  for j := 1 step 1 until δ[i, COL δ dimension] + 2 do
                     generate variable of type ('integer')
               end ;
```

END HANDLING A NEW IDENTIFIER:
        end ;

END STORAGE ALLOCATION:
        max address := *max* (*max address, relative address counter*)
     end storage allocation for blocks ;

    procedure *storage allocation for procedures* ;
      begin
        integer i, j, *index number, entry of proc id* ;
        for *delta* := *delta* $+ 1$ **step** 1 **until** *end identifier list* **do**
          if δ[*delta, COL δ program structure symbol*] =
              ' **procedure begin**' **then**
              go to *NEW PROCEDURE FOUND* ;

NEW PROCEDURE FOUND:
      entry of proc id := *delta* $- 1$;

PROCEDURE LINKAGE:
      relative address counter := 4 ;

VECTOR FOR LOCAL ARRAY BLOCKS:
      entry number of free storage cell := *end delta* $+ 1$;
      for i := 0 **step** 1 **until**
        δ[*delta* $- 1$, *COL δ maximum array block level*] **do**
        generate variable of type ('**integer**') ;

PARAMETER BLOCK:
      for i := *delta* $+ 1$ **step** 1 **until**
        delta $+ \delta$[*delta* $- 1$, *COL δ dimension*] **do**
        begin
          relative address counter := *relative address counter* $+ 1$;
          δ[i, *COL δ relative address*] := *relative address counter*
        end ;

INFORMATION VECTORS FOR FORMAL ARRAYS:
      for i := *delta* $+ 1$ **step** 1 **until**
        delta $+ \delta$[*delta* $- 1$, *COL δ dimension*] **do**
        begin
          if δ[i, *COL δ dimension*] $\neq$ '**empty**' $\wedge$
          δ[i, *COL δ kind*] $\neq$ '**variable**' $\wedge$
          δ[i, *COL δ kind*] $\neq$ '**procedure**' **then**
          begin
            index number := *abs* (δ[i, *COL δ dimension*]) ;
            δ[i, *COL δ associated internal identifier*] :=
              end delta $+ 1$;

for $j := 1$ **step** 1 **until** *index number* $+2$ **do**
    generate variable of type ('**integer**') ;
if $\delta[i, COL\ \delta\ mode] =$ '**value**' **then**
    compile ('**value array**' $\oplus\ i$) ;
if $\delta[i, COL\ \delta\ dimension] \geqq 0$ **then**
    $\delta[i, COL\ \delta\ dimension] :=$ '**empty**'
    end
  end ;

VALUE OF A TYPE PROCEDURE:
    if $\delta[entry\ of\ proc\ id, COL\ \delta\ type] \neq$ '**empty**' **then**
      begin
        generate variable of type ($\delta[entry\ of\ proc\ id, COL\ \delta\ type]$) ;
        $\delta[entry\ of\ proc\ id, COL\ \delta\ associated\ internal\ identifier] :=$
        end delta
      end value of a type procedure ;

VALUE PARAMETERS:
    for $i := delta + 1$ **step** 1 **until**
    $delta + \delta[delta - 1, COL\ \delta\ dimension]$ **do**
    if $\delta[i, COL\ \delta\ mode] =$ '**value**' $\wedge$
    $\delta[i, COL\ \delta\ kind] =$ '**variable**' **then**
      begin
        generate variable of type ($\delta[i, COL\ \delta\ type]$) ;
        $\delta[i, COL\ \delta\ associated\ internal\ identifier] :=$ *end delta* ;
        compile ($\delta[i, COL\ \delta\ type] \oplus$ '**name call**' $\oplus\ i$) ;
        compile (*end delta* $\oplus$ ':$=$ **AC**')
      end
    end storage allocation for procedures ;

  procedure *bookkeeping block end or procedure end* ;
    begin
      for *delta* $:= delta + 1$ **step** 1 **until**
      $\delta[1, COL\ \delta\ associated\ internal\ identifier]$ **do**
      if $\delta[delta, COL\ \delta\ program\ structure\ symbol] =$ '**block end**' $\vee$
      $\delta[delta, COL\ \delta\ program\ structure\ symbol] =$
        '**procedure end**' **then**
      go to *END BOOKKEEPING* ;

END BOOKKEEPING:
    end ;

  procedure *storage allocation for generated variables* ;

comment Within loops suitable for recursive address calculation certain variables are generated which are listed at the end of the identifier list. If the delimiter '**do suitable**' appears the variables associated with this loop are supplied with fixed storage addresses. The variables marked by '**identical**' are given the same storage location address. ;

 begin
  integer i ;
  for *delta add* := *delta add* + 1 **step** 1 **until** *end identifier list* **do**
   if δ[*delta add, COL δ program structure symbol*] =
    '**for begin**' **then**
    go to *NEW GENERATED BLOCK FOUND* ;

NEW GENERATED BLOCK FOUND:
  for i := δ[*delta add, COL δ associated internal identifier*] − 1
     step − 1 **until** *delta add* + 1 **do**
  begin
   if δ[i, *COL δ program structure symbol*] = '**for end**' ∨
    δ[i, *COL δ program structure symbol*] =
    '**for begin**' **then**
    go to *END STORAGE ALLOCATION* ;
   δ[i, *COL δ relative address*] := *relative address counter* + 1 ;
   if δ[i, *COL δ mode*] = '**identical**' **then**
   relative address counter := *relative address counter* + 1 ;
   δ[i, *COL δ mode*] := '**generated**' ;
   δ[i, *COL δ static procedure level*] :=
   static procedure level counter ;
   δ[i, *COL δ array block level*] := *array block level counter*
  end ;

END STORAGE ALLOCATION:
   max address := *max* (*max address, relative address counter*)
  end storage allocation for generated variables ;

 procedure *search for an identifier in the identifier*
    list (*identifier*) *result*: (*int id*) ;
  integer *int id* ; **string** *identifier* ;
  begin
   integer i, *begin search, end search* ;
   begin search := δ[*delta, COL δ associated internal identifier*] − 1 ;
L: *end search* := *begin identifier list* ;

SEARCH:
    for $i := $ *begin search* **step** -1 **until** *end search* **do**
     begin
      if $\delta[i, COL\ \delta\ program\ structure\ symbol] = $ '**block end**' $\vee$
       $\delta[i, COL\ \delta\ program\ structure\ symbol] = $
        '**procedure end**' **then**
       begin
        $i := \delta[i, COL\ \delta\ associated\ internal\ identifier]$;
        go to *END SEARCH LOOP*
       end ;
      if $\delta[i, COL\ \delta\ program\ structure\ symbol] = $
       '**block begin**' $\vee$
       $\delta[i, COL\ \delta\ program\ structure\ symbol] = $
       '**procedure begin**' **then**
       begin
        end search := *begin search* $+2$;
        begin search :=
         $\delta[i, COL\ \delta\ associated\ internal\ identifier] - 1$;
        go to *SEARCH*
       end ;
      if $\delta[i, COL\ \delta\ external\ identifier] = $ *identifier* **then**
       go to *END SEARCH FOR AN IDENTIFIER* ;

END SEARCH LOOP:
     end ;
    if *end search* $= $ *begin identifier list* **then**
     begin
      error identifier is not declared ;
      int id := *begin identifier list* ;
      go to *END PROCEDURE*
     end ;
    begin search :=
     $\delta[begin\ search + 1, COL\ \delta\ associated\ internal\ identifier] - 1$;
    go to L ;

END SEARCH FOR AN IDENTIFIER:
    int id := i ;
    if *state* $= $ '**expression**' $\wedge \delta[i, COL\ \delta\ kind] = $ '**label**' $\wedge$
     $\delta[i, COL\ \delta\ type] = $ '**integer**' **then**

begin
 for $i := -1$ **step** -1 **until** *begin identifier list* **do**
  if $\delta[i, COL\ \delta\ external\ identifier] = identifier$ **then**
   go to *INTEGER NUMBER* ;

INTEGER NUMBER:
 int id $:= i$
end integer number ;

END PROCEDURE:
 end search for an identifier in the identifier list ;

procedure *search for a generated identifier (identifier)*
   result: (*int id*) ;
string *identifier* ; **integer** *int id* ;
begin
 integer i ;
 for $i :=$ *end identifier list* -1 **step** -1 **until**
  $\delta[1, COL\ \delta\ associated\ internal\ identifier] + 1$ **do**
  if $\delta[i, COL\ \delta\ external\ identifier] = identifier$ **then**
   go to *IDENTIFIER FOUND* ;

IDENTIFIER FOUND:
 int id $:= i$
end search for a generated identifier ;

comment The counter *delta* for the identifier list δ contains the entry number of that program structure symbol in δ corresponding to the last processed '**block begin**', '**procedure begin**', '**block end**', or '**procedure end**'. Thus, *delta* contains the beginning corresponding to the innermost relevant block or procedure if no local block or procedure has already been processed, and it contains, otherwise, the entry corresponding to the end of the local block or procedure. The search starts always with the entry associated with *delta* (given by the associated internal identifier of the content of *delta*).

Thus the search for an identifier starts at the end of the innermost relevant block or procedure, or at the beginning of the local block, respectively. Then the next smallest block or procedure in the nesting is tested: from the beginning of the innermost block or procedure to the beginning of the block or procedure to be checked, and from its end to the end of the innermost block or procedure. For outer blocks, the process runs analoguously. Local blocks are omitted.

The search mechanism:

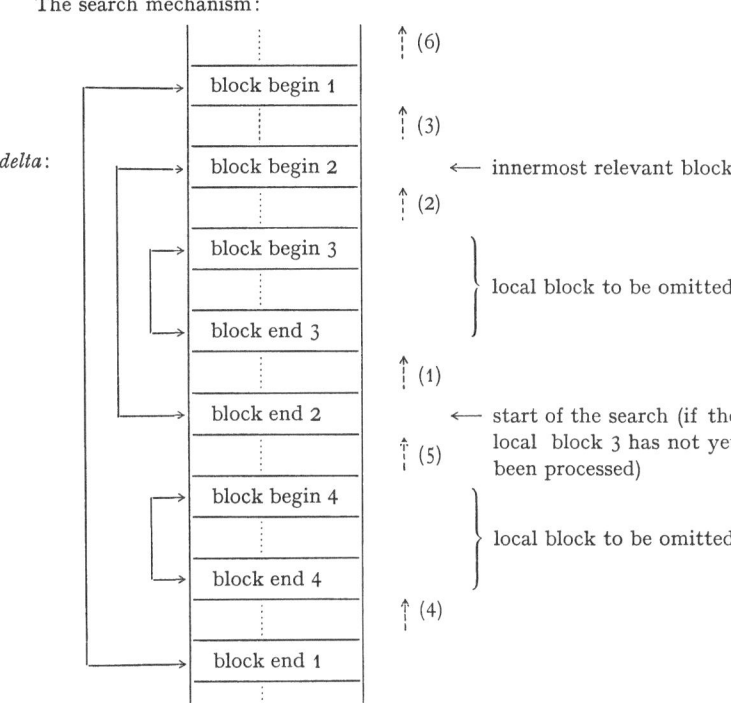

The broken arrows give the direction, and the numbers show the succession of searching. In order to verify this the initial and end variables of the search loop must always be supplied with the proper values.

(A simpler mechanism would require a rearrangement of the identifer list in a prepass and/or references to beginnings or ends of embracing blocks or procedures. For the innermost relevant block it would be enough to push down *delta* together with the '**block begin**' s.)

If an identifier is not found an error message is given. A declaration of this identifier is inserted before the constants at a new beginning of the identifier list, so that this identifier now is declared global to the whole program. In order to avoid further error messages, the identifier is declared as '**unspecified formal parameter**' and handled like these.

Remark: The search for generated identifiers is much simpler. One needs not make allowance for the block structure since all generated identifiers differ. ;

    procedure *search for specification (entry number,*
                number of parameter) result: (state, specification,
                int id of parameter) ;

```
value entry number, number of parameter ;
integer entry number, number of parameter, int id of parameter ;
string state, specification ;
begin
  integer i, int id of procedure ;
  int id of procedure := σ[entry number, internal identifier] ;
  if δ[int id of procedure, COL δ mode] = 'formal' ∨
    σ[entry number, information error] = 'inserted' then
    go to L ;
  if δ[int id of procedure, COL δ dimension] <
      number of parameter then
    go to L ;
  int id of parameter := i := int id of procedure +
                            number of parameter + 1 ;
  if δ[i, COL δ kind] = 'procedure' ∨
    δ[i, COL δ kind] = 'array' ∨
    δ[i, COL δ kind] = 'switch' ∨
    δ[i, COL δ kind] = 'array or switch' ∨
    δ[i, COL δ kind] = 'string' then
    begin
      state := 'statement' ;
      specification := 'otherwise'
    end
  else
    if δ[i, COL δ kind] = 'variable' then
      state := specification := 'expression'
    else
      if δ[i, COL δ kind] = 'label' then
        begin
          state := 'statement' ;
          specification := 'designational'
        end
      else
        begin
L:        state := 'expression' ;
          if σ[int id of procedure, type] ≠ 'empty' then
            specification := 'expression'
          else specification := 'undefined'
        end
end search for specification ;

procedure generate variable of type (x) ;
  string x ;
```

begin
  integer i ;
  end delta $:=$ *end delta* $+ 1$;
  clear end delta ;
  relative address counter $:=$ *relative address counter* $+ 1$;
  max address $:=$ *max* (*max address, relative address counter*) ;
  δ [*end delta, COL* δ *internal identifier*] $:=$ *end delta* ;
  δ [*end delta, COL* δ *mode*] $:=$ '**normal**' ;
  δ [*end delta, COL* δ *type*] $:=$ x ;
  δ [*end delta, COL* δ *kind*] $:=$ '**variable**' ;
  δ [*end delta, COL* δ *relative address*] $:=$ *relative address counter* ;
  δ [*end delta, COL* δ *static procedure level*] $:=$
    static procedure level counter
end generate variable ;

procedure *reduce number cellar* ;
  relative address counter $:=$ *relative address counter* $- 1$;

procedure *bring to AC with prescribed type* ($i, j, x, insert\ AC$) ;
  value $i, j, insert\ AC$;
  integer i, j ; **Boolean** *insert AC* ; **string** x ;

comment The procedure *bring to AC with prescribed type* ($i, j, x,$ *insert AC*) generates the instructions which, at run time, deliver the value of the entity stored in σ_i into the **AC** and, if necessary, perform a type transfer to the type desired, normally given by x. x may have the values '**real**', '**integer**', and '**Boolean**', or '**arithmetic**' and '**type**' which have the meaning of '**undefined**'. Type checks have been made before entering this procedure, so that the actual type of σ_i and the type desired are compatible.

As described in section 4.3 different transfer instructions have to be generated depending on whether the entity stored as σ_i is (1) constituent of an expression or (2) the right part of an assignment statement:

(1) In this case the type desired depends on the type of the entity σ_i or on the immediate companions (e.g. an operator requiring a definite type). The parameter j is irrelevant and is set equal to i ($i = j$).

(2) The type desired depends in general on the type of the left part variables. If this type is undefined additional type check instructions must be generated. Therefore the entry number in σ of the rightmost left part variable is delivered by the parameter j. Thus, the case (2) is given by $i \neq j$ (actually $i = j + 2$).

If the last parameter *insert AC* has the value **true** then the entity σ_i is replaced by the entity associated with the accumulator.

If the content of the accumulator is still a relevant value this is saved in the number cellar. ;

```
begin
    integer int id ;  string type desired, type i ;
    int id := σ[i, internal identifier] ;
    type i := σ[i, type] ;
    type desired := 'empty' ;
    if x = 'real' ∨ x = 'integer' ∨ x = 'Boolean' then
        type desired := x
    else
        if type i = 'real' ∨ type i = 'integer' ∨
            type i = 'Boolean' then
            type desired := type i ;
```

CHECK WHETHER AC IS LOADED:
```
    if accumulator loaded = i then
        begin
            if σ[i, supplement information] ≠ 'content of' then
                begin
                    if type i = x then go to CHECK LEFT PART ;
                    go to TYPE TRANSFER
                end
        end
    else
        if accumulator loaded ≠ 0 then
            take care of AC (accumulator loaded) ;
    if σ[i, kind] = 'number' ∧ type i ≠ type desired then
        begin
            type i := type desired ;
            compile ('AC :=' ⊕ δ[int id, COL δ associated internal
                identifier]) ;
            go to CHECK LEFT PART
        end ;
    if σ[i, kind] = 'procedure' then
        begin
            if σ[i, mode] = 'normal' then
                compile (type desired ⊕ 'procedure call' ⊕ int id)
            else
                if type desired ≠ 'empty' then
                    compile (type desired ⊕ 'formal procedure call' ⊕
                        int id)
                else compile ('type formal procedure call' ⊕ int id)
        end function call
```

else
  if $\sigma[i, mode] =$ '**formal**' **then**
    begin
      if $\sigma[i, information\ error] =$ '**empty**' **then**
        $\delta[int\ id,\ COL\ \delta\ input] :=$ '**input**' ;
      if *type desired* $\neq$ '**empty**' **then**
        compile (*type desired* $\oplus$ '**name call**' $\oplus$ *int id*)
      else *compile* ('**name call**' $\oplus$ *int id*)
    end name call
  else
    begin
      if $\sigma[i, supplement\ information] =$ '**content of**' **then**
        begin
          if $\sigma[i, kind] =$ '**AC**' **then**
            begin
              compile ('**ADDR AC := AC**') ;
              compile ('**AC := content of ADDR AC**')
            end
          else *compile* ('**AC := content of**' $\oplus$ *int id*)
        end subscripted variable
      else *compile* ('**AC :=**' $\oplus$ *int id*) ;
      if $\sigma[i, kind] =$ '**number cellar**' **then**
        reduce number cellar ;
      if *type i* $\neq$ *type desired* **then**

TYPE TRANSFER:
        begin
          if *type desired* $=$ '**Boolean**' **then**
            begin
               if $\sigma[i, kind] =$ '**number cellar**' **then**
                 compile ('**compare Boolean to**' $\oplus$ *int id*)
               else *compile*
                 ('**compare Boolean to ADDR AC**')
            end transfer to **Boolean**
          else
            if *type desired* $=$ '**real**' $\lor$
              type desired $=$ '**integer**' **then**
              begin
                if *type i* $=$ '**real**' $\lor$ *type i* $=$ '**integer**' **then**
                  compile ('**transfer to type**' $\oplus$
                    type desired)
                else

if $\sigma[i, kind] = $ 'number cellar' then
    compile ('possible transfer to' $\oplus$
      type desired $\oplus$ 'depending on' $\oplus$
      int id)
else
    compile ('possible transfer to' $\oplus$
      type desired $\oplus$
      'depending on ADDR AC')
end

end type transfer

end numbers, variables, and expressions ;

CHECK LEFT PART:
    if $i \neq j \wedge$ *undefined left part variable* $\neq 0$ **then**
    begin
      if $x = $ 'Boolean' **then**
      compile ('compare Boolean to' $\oplus$
      $\sigma[j, $ *internal identifier*])
      else
        if *type* $i = $ 'real' $\vee$ *type* $i = $ 'integer' **then**
        compile ('possible transfer from' $\oplus$ *type* $i \oplus$
          'depending on' $\oplus \sigma[j, $ *internal identifier*])
        else
        compile ('possible transfer from ADDR AC to' $\oplus$
          $\sigma[j, $ *internal identifier*])
    end type transfers for the right part and the rightmost
    left part variable ;

CHANGE PUSH DOWN ENTRY:
    if *insert AC* **then**
    begin
      $\sigma[i, mode] := $ 'normal' ;
      if *type desired* $\neq$ 'empty' **then**
      $\sigma[i, type] := $ *type desired* ;
      $\sigma[i, kind] := $ 'AC' ;
      $\sigma[i, $ *internal identifier*$] := $
        $\sigma[i, $ *supplement information*$] := $
        $\sigma[i, $ *information error*$] := $ 'empty'
    end change push down entry ;
    accumulator loaded $:= i$;

END BRING TO AC:
    end bring to AC with prescribed type ;

procedure *take care of AC (j)* ;
  value *j* ; **integer** *j* ;
  begin
    generate variable of type $(\sigma[j, type])$;
    compile (end delta $\oplus$ ':= **AC**') ;
    $\sigma[j, internal\ identifier] := end\ delta$;
    $\sigma[j, kind] :=$ '**number cellar**' ;
    accumulator loaded := 0
  end take care of AC ;

procedure *exchange operands* ;
  begin
    string *x* ; **integer** *i* ;
    for *i* := 1 **step** 1 **until** 7 **do**
      begin
        $x := \sigma[s, i]$;
        $\sigma[s, i] := \sigma[s-2, i]$;
        $\sigma[s-2, i] := x$
      end ;
    if *accumulator loaded* = *s* **then**
      accumulator loaded := *s* − 2
    else
      if *accumulator loaded* = *s* − 2 **then**
        accumulator loaded := *s* ;
    $\sigma[s-1, symbol] := symmetric\ operator\ (\sigma[s-1, symbol])$
  end exchange operands ;

procedure *transfer type of number (i)* ;
  value *i* ; **integer** *i* ;
  begin
    $\sigma[i, internal\ identifier] := \delta[\sigma[i, internal\ identifier],$
      $COL\ \delta\ associated\ internal\ identifier]$;
    $\sigma[i, type] := \delta[\sigma[i, internal\ identifier], COL\ \delta\ type]$
  end transfer type of number ;

comment The following four procedures are used for translating Boolean expressions ;

procedure *bring truth value* ;
  begin
    if $\sigma[s, kind] =$ '**open jump**' **then**
      begin
        compile label $(\sigma[s, internal\ identifier])$;
        if *jump or* ≠ 0 **then**
          begin *compile label (jump or)* ; *jump or* := 0 **end** ;

        if *jump and* $\neq 0$ **then**
          begin *compile label (jump and)* ; *jump and* $:= 0$ **end**
      end no relation : the accumulator contains the truth value
      else
        begin
          deliver label entry number ;
          compile conditional order and address open jumps ;
          compile ('**AC** :='*$\oplus$ internal identifier of false)* ;
          compile simple order and address open jump ;
          compile ('**AC** :='*$\oplus$ internal identifier of true)* ;
          compile label (new label entry number)
        end a relation is found ;
      $\sigma\,[s,\ kind] := $ '**AC**' ;
      $\sigma\,[s,\ internal\ identifier] := \sigma\,[s,\ supplement\ information] :=$
        '**empty**' ;
      accumulator loaded $:= s$
    end bring truth value ;

procedure *deliver label entry number* ;
  if $\sigma\,[s,\ supplement\ information] = $ '**result true**' **then**
    label entry number $:= \sigma\,[s,\ internal\ identifier]$
  else
    if *jump or* $\neq 0$ **then**
      label entry number $:= jump\ or$
    else *generate label (label entry number)* ;

procedure *compile conditional order and address open jumps* ;
  begin
    conditional order $:= instruction\ function\ (relational\ operator,$ '∨'$)$;
    compile (conditional order $\oplus$ label entry number) ;
    relational operator $:= $ '**empty**' ;
    if $\sigma\,[s,\ kind] = $ '**open jump**' ∨
      $\sigma\,[s,\ kind] = $ '**open jump relation**' **then**
      begin
        if $\sigma\,[s,\ supplement\ information] = $ '**result false**' **then**
          compile label ($\sigma\,[s,\ internal\ identifier]$) ;
        if *jump and* $\neq 0$ **then**
          begin *compile label (jump and)* ; *jump and* $:= 0$ **end**
      end open jump
    end compile conditional order and address open jumps ;

procedure *compile simple order and address open jump* ;
  begin
    generate label (new label entry number) ;
    compile ('**local jump to**' *$\oplus$ new label entry number)* ;

compile label (label entry number) ;
if $\sigma[s, supplement\ information] = $ '**result true**' $\wedge$
jump or $\neq 0$ **then**
    begin *compile label (jump or)* ; *jump or* $:= 0$ **end**
end compile simple order and address open jump ;

procedure *generate label (label entry number)* ;
  integer *label entry number* ;
  begin
    end delta $:= label\ entry\ number := end\ delta + 1$;
    clear end delta ;
    $\delta[end\ delta,\ COL\ \delta\ internal\ identifier] := end\ delta$;
    $\delta[end\ delta,\ COL\ \delta\ mode] := $ '**generated**' ;
    $\delta[end\ delta,\ COL\ \delta\ kind] := $ '**label**'
  end generate label ;

procedure *compile jump* ;
  begin
    check designational expression ;
    if $\sigma[s, kind] = $ '**label**' $\vee$
      $\sigma[s, kind] = $ '**switch designator**' **then**
      begin
        if $\sigma[s, mode] = $ '**normal**' **then**
          begin
            if *static procedure level counter* $=$
                $\delta[\sigma[s, internal\ identifier],$
                $COL\ \delta\ static\ procedure\ level] \wedge$
             array block level counter $=$
                $\delta[\sigma[s, internal\ identifier],$
                $COL\ \delta\ array\ block\ level]$ **then**
             compile ('**local jump to**' $\oplus$
                $\sigma[s, internal\ identifier])$
               else *compile* ('**jump to**' $\oplus \sigma[s, internal\ identifier])$
          end
        else *compile* ('**formal procedure exit**' $\oplus$
          $\sigma[s, internal\ identifier])$
      end
  end compile jump ;

procedure *jump over declarations* ;
  if *label jump over* $= 0$ **then**
    begin
      generate label (label jump over) ;
      compile ('**local jump to**' $\oplus$ *label jump over*)
    end jump over declarations ;

procedure *address label jump over* ;
  if ¬ ($\chi[k+1] = $ '**switch**' ∨ $\chi[k+1] = $ '**procedure begin**' ∨
     $\chi[k+2] = $ '**switch**' ∨ $\chi[k+2] = $ '**procedure begin**') **then**
  begin
    compile label (*label jump over*) ;
    label jump over := 0
  end address label jump over ;

comment The following two procedures are used when handling actual parameters in parameter positions whose specifications are not known at first. From those actual parameters which may be arithmetic/Boolean expressions or designational expressions (see section 4.10.1.1) two programs are produced:

a) A program to deliver an arithmetic or Boolean value is generated. If an entity occurs which is only allowed in designational expressions then the specification is changed to "designational", the expression program part already produced must be eliminated, and the decomposition is started again. This is handled by the procedure *change situation to specification label*.

b) If the specification has not yet been fixed a second program part (a designational expression) is produced. If one finds out that the actual parameter may not be a designational expression the instructions in question are eliminated by means of the procedure *change situation to specification expression*. This is only possible if the actual parameter has a form like e.g.

      if ... **then** ... v ... **else** ... **if** ... v ... **then** ...

or

      if ... **then** ... v ... **else** ... $a[... v ...]$ ...

or

    $v[... v[...] ...]$

where v is an unspecified formal parameter. From the second if clause, or the following subscript position resp., v is known to be an arithmetic or Boolean entity. But this knowledge can be utilized only in the second pass through the expression. ;

procedure *change situation to specification expression* ;
  begin
    for $s := s$ **step** -1 **until** 1 **do**
      if $\sigma[s-1, symbol] = $ '**procedure call**' **then** **go to** L ;
$L:$ $p := p1$;
    $k := k1$; *incoming symbol* := $\chi[k]$;
    specification := *state* := '**expression**'
  end change situation to specification expression ;

procedure *change situation to specification label* ;
  begin
    state := '**statement**' ;
    specification := '**designational**' ;
    for $s := s-1$ **step** -1 **until** 1 **do**
      if $\sigma[s, symbol]$ = '**procedure call**' **then**
        go to L ;
$L:$ $p := \sigma[s, information\ 3]$;
       $\pi[p+1, mark] :=$ '**empty**' ;
       $k := \sigma[s, information\ 2]$
  end ;

procedure *check undefined specification* ;
  if *specification* = '**undefined**' **then**
    begin
      if $(\sigma[s, kind]$ = '**number**' $\wedge$ $\sigma[s, type] \neq$ '**integer**') $\vee$
        $\sigma[s, kind]$ = '**variable**' $\vee$
        $\sigma[s, kind]$ = '**procedure**' **then**
        specification := *state* := '**expression**'
    end ;

comment The remaining procedures serve for error checking, error messages, and inserting missing specifications which operation is completely included in the check procedures. The check procedures consist of

(1) a procedure *check delimiter pair* which checks the formally described part of the syntax and tries to find proper corrections.

(2) procedures which test whether the use of entities is according to the declaration or specification. Suitable corrections may simply be made within the push down σ. ;

procedure *error message* (x) ;
  string x ;
  code ;

string procedure *text* (x, y) ;
  string x, y ;
  code ;

comment The following table gives the values of the string procedure *text* depending on the parameters x and y.

y \ x	'empty' | 'normal'	'value' | 'formal'	'generated'
'variable'	'a variable'	'a formal variable'	'a generated variable (possibly a subscripted variable)'
'number'	'a number'	./.	./.
'array'	'an array identifier'	'a formal array identifier'	./.
'label'	'a label identifier'	'a formal label identifier'	./.
'switch'	'a switch identifier'	'a formal switch identifier'	./.
'string'	'a string'	'a formal string identifier'	./.
'procedure'	'a procedure identifier'	'a formal procedure identifier'	./.
'unspecified formal parameter'	./.	'a formal parameter'	./.
'AC'	'an expression'	./.	./.
'number cellar'	'a subscripted variable'	./.	./.
'relation'	'a relation'	./.	./.
'open jump' | 'open jump relation'	'a Boolean expression'	./.	./.
'designational'	'a designational expression'	./.	./.
'switch designator'	'a switch designator'	./.	./.
'variable or label'	./.	'a formal parameter'	./.
'array or switch'	./.	'a formal array or switch identifier'	./.

procedure *check delimiter pair* ;
  begin
    Boolean procedure *symbol may follow an entity* (x) ;
      string x ;
      code ;

    Boolean procedure *symbol may follow a delimiter* (x) ;
      string x ;
      code ;

    Boolean procedure *triple delimiter entity delimiter is*
      allowed (x, y) ;
      string x, y ;
      code ;

    Boolean procedure *pair delimiter delimiter is allowed* (x, y) ;
      string x, y ;
      code ;

    procedure *exit for delimiter entity delimiter* (x, y) ;
      string x, y ;
      code ;

    procedure *exit for delimiter delimiter* (x, y) ;
      string x, y ;
      code ;

    procedure *insert an entity* ;
      begin
        push down entity (0, '**formal**', '**empty**',
          '**unspecified formal parameter**', '**empty**') ;
        $\sigma[s, \textit{information error}] := $ '**inserted**'
      end ;

    procedure *delete uppermost state* ;
      begin
        if $\sigma[s, \textit{symbol}] = $ '[' $\lor$ $\sigma[s, \textit{symbol}] = $ '**if**' **then**
          begin
            specification := $\sigma[s, \textit{information 2}]$;
            state := $\sigma[s, \textit{information state}]$;
            if $\sigma[s-1, \textit{del}] = $ '**entity**' **then**
              $\sigma[s-1, \textit{information error}] := $ '**inserted**'
          end
          else
            if $\sigma[s, \textit{symbol}] = $ ':=' **then**

begin
    state := $\sigma[s,$ *information state*$]$;
    type of left part := '**type**' ;
    undefined left part variable := 0 ;
    if $\sigma[s-1,$ *del*$]$ = '**entity**' **then**
      $\sigma[s-1,$ *information error*$]$:= '**inserted**'
end
else
  if $\sigma[s,$ *symbol*$]$ = '**for**' ∨
    $\sigma[s,$ *symbol*$]$ = '**for n**' ∨
    $\sigma[s,$ *symbol*$]$ = '**for suitable**' ∨
    $\sigma[s,$ *symbol*$]$ = '**for inserted 1**' ∨
    $\sigma[s,$ *symbol*$]$ = '**for inserted 2**' ∨
    $\sigma[s,$ *symbol*$]$ = '**for n inserted**' **then**
    state := '**statement**'
  else
    if $\sigma[s,$ *symbol*$]$ = '**procedure call**' **then**
      begin
        s := *entry number of procedure identifier* ;
        state := $\sigma[s+1,$ *information state*$]$;
        specification := $\sigma[s+1,$ *information 2*$]$;
        entry number of procedure identifier :=
          $\sigma[s+1,$ *internal identifier*$]$;
      end
    else
      if $\sigma[s,$ *symbol*$]$ = '**array [**' ∨
        $\sigma[s,$ *symbol*$]$ = '**array :**' **then**
        begin
          state := '**statement**' ;
          delete $(s -$ *begin subscript bounds* $+ 2)$
        end
      else
        if $\sigma[s,$ *symbol*$]$ = '**switch :=**' **then**
          delete $(s-$ *begin switch list* $+ 2)$
end delete uppermost state ;

TEST AND POSSIBLE ERROR MESSAGE:
    if *delimiter* $(\sigma[s,$ *symbol*$])$ **then**
      begin
        if *symbol may follow a delimiter* (*incoming symbol*) **then**
          begin
            if *pair delimiter delimiter is allowed*
              $(\sigma[s,$ *symbol*$],$ *incoming symbol*) **then**

begin
$\sigma[s, \text{ information error}] := \text{'empty'}$;
go to *END CHECK DELIMITER PAIR*
end ;

MESSAGE DD:

 if *incoming symbol correct* $\wedge$
 $\sigma[s, \text{ information error}] = \text{'empty'}$ **then**
 error message ('error: the incoming symbol' $\oplus$
  incoming symbol $\oplus$ 'finds' $\oplus$ $\sigma[s, \text{ symbol}]$ $\oplus$
  'as uppermost state.') ;
 $\sigma[s, \text{ information error}] := \text{'error announced'}$;
 incoming symbol correct := **false** ;
 error := **true** ;
 exit for delimiter delimiter
  $(\sigma[s, \text{ symbol}], \text{ incoming symbol})$
 end ;

INSERT AN ENTITY:

 error := **true** ;
 insert an entity ;
 if *triple delimiter entity delimiter is allowed*
  $(\sigma[s-1, \text{ symbol}], \text{ incoming symbol})$ **then**
  begin
  error message ('The incoming symbol' $\oplus$
   incoming symbol $\oplus$ 'finds' $\oplus$ $\sigma[s-1, \text{ symbol}]$ $\oplus$
   'as uppermost entry in σ. There is missing an
   entity.') ;
  $\sigma[s-1, \text{ information error}] := \text{'empty'}$;
  go to *END CHECK DELIMITER PAIR*
  end ;
 if *incoming symbol correct* $\wedge$
  $\sigma[s-1, \text{ information error}] = \text{'empty'}$ **then**
  error message ('error: the incoming symbol' $\oplus$
   incoming symbol $\oplus$ 'finds' $\oplus$ $\sigma[s-1, \text{ symbol}]$ $\oplus$
   'as uppermost state.') ;

1: *incoming symbol correct* := **false** ;
  $\sigma[s-1, \text{ information error}] := \text{'error announced'}$;
  exit for delimiter entity delimiter
   $(\sigma[s-1, \text{ symbol}], \text{ incoming symbol})$
 end uppermost entry in σ is a delimiter ;

UPPERMOST ENTRY IS AN ENTITY:
 if *symbol may follow an entity* (*incoming symbol*) **then**

18*

begin
 if *triple delimiter entity delimiter is allowed*
   ($\sigma[s-1, symbol]$, *incoming symbol*) **then**
  begin
   $\sigma[s-1, information\ error] := $ '**empty**' ;
   go to *END CHECK DELIMITER PAIR*
  end ;
 if *incoming symbol correct* $\wedge$
  $\sigma[s-1, information\ error] = $ '**empty**' **then**
  begin
   if $\sigma[s, information\ error] \neq $ '**inserted**' **then**
    error message ('error: the incoming symbol' $\oplus$
     incoming symbol $\oplus$ 'finds' $\oplus$ $\sigma[s-1, symbol]$ $\oplus$
     'and' $\oplus$ *text* ($\sigma[s, mode]$, $\sigma[s, kind]$) $\oplus$
     'as uppermost entries in σ.')
   else
    error message ('error: the incoming symbol' $\oplus$
     incoming symbol $\oplus$ 'finds' $\oplus$ $\sigma[s-1, symbol]$ $\oplus$
     'as uppermost state in σ.')
  end triple is not allowed ;
  error := **true** ;
  go to *1*
 end symbol may follow an entity ;
DELETE AN ENTITY:
 if *pair delimiter delimiter is allowed*
   ($\sigma[s-1, symbol]$, *incoming symbol*) **then**
  begin
   if $\sigma[s, information\ error] \neq $ '**inserted**' **then**
    begin
     error message ('error: the incoming symbol' $\oplus$
      incoming symbol $\oplus$ 'finds' $\oplus$ *text* ($\sigma[s, mode]$,
      $\sigma[s, kind]$) $\oplus$ 'as uppermost entry in σ.') ;
     error := **true**
    end ;
   $\sigma[s-1, information\ error] := $ '**empty**' ;
   delete (1) ;
   go to *END CHECK DELIMITER PAIR*
  end ;
 if *incoming symbol correct* $\wedge$
  $\sigma[s-1, information\ error] = $ '**empty**' **then**

begin
    if $\sigma[s, information\ error] \neq$ '**inserted**' **then**
        error message ('error: the incoming symbol' $\oplus$
            incoming symbol $\oplus$ 'finds' $\oplus$ $\sigma[s-1, symbol]$ $\oplus$
            'and' $\oplus$ *text* $(\sigma[s, mode], \sigma[s, kind]) \oplus$
            'as uppermost entries in σ.')
    else
        error message ('error: the incoming symbol' $\oplus$
            incoming symbol $\oplus$ 'finds' $\oplus$ $\sigma[s-1, symbol]$ $\oplus$
            'as uppermost entries in σ.')
    end ;
$error :=$ **true** ;
incoming symbol correct := **false** ;
delete (1) ;
$\sigma[s, information\ error] :=$ '**error announced**' ;
exit for delimiter delimiter $(\sigma[s, symbol], incoming\ symbol)$;

EXITS FOR AN INCORRECT INCOMING DELIMITER
  FINDING A DELIMITER AND AN ENTITY AS
  UPPERMOST ENTRIES IN σ:
INCORRECT INCOMING COLON EQUAL:
A:
    for $i := s-2$ **step** -1 **until** 1 **do**
      if $\sigma[i, del] =$ '**delimiter**' **then go to** $A0$;
$A0$:
    if *triple delimiter entity delimiter is allowed*
        $(\sigma[i, symbol], incoming\ symbol)$ **then**
      begin
        delete (1) ; *delete uppermost state* ;
        $s := i$;
        insert an entity ;
        go to *INCOMING COLON EQUAL*
      end push down contains an admissible entry next to the
      uppermost state ;
    push down (incoming symbol) ;
    $\sigma[s, information\ state] :=$ *state* ; *state* := '**expression**' ;
    go to *READ NEXT SYMBOL* ;

INCORRECT INCOMING LABEL COLON:
B:
    go to *INCOMING LABEL COLON* ;

INCORRECT INCOMING CLOSING ROUND BRACKET:
INCORRECT INCOMING CLOSING SQUARE BRACKET:
D:
    go to *READ NEXT SYMBOL* ;

INCORRECT INCOMING COMMA:
INCORRECT INCOMING COLON:
E:

      delete (1) ; **go to** *READ NEXT SYMBOL* ;

INCORRECT INCOMING THEN:
F:

      $s := s-1$;

      if $\sigma[s, del] =$ '**delimiter**' **then**

        begin

          if *state order* $(\sigma[s, symbol]) <$ *order expression bracket* **then**

            go to *PROCESS* ;

          if $\sigma[s, symbol] =$ '**if**' **then**

            begin *insert an entity* ; **go to** *IF* **end**

          else

            if $\sigma[s, symbol] \neq$ '(' $\wedge$ $\sigma[s, symbol] \neq$ '(**not**' $\wedge$

              $\sigma[s, symbol] \neq$ '**procedure call**' $\wedge$

              $\sigma[s, symbol] \neq$ '[' **then**

              begin

G: *push down* (*incoming symbol*) ;

                go to *READ NEXT SYMBOL*

              end

            else *delete uppermost state*

        end ;

      go to *F* ;

INCORRECT INCOMING ELSE:
H:

      $s := s-1$;

      if $\sigma[s, del] =$ '**delimiter**' **then**

        begin

          if *state order* $(\sigma[s, symbol]) <$ *order expression bracket* **then**

            begin *insert an entity* ; **go to** *PROCESS 2* **end** ;

          if $\sigma[s, symbol] =$ '**then**' **then**

            begin

              insert an entity ;

              go to *THEN EXPRESSION*

            end

          else

            if $\sigma[s, symbol] =$ '**if**' **then**

              begin

                delete uppermost state ;

I: *delete* (1) ;

H0:   *push down (incoming symbol)* ;
           $\sigma\,[s,\,information\ type] := $ '**type**' ;
           go to *READ NEXT SYMBOL*
        end
      else
        if $\sigma\,[s,\,symbol] \neq$ '(' $\wedge$ $\sigma\,[s,\,symbol] \neq$ '(**not**' $\wedge$
        $\sigma\,[s,\,symbol] \neq$ '**procedure call**' $\wedge$
        $\sigma\,[s,\,symbol] \neq$ '[' **then**
          go to *H0*
        else *delete uppermost state*
    end incoming else ;
   go to *H* ;

INCORRECT INCOMING FOR COMMA:
J:
      for $s := s - 1$ **step** -1 **until** 1 **do**
      if $\sigma\,[s,\,del] = $ '**delimiter**' **then**
        begin
          if *state order* $(\sigma\,[s,\,symbol]) <$ *order expression bracket* **then**
            begin *insert an entity* ;   **go to** *PROCESS 2* **end** ;
          if *state order* $(\sigma\,[s,\,symbol]) =$ *order statement bracket* **then**
            go to *READ NEXT SYMBOL*
          else
            if *triple delimiter entity delimiter is allowed*
                $(\sigma\,[s,\,symbol],\ incoming\ symbol)$ **then**
              begin *insert an entity* ;   **go to** *PROCESS 2* **end**
            else *delete uppermost state*
          end incoming for comma ;

INCORRECT INCOMING DO:
K:
      for $s := s - 1$ **step** -1 **until** 1 **do**
      if $\sigma\,[s,\,del] = $ '**delimiter**' **then**
        begin
          if *state order* $(\sigma\,[s,\,symbol]) <$ *order expression bracket* **then**
            begin *insert an entity* ;   **go to** *PROCESS 2* **end** ;
          if *state order* $(\sigma\,[s,\,symbol]) =$ *order statement bracket* **then**
            begin
              push down ('**do**') ;
              go to *READ NEXT SYMBOL*
            end
          else
            if *triple delimiter entity delimiter is allowed*
                $(\sigma\,[s,\,symbol],\ incoming\ symbol)$ **then**

begin *insert an entity* ; **go to** *PROCESS 2* **end**
else *delete uppermost state*
end incoming **do** ;

INCORRECT INCOMING STEP OR WHILE:
INCORRECT INCOMING UNTIL:
L:
    for $s := s-1$ **step** -1 **until** 1 **do**
    if $\sigma[s, del] = $ '**delimiter**' **then**
      begin
        if *state order* $(\sigma[s, symbol]) <$ *order expression bracket* **then**
          begin *insert an entity* ; **go to** *PROCESS 2* **end** ;
        if $\sigma[s, symbol] = $ '**for**' $\vee$ $\sigma[s, symbol] = $ '**for n**' $\vee$
          $\sigma[s, symbol] = $ '**for suitable**' $\vee$
          $\sigma[s, symbol] = $ '**for inserted 1**' $\vee$
          $\sigma[s, symbol] = $ '**for inserted 2**' $\vee$
          $\sigma[s, symbol] = $ '**for n inserted**' **then**
        begin
          insert an entity ;
          if *incoming symbol* $\neq$ '**until**' **then**
            push down ('**for** $:=$')
          else *push down* ('**step**') ;
          insert an entity ;
          type of loop var $:= $ '**integer**' ;
          kind of loop var $:= $ '**simple**' ;
          go to *PROCESS 2*
        end
        else
          if *state order* $(\sigma[s, symbol])$
            $= $ *order statement bracket* **then**
          begin
            push down ('**for**') ;
            insert an entity ;
            if *incoming symbol* $= $ '**until**' $\vee$
              incoming symbol $= $ '**until constant**' **then**
              push down ('**step**')
            else *push down* ('**for** $:=$') ;
            insert an entity ;
            state $:= $ '**expression**' ;
            type of loop var $:= $ '**integer**' ;
            kind of loop var $:= $ '**simple**' ;
            go to *PROCESS 2*
          end

else
    if *triple delimiter entity delimiter is allowed*
        $(\sigma[s, symbol], incoming\ symbol)$ **then**
        begin *insert an entity* ; **go to** $PROCESS\ 2$ **end**
    else *delete uppermost state*
end incoming **step**, **until**, or **while** ;

INCORRECT INCOMING THEN STATEMENT:

INCORRECT INCOMING ELSE STATEMENT:

$M: \quad N:$

    for $s := s - 1$ **step** -1 **until** 1 **do**
    if $\sigma[s, del] = $ '**delimiter**' **then**
    begin
        if *triple delimiter entity delimiter is allowed*
            $(\sigma[s, symbol], incoming\ symbol)$ **then**
        begin *insert an entity* ; **go to** $PROCESS\ 2$ **end** ;
        if *state order* $(\sigma[s, symbol]) = order\ statement\ bracket$ **then**
        begin
            push down (incoming symbol) ;
            go to $READ\ NEXT\ SYMBOL$
        end
        else *delete uppermost state*
    end ;

INCOMING SEMICOLON:

INCOMING END OR BLOCK END:

$O: \quad$ *delete* (1) ;

$T: \quad$ *delete uppermost state* ; *delete* (1) ;
    go to $PROCESS$;

EXITS FOR AN INCORRECT INCOMING DELIMITER
  FINDING A DELIMITER AS UPPERMOST ENTRY IN σ:

INCORRECT INCOMING OPENING BRACKET:

$P:$

    go to $INCOMING\ OPENING\ BRACKET$
      [*opening bracket nr (incoming symbol)*] ;

$Q:$

    delete uppermost state ; *delete* (1) ;
    go to $PROCESS$;

INCORRECT INCOMING STATEMENT END:

$R:$

    go to $READ\ NEXT\ SYMBOL$;

$S: \quad$ **go to** $PROCESS\ 2$;

$C'': \quad$ **go to** $ARITHMETIC\ OR\ BOOLEAN\ OPERATOR$;

$C':$

END CHECK DELIMITER PAIR:
    end ;

comment The procedure *check delimiter pair* tests whether the pair (uppermost state in σ, incoming delimiter) is permitted. At first is tested whether the incoming delimiter may follow immediately an entity or whether it may follow immediately a delimiter. For this check two code procedures of type **Boolean** are used:

  symbol may follow an entity

and   *symbol may follow a delimiter.*

The function values of these procedures are not listed as they are obvious.

Depending on the result of this test and possibly after having deleted a superfluous entity or inserted a missing one, the code procedure

  triple delimiter entity delimiter is allowed

or   *pair delimiter delimiter is allowed*   resp.

is called with the uppermost state and the incoming delimiter as parameters. These code procedures deliver the result **true** if in the associated check matrix the field corresponding to the delimiters is empty, otherwise **false** which means a syntactical error. A star ∗ on a field of the matrices means the respective delimiter pair can not occur (if the translator works correct).

The two check matrices are also used for correcting the error announced. The letters denote labels within the procedure *check delimiter pair* which mark certain correcting program parts. Thus, the matrices are used as generalized switches called by the code procedures

  exit for delimiter entity delimiter

or   *exit for delimiter delimiter*   respectively.

Corrections are tried in a rather simple way and could be made in more detail, that means taking more special situations into consideration. Generally the assumption is made that the new incoming delimiter is correct because, in order to avoid redundant error messages, one does not want to change the original program by inserting or changing delimiters.

Entities may be deleted together with incompatible states without any testing. Such tests could easily be added to the program. A mark '**error announced**' and a variable *incoming symbol correct* are used to give error messages not more than once with the same state or incoming symbol.

Remarks: The check matrices can be stored more densely if storage place requires this, if one inserts a prechoice according to the state order or character order. Then the matrices can be divided as suggested by the broken lines. One should associate with each state or delimiter the row or column numbers by function tables so that several states or delimiters can use the same row or column. The code procedures *symbol may follow an entity* and *symbol may follow a delimiter* could be realized in such a way, that they deliver the respective column number in the case **true** and the number 0 in the case **false**.

The labels C', C'' in the matrices mean: the pair could be handled as if it were correct for the present since the error will be detected later on when checking entities (this may not be true if the immediate neighbourhood contains another syntactical error).

syntax check and correcting matrix for pair (delimiter, delimiter)

state			content real of' / content integer of' / content Boolean of'	'('	'if'	'go to' / 'for' / 'do suitable' / 'for suitable'	'array' / 'switch' / procedure begin'	'begin' / 'block begin' / begin initial value' / begin increment'	'else Σ'	'for end'	'procedure end'	'Σ' / 'end' / 'block end' / end initial value' / end increment'
⟨arithmetic operator⟩	S	R			P	Q	Q	Q	Q	*	*	Q
⟨relational operator⟩		R			P	Q	Q	Q	Q	*	*	Q
⟨Boolean operator⟩					P	Q	Q	Q	Q	*	*	Q
'(' / '(not' / 'procedure call' / 'if' / 'else' / ':=' / 'while'						Q	Q	Q	N	*	*	T
'[' / '[̄' / 'array [' / 'array :'		R				Q	Q	Q	N	*	*	T
'step' / 'step inserted' / 'step suitable' / 'until' / 'until suitable' / 'until constant' / 'for :=' / 'for := suitable'		R				Q	Q	Q	N	*	*	T
'then'					P	Q	Q	Q	N	*	*	T
'go to' / 'switch :='	R	R				Q	*	Q	N	*	*	T
'array' / 'switch' / 'procedure begin'	*	*	*	*	*	*	*	*	*	*	*	
'for' / 'for suitable' / 'for n' / 'for inserted 1' / 'for inserted 2' / 'for n inserted'	R	R		P	P	Q	Q	Q	N	*	*	T
'then Σ'	R	R		P	P			P		*	*	
'else Σ'	R	R		P				P	N	*	*	
'begin' / 'block begin' / 'array block begin' / 'begin initial value' / 'begin increment'	R	R		P					N	*	*	
'do' / 'do n' / 'do suitable' / 'do n suitable'	*	*	*	*	*	*	*			*		*
'procedure body begin'	*	*	*	*	*	*	*			*	*	*

syntax check and correcting matrix for triple (delimiter, entity, delimiter)

state	⟨binary arithmetic operators⟩	⟨Boolean operators⟩	⟨relational operators⟩	'red init'	'('	'[' | '[̄'	':='	'label :'	•	')'	']'	';'	':'	'then'	'else'	'step' | 'while'	'until' | 'until constant'	'for,' | 'begin increment'	'do,' | 'begin initial value'	'then Σ'	'else Σ'	';'	'end' | 'block end'
⟨arithmetic operators⟩								C''	B														
⟨relational operators⟩			C'					C''	B			C'	C'				C'	C'					
⟨Boolean operators⟩								C''	B			C'	C'				C'	C'					
'(' | '(not)'								A	•	D	E	E		F	H	L	L	J	K	M	N	O	O
'procedure call'								A	•	D	E			F	H	L	L	J	K	M	N	O	O
'[' | '[̄'		C'	C'					A	•	D	E			F	H	L	L	J	K	M	N	O	O
'if'								A	•	D	D	E	E		H	L	L	J	K		N	O	O
'then'								A	•	D	D	E	E	F		L	L	J	K	M	N	O	O
'else'								A	•						I								
':='									•	D	D	E	E	G	I	L	L	J	K	M			
'for'|'for n'|'for suitable'|'for inserted 1'|'for inserted 2'|'for n inserted'	C'	C'	C'	C'					•	D	D	E	E	G	I	L	L			M	N	O	O
'for:=' | 'for:= suitable'		C'	C'					A	•	D	D	E	E	G	I	L				M	N	O	O
'step' | 'step suitable' | 'step inserted'		C'	C'					A	•	D	D	E	E	G	I	L		J	K	M	N	O	O
'until' | 'until suitable' | 'until constant'		C'	C'					A	•	D	D	E	E	G	I	L	L			M	N	O	O
'while'								A	•	D	D	E	E	G	I	L	L			M	N	O	O
'go to'	C'	C'	C'		C'			A	•	D	D	E	E	G	I	L	L	J	K	M			
'array'	*	*	*	*	*	*	*	*	*	*	*	*	*	*	*	*	*	*	*	*	*	*	*
'array ['		C'	C'					A	•	D	D	E		G	I	L	L	J	K	M	N	O	O
'array :'		C'	C'					A	•	D		E		G	I	L	L	J	K	M	N	O	O
'switch'	*	*	*	*	*	*	*	*	*	*	*	*	*	*	*	*	*	*	*	*	*	*	*
'switch:='	C'	C'	C'	C'	C'			A	•	D	D	E		G	I	L	L	J	K	M	N		O
'procedure begin'	*	*	*	*	*	*	*	*	*	*	*	*	*	*	*	*	*	*	*	*	*	*	*
'then Σ'	C'	C'	C'							D	D	E	E	G	I	L	L	J	K	M			
'else Σ'	C'	C'	C'							D	D	E	E	G	I	L	L	J	K	M	N		
'begin' | 'block begin' | 'array block begin' | 'begin initial value' | 'begin increment'	C'	C'	C'							D	D	E	E	G	I	L	L	J	K	M	N		
'do' | 'do n' | 'do suitable' | 'do n suitable'	*	*	*	*	*	*	*	*	*	*	*	*	*	*	*	*	*	*	*	*	*	*	*
'procedure body begin'	*	*	*	*	*	*	*	*	*	*	*	*	*	*	*	*	*	*	*	*	*	*	*

:

    procedure *error identifier is not declared* ;
        begin
            integer i, j ;
            error message ('the identifier' $\oplus$ *incoming symbol* $\oplus$
                'is not declared') ;
            error := **true** ;
            $i :=$ *begin identifier list* $:=$ *begin identifier list* $- 1$;

for $j := 1$ **step** 1 **until** 16 **do** $\delta[i, j] := $ '**empty**' ;
$\delta[i, COL\ \delta\ external\ identifier] := incoming\ symbol$;
$\delta[i, COL\ \delta\ internal\ identifier] := i$;
$\delta[i, COL\ \delta\ mode] := $ '**formal**' ;
$\delta[i, COL\ \delta\ kind] := $ '**unspecified formal parameter**' ;
$\delta[i, COL\ \delta\ array\ block\ level] := array\ block\ level\ counter$;
$\delta[i, COL\ \delta\ static\ procedure\ level] := $
  static procedure level counter ;
$\delta[i, COL\ \delta\ maximum\ array\ block\ level] := 0$
end error identifier is not declared ;

procedure *check identifier* $(int\ id)$;
  value *int id* ; **integer** *int id* ;

comment The procedure *check identifier* is called if a number or an identifier is read, it checks whether the uppermost entry in σ is an entity which is not admissible. In certain cases then a '(' or a '[' is inserted, otherwise the superfluous entity is deleted ;

if $\sigma[s, del] = $ '**entity**' **then**
  begin
    if $\sigma[s, information\ error] \neq $ '**inserted**' **then**
      begin
        error message $(text\ (\sigma[s, mode], \sigma[s, kind]) \oplus$
          'is followed by' $\oplus$
          $text\ (\delta[int\ id, COL\ \delta\ mode], \delta[int\ id, COL\ \delta\ kind]) \oplus$
          ', possibly there is missing a delimiter') ;
        $error := $ **true**
      end ;
    if $\sigma[s, kind] = $ '**procedure**' **then**
      begin
        if $\delta[\sigma[s, internal\ identifier], COL\ \delta\ dimension] =$
          '**empty**' $\vee$
          $\delta[\sigma[s, internal\ identifier], COL\ \delta\ dimension] = 0 \vee$
          $specification = $ '**otherwise**' $\vee$
          $specification = $ '**undefined**' **then**
          go to $DELETE\ ENTITY$;
        $incoming\ symbol := $ '(' ;
        if *normal program* **then**
          $k := k - 1$;
        else $k\ add := k\ add - 1$;
        go to $FUNCTION\ OR\ PROCEDURE\ CALL$
      end ;

if $\sigma[s, kind] = $ '**array**' $\lor$ $\sigma[s, kind] = $ '**switch**' $\lor$
$\sigma[s, kind] = $ '**array or switch**' **then**
begin
   if *specification* $= $ '**otherwise**' $\lor$
     specification $= $ '**undefined**' **then**
    go to *DELETE ENTITY* ;
   incoming symbol $:= $ '[' ;
   if *normal program* **then**
    $k := k - 1$
   else $k\ add := k\ add - 1$;
   go to *INCOMING SQUARE BRACKET*
end ;

DELETE ENTITY:
   if *accumulator loaded* $= s$ **then** *accumulator loaded* $:= 0$;
   delete (1)
  end check identifier ;

  procedure *check compatibility* (i, x, y) ;
   value i ; **integer** i ; **string** x, y ;

comment The procedure *check compatibility* (i, x, y) tests whether the entity stored in σ_i is compatible with the entity required by the translation process with kind x and type y. The actual check is performed by two code procedures. ;

  begin
   integer *int id* ;

   procedure *kind test matrix* (x, y) ;
    string x, y ;
    code ;

   procedure *type test matrix* (x, y) ;
    string x, y ;
    code ;

comment The procedures *kind test matrix* and *type test matrix* represent generalized switches. Depending on the actual values of type **string** of the parameters x and y a certain global label is selected and a global jump to this label is executed. The labels are listed in the following tables.

a) kind test matrix (x, y)

x \ y	'entity'	'variable'	'left part variable'
'number'	*L1*	*INCORRECT*	*INCORRECT*
'variable'	*L1*	*L1*	*L1*
'number cellar'	*L1*	*L1*	*L1*
'AC'	*L1*	*L5*	*L5*
'relation'	*L1*	*INCORRECT*	*INCORRECT*
'open jump' | **'open jump relation'**	*L1*	*INCORRECT*	*INCORRECT*
'procedure'	*L2*	*INCORRECT*	*L1*
'variable or label'	*L3*	*L3*	*L3*
'index register'	*L1*	*L1*	*L1*
'unspecified formal parameter'	*L4*	*L4*	*L4*
⟨otherwise⟩	*INCORRECT*	*INCORRECT*	*INCORRECT*

b) type test matrix (x, y)

x \ y	'real' | 'integer' | 'arithmetic'	'Boolean'	'type'
'real'	*M1*	*TYPE INCORRECT*	*M1*
'integer'	*M1*	*TYPE INCORRECT*	*M1*
'Boolean' | **'Boolean*'**	*TYPE INCORRECT*	*M5*	*M5*
'arithmetic'	*M1*	*TYPE INCORRECT*	*M1*
'type'	*M2*	*M3*	*M1*
'empty'	*M2*	*M3*	*M4*

$;$

$\quad int\ id := \sigma[i, internal\ identifier]\ ;$

$\quad kind\ test\ matrix\ (\sigma[i, kind], x)\ ;$

L2: $\quad$ check procedure without parameters (i) ;

$\quad$ **go to** *CHECK TYPE* ;

L3: $\quad$ **if** $\sigma[s, information\ error] \neq$ **'inserted'** $\wedge$

$\quad\quad$ *specification* $\neq$ **'undefined'** **then**

$\quad\quad$ $\delta[int\ id, COL\ \delta\ kind] :=$ **'variable'** ;

$\quad$ $\sigma[i, kind] :=$ **'variable'** ;

$\quad$ **go to** *CHECK TYPE* ;

L4: $\quad$ **if** $\sigma[s, information\ error] \neq$ **'inserted'** $\wedge$

$\quad\quad$ *specification* $\neq$ **'undefined'** **then**

$\quad\quad$ $\delta[int\ id, COL\ \delta\ kind] :=$ **'variable'**

$\quad$ **else**

$\quad\quad$ **if** $\sigma[s, information\ error] \neq$ **'inserted'** $\wedge$

$\quad\quad$ *specification* $=$ **'undefined'** **then**

$\quad\quad\quad$ $\delta[int\ id, COL\ \delta\ kind] :=$ **'variable or label'** ;

$\quad$ $\sigma[i, kind] :=$ **'variable'** ;

$\quad$ **go to** *CHECK TYPE* ;

L5: $\quad$ **if** $\sigma[s, supplement\ information] =$ **'content of'** **then**

$\quad\quad$ **go to** *CHECK TYPE* ;

INCORRECT:

> **if** $\sigma[i,\ \textit{information error}] \neq$ '**inserted**' **then**
>> **begin**
>>> $\sigma[i,\ \textit{information error}] :=$ '**inserted**' ;
>>> $\textit{error} :=$ **true** ;
>>> $\textit{error message}\ (\textit{text}\ (\sigma[i,\ \textit{mode}],\ \sigma[i,\ \textit{kind}]) \oplus$
>>>> '`occurs instead of`' $\oplus\ \textit{text}\ ($ '**empty**', $x))$
>>
>> **end** ;
>> $\sigma[i,\ \textit{internal identifier}] := 0$;
>> $\sigma[i,\ \textit{mode}] :=$ '**normal**' ;
>> $\sigma[i,\ \textit{kind}] :=$ '**variable**' ;
>> $\sigma[i,\ \textit{supplement information}] :=$ '**empty**' ;

L1:

CHECK TYPE:

> $\textit{type test matrix}\ (\sigma[i,\ \textit{type}],\ y)$;

M2:   **if** $\sigma[i,\ \textit{information error}] \neq$ '**inserted**' $\wedge$
      $\sigma[i,\ \textit{mode}] =$ '**formal**' **then**
      $\delta[\textit{int id, COL}\ \delta\ \textit{type}] :=$ '**arithmetic**' ;
      $\sigma[i,\ \textit{type}] :=$ '**arithmetic**' ;
      go to *END CHECK COMPATIBILITY* ;

M3:   **if** $\sigma[i,\ \textit{information error}] \neq$ '**inserted**' $\wedge$
      $\sigma[i,\ \textit{mode}] =$ '**formal**' **then**
      begin
        $\delta[\textit{int id, COL}\ \delta\ \textit{type}] :=$ '**Boolean***' ;
        $\sigma[i,\ \textit{type}] :=$ '**Boolean**'
      end
      else $\sigma[i,\ \textit{type}] :=$ '**type**' ;
      go to *END CHECK COMPATIBILITY* ;

M4:   **if** $\sigma[i,\ \textit{information error}] \neq$ '**inserted**' $\wedge$
      $\sigma[i,\ \textit{mode}] =$ '**formal**' **then**
      $\delta[\textit{int id, COL}\ \delta\ \textit{type}] :=$ '**type**' ;
      $\sigma[i,\ \textit{type}] :=$ '**type**' ;
      go to *END CHECK COMPATIBILITY* ;

M5:   **if** $\sigma[i,\ \textit{type}] =$ '**Boolean***' **then** $\sigma[i,\ \textit{type}] :=$ '**Boolean**' ;
      go to *END CHECK COMPATIBILITY* ;

TYPE INCORRECT:

> **if** $\sigma[i,\ \textit{information error}] \neq$ '**inserted**' **then**
>> **begin**
>>> $\sigma[i,\ \textit{information error}] :=$ '**inserted**' ;
>>> $\textit{error} :=$ **true** ;

error message ('type error: the type is' $\oplus$ $\sigma[i, type]$ $\oplus$
 'instead of' $\oplus$ y)
 end ;
 $\sigma[i, type] := y$;
M1: **if** $x=$ '**left part variable**' $\wedge$ ($\sigma[i, type] =$ '**real**' $\wedge$ $y=$
  '**integer**' $\vee$ $\sigma[i, type] =$ '**integer**' $\wedge$ $y =$ '**real**') **then**
  go to *TYPE INCORRECT* ;
END CHECK COMPATIBILITY:
 end ;

 comment The type **Boolean*** is used for unspecified formal
parameters which are found to be variables or procedures of type
Boolean. They must be distinguished from parameters specified to
be variables or procedures for the purpose of checking actual-formal parameter correspondence. Compare with the check matrix in the appendix ;

 procedure *check array or switch identifier* ;

 comment The procedure *check array or switch identifier* is called if
'[' is coming in. As in the former procedure *check compatibility* a code
procedure is used for the proper check. ;
  begin
   procedure *switch matrix* (x, y, z) ;
   string x, y, z ;
   code ;

 comment The procedure *switch matrix* represents a generalized
switch. Depending on the three parameters x, y, z a global label is
selected and a global jump to this label is executed. The labels are
listed in the following table as function of the parameters x, y, z.

			z			
x	y	'array'	'array or switch'	'unspecified formal parameter'	'switch'	⟨otherwise⟩
'statement'	'empty'	*L1*	*L6*	*L6*	*L1*	*INCORRECT 3*
'statement'	'designational'	*INCORRECT 4*	*L5*	*L5*	*L1*	*INCORRECT 4*
'statement'	'undefined'	*L7*	*L1*	*L8*	*L1*	*INCORRECT 2*
'expression'	'empty'	*L1*	*L4*	*L4*	*INCORRECT 1*	*INCORRECT 1*
'expression'	'expression'	*L1*	*L4*	*L4*	*INCORRECT 1*	*INCORRECT 1*
'expression'	'undefined'	*L2*	*L1*	*L8*	*L3*	*INCORRECT 2*
'statement'	'otherwise'	*L1*	*L1*	*L8*	*L1*	*INCORRECT 2*

 The parameter x is the *state*, y is the *specification*, and z is the kind
of the uppermost entity in σ. ;
  switch matrix (*state, specification,* $\sigma[s, kind]$) ;
L1: $\sigma[s,$ *supplement information*$] := 0$;
  go to *END CHECK ARRAY OR SWITCH IDENTIFIER* ;

L2: $\sigma[s,$ *supplement information* $] := 0$;
        specification $:=$ '**expression**' ;
        go to *END CHECK ARRAY OR SWITCH IDENTIFIER* ;
L3:   *change situation to specification label* ;
        go to *READ NEXT SYMBOL* ;
L4: $\sigma[s,$ *supplement information* $] := 0$;
        if $\sigma[s,$ *information error* $] \neq$ '**inserted**' **then**
            begin
                $\delta[\sigma[s,$ *internal identifier*$],$ *COL* δ *kind*$] :=$ '**array**' ;
                $\delta[\sigma[s,$ *internal identifier*$],$ *COL* δ *type*$] :=$ '**type**'
            end ;
        $\sigma[s,$ *type*$] :=$ '**type**' ;
        $\sigma[s,$ *kind*$] :=$ '**array**' ;
        go to *END CHECK ARRAY OR SWITCH IDENTIFIER* ;
L5:   **if** $\sigma[s,$ *information error* $] \neq$ '**inserted**' **then**
            $\delta[\sigma[s,$ *internal identifier*$],$ *COL* δ *kind*$] :=$ '**switch**' ;
        $\sigma[s,$ *kind*$] :=$ '**switch**' ;
        go to *END CHECK ARRAY OR SWITCH IDENTIFIER* ;
L6:   **if** *state order* $(\sigma[s-1,$ *symbol*$]) =$ *order statement bracket* **then**
            go to *L4* ;
        go to *L5* ;
L7:   *change situation to specification expression* ;
        go to *END PROCEDURE CALL* ;
L8: $\sigma[s,$ *supplement information* $] := 0$;
        if $\sigma[s,$ *information error* $] \neq$ '**inserted**' **then**
            begin
                $\delta[\sigma[s,$ *internal identifier*$],$ *COL* δ *kind*$] :=$
                    '**array or switch**' ;
                $\delta[\sigma[s,$ *internal identifier*$],$ *COL* δ *type*$] :=$ '**type**'
            end ;
        $\sigma[s,$ *kind*$] :=$ '**array or switch**' ;
        $\sigma[s,$ *type*$] :=$ '**type**' ;
        go to *END CHECK ARRAY OR SWITCH IDENTIFIER* ;
INCORRECT 1:
        $\sigma[s,$ *type*$] :=$ '**type**' ;
        $\sigma[s,$ *kind*$] :=$ '**array**' ;
        if $\sigma[s,$ *information error* $] \neq$ '**inserted**' **then**
            error message ('error: [does not follow an array identifier') ;
        go to *INCORRECT* ;

INCORRECT 2:
>$\sigma[s, type] := $ '**type**' ;
>$\sigma[s, kind] := $ '**array or switch**' ;
>**if** $\sigma[s, \text{ }information\text{ }error] \neq $ '**inserted**' **then**
>>*error message* ('error: [does not follow an array or
>>switch identifier') ;
>
>**go to** *INCORRECT* ;

INCORRECT 3:
>**if** *state order* $(\sigma[s-1, symbol]) = $ *order statement bracket* **then**
>>**go to** *INCORRECT 1* ;

INCORRECT 4:
>$\sigma[s, type] := $ '**empty**' ;
>$\sigma[s, kind] := $ '**switch**' ;
>**if** $\sigma[s, \text{ }information\text{ }error] \neq $ '**inserted**' **then**
>>*error message* ('error: [does not follow a switch identifier') ;

INCORRECT:
>$\sigma[s, \text{ }information\text{ }error] := $ '**inserted**' ;
>$error := $ **true** ;
>$\sigma[s, \text{ }internal\text{ }identifier] := 0$;
>$\sigma[s, mode] := $ '**normal**' ;
>$\sigma[s, \text{ }supplement\text{ }information] := 0$;

END CHECK ARRAY OR SWITCH IDENTIFIER:
>**end** ;

>**procedure** *check subscript position* ;

comment If a ']' is read (which does not terminate an array declaration) then it is necessary to check whether the number of index positions is correct. This is done by *check subscript position* which also performs the appropriate preparations. The procedure counts index positions, checks index position numbers for index vectors, subscripted variables, and switch designators, and if necessary gives an error message, corrects, if possible, entries in σ or in the identifier list, and changes the undefined specification of parameters if more than one index position is found. (Compare also with the comment in pass two on page 209.) ;

>**begin**
>>**integer** *int id, index number* ;
>>**if** $\sigma[s-2, kind] = $ '**switch**' $\wedge$ *incoming symbol* = ',' **then**
>>>**begin**
>>>>**if** $\sigma[s-2, \text{ }information\text{ }error] \neq $ '**inserted**' **then**

19*

begin
   error message ('switch designator has more than one
      subscript position') ;
   $\sigma[s-2, \textit{supplement information}] := \text{'\textbf{inserted}'}$;
   error := **true**
end ;
delete (1) ;
go to *READ NEXT SYMBOL*
end switch ;
if $\sigma[s-2, \textit{kind}] = \text{'\textbf{switch}'}$ **then go to** *END CHECK* ;
index number := $\sigma[s-2, \textit{supplement information}] :=$
   $\sigma[s-2, \textit{supplement information}] + 1$;
if *incoming symbol* $\neq$ ']' **then go to** *END CHECK* ;
if $\sigma[s-2, \textit{information error}] \neq \text{'\textbf{inserted}'}$ **then**
   begin
      int id := $\sigma[s-2, \textit{internal identifier}]$;
      if $\sigma[s-2, \textit{kind}] = \text{'\textbf{array or switch}'} \wedge$
         index number > 1 **then**
         begin
            $\sigma[s-2, \textit{kind}] := \delta[\textit{int id}, COL\ \delta\ \textit{kind}] := \text{'\textbf{array}'}$;
            $\delta[\textit{int id}, COL\ \delta\ \textit{dimension}] := \textit{index number}$;
            if $\sigma[s-1, \textit{information 2}] = \text{'\textbf{undefined}'}$ **then**
               $\sigma[s-1, \textit{information 2}] := \text{'\textbf{expression}'}$
         end
      else
         if $\sigma[s-2, \textit{mode}] \neq \text{'\textbf{normal}'} \wedge$
            $\delta[\textit{int id}, COL\ \delta\ \textit{dimension}] = \text{'\textbf{empty}'}$ **then**
            $\delta[\textit{int id}, COL\ \delta\ \textit{dimension}] := \textit{index number}$
         else
            if *index number* $\neq$
               $abs(\delta[\textit{int id}, COL\ \delta\ \textit{dimension}])$ **then**
               begin
                  error message ('incorrect number of subscript
                     positions') ;
                  error: = **true** ;
                  $\sigma[s-2, \textit{information error}] := \text{'\textbf{inserted}'}$
               end
   end
else
   if $\sigma[s-2, \textit{kind}] = \text{'\textbf{array or switch}'} \wedge$
      index number > 1 **then**

begin
    $\sigma[s-2,\ kind] := \text{'}$**array**$\text{'}$;
    if $\sigma[s-1,\ information\ 2] = \text{'}$**undefined**$\text{'}$ **then**
        $\sigma[s-1,\ information\ 2] := \text{'}$**expression**$\text{'}$
end inserted array or switch identifier ;

END CHECK:
    end check subscript position ;

  procedure *check designational expression* ;
  begin
    if $\sigma[s,\ information\ error] \neq \text{'}$**inserted**$\text{'}$ **then**
      begin
        if $\sigma[s,\ kind] \neq \text{'}$**label**$\text{'} \wedge$
         $\sigma[s,\ kind] \neq \text{'}$**switch designator**$\text{'} \wedge$
         $\sigma[s,\ kind] \neq \text{'}$**designational**$\text{'} \wedge$
         $\sigma[s,\ kind] \neq \text{'}$**variable or label**$\text{'} \wedge$
         $\sigma[s,\ kind] \neq \text{'}$**unspecified formal parameter**' **then**
        begin
         error message ('incorrect use of$\text{'} \oplus$
          text ($\sigma[s,\ mode],\ \sigma[s,\ kind]) \oplus$
          'instead of a designational expression') ;
         error := **true**
        end error found
      else
        if *specification* $\neq \text{'}$**undefined**' **then**
         begin
          if $\sigma[s,\ kind] = \text{'}$**variable or label**$\text{'} \vee$
           $\sigma[s,\ kind] =$
            '**unspecified formal parameter**' **then**
          begin
           $\delta[\sigma[s,\ internal\ identifier],\ COL\ \delta\ kind] :=$
            '**label**' ;
           $\delta[\sigma[s,\ internal\ identifier],\ COL\ \delta\ type] :=$
            '**empty**'
          end
         end
        else
         if $\sigma[s,\ kind] =$
          '**unspecified formal parameter**' **then**

```
            begin
              δ[σ[s, internal identifier], COL δ kind] :=
                'variable or label' ;
              δ[σ[s, internal identifier], COL δ type] :=
                'type'
            end
        end
    end check designational expression ;

  procedure check procedure call ;
```

comment The procedure *check procedure call* is called if a simple identifier which is not a label is found on statement level. This then must be the call of a procedure without parameters. ;

```
    begin
      if σ[s, information error] ≠ 'inserted' then
        begin
          if σ[s, kind] = 'procedure' then
            begin
              if σ[s, type] ≠ 'empty' then
                begin
                  error message ('incorrect use of a function as
                    procedure statement') ;
                  error := true
                end ;
              check procedure without parameters (s)
            end
          else
            if σ[s, kind] = 'unspecified formal parameter' then
              begin
                δ[σ[s, internal identifier], COL δ kind] :=
                  'procedure' ;
                δ[σ[s, internal identifier], COL δ dimension] := 0
              end
            else
              begin
                error message ('incorrect use of' ⊕
                  text (σ[s, mode], σ[s, kind]) ⊕
                  'as procedure statement') ;
                error := true
              end
        end
    end check procedure call ;
```

procedure *check procedure without parameters* (i) ;
  value i ; **integer** i ;

comment The procedure *check procedure without parameters* (i) is called if the procedure identifier stored in σ_i represents a procedure or function call without parameters. It checks whether the number of parameters is declared to be 0, or is inserted to be 0 in the case of a formal procedure identifier ;

    begin
      integer *int id* ;
      if $\sigma[i, \textit{information error}] \neq$ '**inserted**' **then**
        begin
          int id $:= \sigma[i, \textit{internal identifier}]$;
          if $\delta[\textit{int id, COL } \delta \textit{ dimension}] \neq 0$ **then**
            begin
              if $\delta[\textit{int id, COL } \delta \textit{ dimension}] =$ '**empty**' $\wedge$
                 $\delta[\textit{int id, COL } \delta \textit{ error}] \neq$ '**incorrect**' **then**
                 $\delta[\textit{int id, COL } \delta \textit{ dimension}] := 0$
              else
                begin
                  error message ('incorrect use of a procedure
                    identifier with parameters as procedure without
                    parameters') ;
                  error $:=$ **true** ;
                  if $\sigma[i, \textit{mode}] =$ '**formal**' **then**
                    begin
                      $\delta[\textit{int id, COL } \delta \textit{ dimension}] :=$ '**empty**' ;
                      $\delta[\textit{int id, COL } \delta \textit{ error}] :=$ '**incorrect**'
                  end
                end
            end
        end
    end check procedure without parameters ;

procedure *check procedure identifier with parameters* ;

comment If a '(' follows an identifier then this identifier must be a procedure identifier. This is checked for in *check procedure identifier with parameters*. The number of parameters is tested later on in connection with the closing of a procedure or function call. ;

    begin
      if $\sigma[s, \textit{information error}] \neq$ '**inserted**' **then**
        begin
          if $\sigma[s, \textit{kind}] =$ ' **unspecified formal parameter**' **then**

```
      begin
        δ[σ[s, internal identifier], COL δ kind] :=
          σ[s, kind] := 'procedure' ;
        if state = 'expression' then
          δ[σ[s, internal identifier], COL δ type] :=
            σ[s, type] := 'type'
      end
    else
      if σ[s, kind] ≠ 'procedure' then
        begin
          error message ('incorrect use of a' ⊕
            text (σ[s, mode], σ[s, kind]) ⊕
            'as procedure identifier') ;
          error := true ;
          σ[s, information error] := 'inserted' ;
          go to L
        end
    end entity is not inserted
  else
L:      begin
        σ[s, kind] := 'procedure' ;
        if state = 'expression' then σ[s, type] := 'type'
      end ;
    if specification = 'undefined' then
      specification := 'expression'
  end check procedure identifier with parameters ;
procedure check number of parameters (i) ;
  value i ; integer i ;
  begin
    integer int id ;
    int id := σ[i, internal identifier] ;
    if σ[i, information error] = 'empty' then
      begin
        if σ[i, mode] = 'normal' then
          begin
            if δ[int id, COL δ dimension] ≠
              number of parameter − 1 then
              begin
                error message ('the procedure call has an incorrect
                  number of parameters') ;
                error := true
              end
          end normal procedure
```

else
    if $\delta[int\ id,\ COL\ \delta\ error] =$ '**empty**' **then**
      begin
        if $\delta[int\ id,\ COL\ \delta\ dimension] =$ '**empty**' **then**
          $\delta[int\ id,\ COL\ \delta\ dimension] :=$
           number of parameter $- 1$
        else
          if $\delta[int\ id,\ COL\ \delta\ dimension] \neq$
           number of parameter $- 1$ **then**
           begin
             error message ('the formal procedure call has
               an incorrect number of parameters') ;
             $\delta[int\ id,\ COL\ \delta\ dimension] :=$ '**empty**' ;
             $\delta[int\ id,\ COL\ \delta\ error] :=$ '**incorrect**' ;
             error $:=$ **true**
           end
      end formal procedure call
    end
  end check number of parameters ;

procedure *check specification otherwise* ;
  if *specification* $=$ '**otherwise**' **then**
    begin
      error message ('the actual parameter on a parameter
        position specified '**array**', '**switch**', '**string**', or
        '**procedure**' contains a delimiter') ;
      error $:=$ **true**
    end check ;

procedure *test compatibility of actual and formal parameter* $(ap,\ fp)$;
  value $ap,\ fp$; **integer** $ap,\ fp$;

comment The procedure *test compatibility of actual and formal parameter* makes use of the global procedure *check agreeability of actual parameter and specification* p. 356 which delivers the result **true** or **false** as the case may be. In the case **false** an error message is given. Otherwise one may try to complete specifications by calling the code procedure *switch matrix for correcting inserted specifications*. This code procedure employs the correspondence matrix given in the appendix. If the respective field contains a number then a jump to the label given by it is executed. The procedure *test compatibility of actual and formal parameter* contains the marks *1, ..., 25* which denote the different possible supplementary entries into the identifier list. If the actual parameter itself is a formal parameter then the code procedure *switch matrix for correcting inserted specifications* is once more called with parameters exchanged, in order to complete eventually the specification of the actual parameter. An empty mark has the meaning of 0. ;

    begin
        integer a ;

        Boolean *agreeable, correct fp* ;

        procedure *switch matrix for correcting inserted*
                          specifications (mode a, type a, kind a,
                          mode f, type f, kind f) ;
          string *mode a, type a, kind a, mode f, type f, kind f* ;
          code ;
        check agreeability of actual parameter and specification (ap, fp)
        result: *(agreeable)* ;
        if *agreeable* **then**
        begin
          correct fp := **true** ;

L: *switch matrix for correcting inserted specifications*
            ($\delta[ap, COL\ \delta\ mode]$, $\delta[ap, COL\ \delta\ type]$,
            $\delta[ap, COL\ \delta\ kind]$, $\delta[fp, COL\ \delta\ mode]$,
            $\delta[fp, COL\ \delta\ type]$, $\delta[fp, COL\ \delta\ kind]$) ;

1: $\delta[fp, COL\ \delta\ type]$:= '**arithmetic**' ; **go to** 0 ;
2: $\delta[fp, COL\ \delta\ type]$:= $\delta[ap, COL\ \delta\ type]$; **go to** 0 ;
3: $\delta[fp, COL\ \delta\ type]$:= '**Boolean***' ; **go to** 0 ;
4: $\delta[fp, COL\ \delta\ kind]$:= $\delta[ap, COL\ \delta\ kind]$; **go to** 0 ;
5: $\delta[fp, COL\ \delta\ type]$:= '**arithmetic**' ;
          $\delta[fp, COL\ \delta\ kind]$:= $\delta[ap, COL\ \delta\ kind]$; **go to** 0 ;
6: $\delta[fp, COL\ \delta\ type]$:= $\delta[ap, COL\ \delta\ type]$;
          $\delta[fp, COL\ \delta\ kind]$:= $\delta[ap, COL\ \delta\ kind]$; **go to** 0 ;
7: $\delta[fp, COL\ \delta\ type]$:= '**Boolean***' ;
          $\delta[fp, COL\ \delta\ kind]$:= $\delta[ap, COL\ \delta\ kind]$; **go to** 0 ;
8: $\delta[fp, COL\ \delta\ type]$:= '**empty**' ;
          $\delta[fp, COL\ \delta\ kind]$:= $\delta[ap, COL\ \delta\ kind]$; **go to** 0 ;
9: $\delta[fp, COL\ \delta\ type]$:= '**arithmetic**' ;
          $\delta[fp, COL\ \delta\ kind]$:= '**variable**' ; **go to** 0 ;
10: $\delta[fp, COL\ \delta\ type]$:= $\delta[ap, COL\ \delta\ type]$;
          $\delta[fp, COL\ \delta\ kind]$:= '**variable**' ; **go to** 0 ;
11: $\delta[fp, COL\ \delta\ type]$:= '**Boolean***' ;
          $\delta[fp, COL\ \delta\ kind]$:= '**variable**' ; **go to** 0 ;
12: $\delta[fp, COL\ \delta\ type]$:= '**type**' ;
          $\delta[fp, COL\ \delta\ kind]$:= '**variable or label**' ; **go to** 0 ;
13: $\delta[fp, COL\ \delta\ output]$:= '**not output**' ; **go to** 0 ;
14: $\delta[fp, COL\ \delta\ type]$:= '**arithmetic**' ;
          $\delta[fp, COL\ \delta\ output]$:= '**not output**' ; **go to** 0 ;

15: $\delta[fp, COL\ \delta\ type] := \text{`} \textbf{Boolean*} \text{'}$;
$\delta[fp, COL\ \delta\ output] := \text{`}\textbf{not output}\text{'}$; **go to** 0 ;

16: $\delta[fp, COL\ \delta\ kind] := \text{`}\textbf{variable}\text{'}$;
$\delta[fp, COL\ \delta\ output] := \text{`}\textbf{not output}\text{'}$; **go to** 0 ;

17: $\delta[fp, COL\ \delta\ type] := \text{`}\textbf{arithmetic}\text{'}$;
$\delta[fp, COL\ \delta\ kind] := \text{`}\textbf{variable}\text{'}$;
$\delta[fp, COL\ \delta\ output] := \text{`}\textbf{not output}\text{'}$; **go to** 0 ;

18: $\delta[fp, COL\ \delta\ type] := \text{`}\textbf{Boolean*}\text{'}$; **go to** 0 ;

19: $\delta[fp, COL\ \delta\ type] := \text{`}\textbf{type}\text{'}$; **go to** 0 ;

20: $\delta[fp, COL\ \delta\ type] := \delta[ap, COL\ \delta\ type]$;
$\delta[fp, COL\ \delta\ kind] := \text{`}\textbf{variable or label}\text{'}$;
$\delta[fp, COL\ \delta\ output] := \text{`}\textbf{not output}\text{'}$; **go to** 0 ;

21: $\delta[fp, COL\ \delta\ type] := \text{`}\textbf{arithmetic}\text{'}$;
$\delta[fp, COL\ \delta\ kind] :=$
   if $\delta[ap, COL\ \delta\ dimension] = 0$ **then**
     'variable'
   else **'procedure'** ;
go to 0 ;

22: $\delta[fp, COL\ \delta\ type] := \text{`}\textbf{Boolean*}\text{'}$;
$\delta[fp, COL\ \delta\ kind] :=$
   if $\delta[ap, COL\ \delta\ dimension] = 0$ **then**
     'variable'
   else **'procedure'** ;
go to 0 ;

23: **if** $\delta[ap, COL\ \delta\ dimension] \neq \text{`}\textbf{empty}\text{'}$ **then**
   begin
     $\delta[fp, COL\ \delta\ type] := \text{`}\textbf{arithmetic}\text{'}$;
     $\delta[fp, COL\ \delta\ kind] :=$
       if $\delta[ap, COL\ \delta\ dimension] = 0$ **then**
         'variable'
       else **'procedure'**
   end ;
go to 0 ;

24: **if** $\delta[ap, COL\ \delta\ dimension] \neq \text{`}\textbf{empty}\text{'}$ **then**
   begin
     $\delta[fp, COL\ \delta\ type] := \text{`}\textbf{Boolean*}\text{'}$;
     $\delta[fp, COL\ \delta\ kind] :=$
       if $\delta[ap, COL\ \delta\ dimension] = 0$ **then**
         'variable'
       else **'procedure'**
   end ;
go to 0 ;

25:              **if** $\delta[ap, COL\ \delta\ dimension] \neq$ '**empty**' **then**
                 begin
                     $\delta[fp, COL\ \delta\ type] :=$ '**type**'
                     $\delta[fp, COL\ \delta\ kind] :=$
                         if $\delta[ap, COL\ \delta\ dimension] = 0$ **then**
                             '**variable**'
                         else '**procedure**'
                 end ;
0:               **if** $correct\ fp \wedge \delta[ap, COL\ \delta\ mode] =$ '**formal**' **then**
                 begin
                     $correct\ fp :=$ **false** ;
                     $a := ap$; $ap := fp$; $fp := a$;
                     go to L
                 end correct formal and actual parameter specification
             end actual parameter is agreeable
         else
           begin
             $error :=$ **true** ;
             $error\ message$ ('the actual parameter:' $\oplus$
                 $\delta[ap, COL\ \delta\ mode] \oplus \delta[ap, COL\ \delta\ type] \oplus$
                 $\delta[ap, COL\ \delta\ kind] \oplus$
                 'is not compatible with the associated specification:' $\oplus$
                 $\delta[fp, COL\ \delta\ mode] \oplus \delta[fp, COL\ \delta\ type] \oplus$
                 $\delta[fp, COL\ \delta\ kind])$
           end error
       end test compatibility of actual and formal parameter ;
     comment This is the end of the declaration part of pass 3 ;

   comment Here the statements of pass 3 begin ;
INITIALIZATIONS FOR THE DECOMPOSITION
 AND GENERATION PASS:
     $del := mark := 1$;
     $internal\ identifier := symbol := information\ 4 := instruction := 2$;
     $mode := information\ label := information\ identifier :=$
         $information\ 0 := 3$;
     $type := information\ state := information\ 1 := 4$;
     $kind := information\ type := information\ 2 := 5$;
     $supplement\ information := information\ 3 := 6$;
     $information\ error := 7$;
     $internal\ identifier\ of\ true := -1$;
     $internal\ identifier\ of\ false := -2$;
     $int\ id\ of\ number\ 0 := -3$;
     $int\ id\ of\ number\ 1 := -5$;

order opening bracket := 2 ;
order expression bracket := 12 ;
order statement bracket := 13 ;
$k := p :=$ *accumulator loaded* :=
  max address := *array block level counter* :=
  static procedure level counter := *delta* := *label jump over* :=
  entry number of procedure identifier := *int id of parameter* :=
  jump and := *jump or* := *undefined left part variable* := $s :=$ 0 ;
normal program := *incoming symbol correct* := **true** ;
content of := *not bool* := **false** ;
state := '**statement**' ;
π [1, *mark*] := *specification* := *relational operator* := '**empty**' ;
type of left part := '**type**' ;
delta add := δ [1, *COL* δ *associated internal identifier*] ;
end delta := *end identifier list* ;
entry number of free storage cell := *end delta* + 1 ;
relative address counter := 4 ;
for $i :=$ 0 **step** 1 **until** δ [0, *COL* δ *maximum array block level*] **do**
  generate variable of type ('**integer**') ;
push down ('**block begin**') ;
σ [*s, information 3*] := *relative address counter* ;
δ [0, *COL* δ *associated internal identifier*] := 0 ;

READ NEXT SYMBOL:

  read ;
  if *delimiter* (*incoming symbol*) **then**
    begin
PROCESS:
      check delimiter pair ;
PROCESS 1:
      if *character order* (*incoming symbol*) $\leq$
          order opening bracket **then**
        go to *INCOMING OPENING BRACKET*
          [*opening bracket nr* (*incoming symbol*)] ;
PROCESS 2:
      if σ [*s, del*] = '**delimiter**' **then**
        begin
          if *state order* (σ [*s, symbol*]) = *order statement bracket* $\wedge$
              character order (*incoming symbol*) $\geq$
                state order (σ [*s, symbol*]) **then**
            go to *STATEMENT BRACKET PAIR*
              [*statement bracket nr* (σ [*s, symbol*])]
        end

else
  if *character order (incoming symbol)* $\geqq$
    state order $(\sigma[s-1, symbol])$ **then**
    begin
      if *state order* $(\sigma[s-1, symbol]) \geqq$
        order expression bracket **then**
        go to *EXPRESSION BRACKET PAIR*
          [expression bracket nr $(\sigma[s-1, symbol])]$;
        go to *ARITHMETIC OR BOOLEAN OPERATOR*
    end ;
  if *Boolean operator (incoming symbol)* **then**
    go to *INCOMING BOOLEAN OPERATOR* ;
  if $\sigma[s, del] = $ '**entity**' **then**
    begin
      if $\sigma[s, kind] = $ '**AC**' $\wedge$ $(\sigma[s, type] = $ '**type**' $\vee$
      $\sigma[s, type] = $ '**arithmetic**') $\wedge$
      $\sigma[s, supplement\ information] \neq $ '**content of**' **then**
      begin
        if *integer operator (incoming symbol)* **then**
          begin
            compile ('**possible transfer to integer**
              depending on ADDR AC') ;
            $\sigma[s, type] := $ '**integer**'
          end
          else
          begin
            compile ('**possible transfer to real**
              depending on ADDR AC') ;
            $\sigma[s, type] := $ '**real**'
          end
        end insert possible transfers for conditional
        expressions with undefined type
    end ;
  push down (incoming symbol) ;
  state := '**expression**' ;
  if *specification* = '**undefined**' **then**
    specification := '**expression**'
  end incoming symbol is a delimiter
else

begin
  if *index register (incoming symbol)* **then**
    push down entity (incoming symbol, '**normal**', '**integer**',
      '**index register**', '**empty**')
  else
    begin
      if *generated identifier (incoming symbol)* **then**
        search for a generated identifier (incoming symbol)
          result: *(int id)*
      else *search for an identifier in the identifier*
        list (incoming symbol) result: *(int id)* ;
      check identifier (int id) ;
      push down entity (
        if $\delta[int\ id,\ COL\ \delta\ mode] =$ '**value**' $\wedge$
        $\delta[int\ id,\ COL\ \delta\ kind] =$ '**variable**' **then**
        $\delta[int\ id,\ COL\ \delta\ associated\ internal\ identifier]$
        else *int id,*
        $\delta[int\ id,\ COL\ \delta\ mode],$
        $\delta[int\ id,\ COL\ \delta\ type],$
        $\delta[int\ id,\ COL\ \delta\ kind],$
        '**empty**') ;
      if *content of* **then**
        begin
          $\sigma[s,\ type] := type\ variable$;
          type variable := '**empty**' ;
          $\sigma[s,\ supplement\ information] :=$ '**content of**' ;
          content of := **false**
        end
    end
  end incoming entity ;
  go to *READ NEXT SYMBOL* ;

ARITHMETIC OR BOOLEAN OPERATOR:
  if $\sigma[s-2,\ del] =$ '**delimiter**' **then**
    begin
      if $\sigma[s-1,\ symbol] =$ '+' **then**
        begin
          check specification otherwise ;
          for $i := 1$ **step** 1 **until** 7 **do**
            $\sigma[s-1,\ i] := \sigma[s,\ i]$;

```
      if accumulator loaded = s then
         accumulator loaded := s−1 ;
      s := s−1 ;
      go to PROCESS
    end unary + ;
  go to UNARY OPERATOR
end unary operator ;
```

if *Boolean operator* $(\sigma[s-1, symbol])$ **then**
  go to *BINARY BOOLEAN OPERATOR* ;
if $\sigma[s-1, symbol] = `\oplus'$ **then go to** *CONCATENATE IR* ;
go to *BINARY ARITHMETIC OR RELATIONAL*
  OPERATOR ;

UNARY OPERATOR:
  if $\sigma[s-1, symbol] = `-'$ **then**
    begin
      $\sigma[s-1, symbol] := `-_u'$;
      check compatibility (s, '**entity**', '**arithmetic**') ;
      type wanted := $\sigma[s, type]$
    end
  else
    begin
      check compatibility (s, '**entity**', '**Boolean**') ;
      type wanted := '**Boolean**'
    end ;
  if $\sigma[s, kind] \ne `\mathbf{AC}' \wedge \sigma[s, kind] \ne `\textbf{procedure}' \wedge$
  $\sigma[s, mode] \ne `\textbf{formal}'$ **then**
    begin
      if $\sigma[s, supplement\ information] = `\textbf{content of}'$ **then**
        compile $(`\mathbf{AC}:=' \oplus \sigma[s-1, symbol] \oplus `\textbf{content of}' \oplus$
          $\sigma[s, internal\ identifier])$
      else *compile* $(`\mathbf{AC}:=' \oplus \sigma[s-1, symbol] \oplus$
        $\sigma[s, internal\ identifier])$
    end number, simple variable, subscripted variable with
      rec addr calc
  else
    if $\sigma[s, kind] = `\mathbf{AC}' \wedge$
    $\sigma[s, supplement\ information] = `\textbf{content of}'$ **then**
      begin
        compile $(`\textbf{ADDR AC}:= \mathbf{AC}')$;
        compile $(`\mathbf{AC}:=' \oplus \sigma[s-1, symbol] \oplus$
          $`\textbf{content of ADDR AC}')$;
```
```

if $\sigma[s-1, symbol] = '\neg' \wedge \sigma[s, type] \neq$ '**Boolean**' **then**
    compile ('**compare Boolean to ADDR AC**')
    end subscripted variables
else
  begin
    bring to AC with prescribed type (s, s, *type wanted*, **false**) ;
    compile ('**AC**:=' $\oplus \sigma[s-1, symbol] \oplus$ '**AC**')
  end ;
delete (1) ;
clear sigma (s) ;
$\sigma[s, del] :=$ '**entity**' ;
$\sigma[s, mode] :=$ '**normal**' ;
$\sigma[s, type] :=$ *type wanted* ;
$\sigma[s, kind] :=$ '**AC**' ;
accumulator loaded := s ;
go to *PROCESS* ;

comment Within the following part of the program the instructions for binary arithmetic and relational operations are generated.

Remarks:

1. We do not distinguish between normalized and nonnormalized operations, as the treatment depends on the special number representation, and thus is machine-dependent. But we note that, if one uses (following the ALCOR suggestions) nonnormalized addition and subtraction the operator '$-$' used for the reduction of a relation $a \varrho b$ to $a - b \varrho 0$ must be different from the usual '$-$'. This new operator must include the normalization, as there are in general different representations for the normalized and nonnormalized numbers 0.

2. The following optimizations are implemented:

a) $E + 0$ is replaced by E. This may easily occur when evaluating the distance function for an index vector.

b) A relation of the special form $E \varrho 0$ or $0 \varrho E$ does not require a reduction. But another warning is necessary here: the implementor must examine whether it is necessary to insert an instruction for normalization in case E is an expression of real type.

Other optimizations may easily be introduced. But one should always consider whether they are of advantage for the user (e.g. the ALGOL programmer may write $E \times 1$ in the program under the assumption of having a certain influence on normalization).

3. The operator $\uparrow$ is used according to adjustment 2 of section 4.3:

$E1 \uparrow E2$ is of type **integer** if $E1$ is an expression of type **integer** and $E2$ is an unsigned integer number. We use the operator $\uparrow_{ii}$.

If *E1* is an expression of type **real** and *E2* is an unsigned integer number, then we use $\uparrow_{r,i}$, and the result is of type **real**. Otherwise the result is always of type **real**: operator $\uparrow$.

4. Side effects of functions are excluded in the sense that we do not make provision for them. In particular we exchange operands of binary operations. ;

BINARY ARITHMETIC OR RELATIONAL OPERATOR:
    check compatibility (*s*, '**entity**', '**arithmetic**') ;
    check compatibility (*s*— 2, '**entity**', '**arithmetic**') ;
    if $\sigma[s-2, kind]$ = '**index register**' **then**
      begin
        compile ($\sigma[s-2$, *internal identifier*] $\oplus$ ':=' $\oplus$
          $\sigma[s-2$, *internal identifier*] $\oplus$ '$+_i$' $\oplus$
          $\sigma[s$, *internal identifier*]) ;
        delete (5) ;
        go to *PROCESS*
      end ;
    if $\sigma[s$, *internal identifier*] = *int id of number 0* **then**
      begin
        if $\sigma[s-1$, *symbol*] = '$+$' **then**
          begin *delete* (2) ; **go to** *PROCESS* **end**
      end ;
    if *relational op* ($\sigma[s-1$, *symbol*]) **then**
      begin
        if $\sigma[s-2$, *internal identifier*] = *int id of number 0* **then**
          begin
            exchange operands ;
RELATION WITH 0:
            bring to AC with prescribed type (*s*— 2, *s*— 2,
              if $\sigma[s-2$, *type*] = '**integer**' **then** '**integer**' **else**
              '**real**', **true**) ;
            relational operator := $\sigma[s-1$, *symbol*] ;
            go to *RESULT RELATIONAL OPERATOR*
          end ;
        if $\sigma[s$, *internal identifier*] = *int id of number 0* **then**
          go to *RELATION WITH 0* ;
        relational operator := $\sigma[s-1$, *symbol*] ;
        $\sigma[s-1$, *symbol*] := '$-$'
      end preparations for relations
    else *relational operator* := '**empty**' ;

if ¬ *integer operator* ($\sigma[s-1, symbol]$) ∧
  $\sigma[s-1, symbol] \neq$ '/' ∧ $\sigma[s-1, symbol] \neq$ '↑' ∧
  $\sigma[s-2, type] =$ '**integer**' ∧ $\sigma[s, type] =$ '**integer**' **then**
  $\sigma[s-1, symbol] := int\ op\ (\sigma[s-1, symbol])$
else
  if $\sigma[s-1, symbol] =$ '↑' **then**
    begin
      if $\sigma[s, kind] =$ '**number**' ∧ $\sigma[s, type] =$ '**integer**' **then**
        $\sigma[s-1, symbol] :=$ **if** $\sigma[s-2, type] =$ '**integer**' **then**
          '↑$_{ii}$' **else** '↑$_{ri}$'
    end determine type of operator ;
if $\sigma[s-2, mode] =$ '**formal**' ∨ $\sigma[s-2, kind] =$ '**procedure**' **then**
  subroutine 1 := **true**
else *subroutine 1* := **false** ;
if $\sigma[s, mode] =$ '**formal**' ∨ $\sigma[s, kind] =$ '**procedure**' **then**
  subroutine 2 := **true**
else *subroutine 2* := **false** ;
if $\sigma[s, type] \neq type\ 2\ (\sigma[s-1, symbol])$ ∨ *subroutine 2* **then**
  begin
    if $\sigma[s, kind] =$ '**number**' **then**
      begin
        transfer type of number (s) ;
        go to *BINARY OPERATOR 3*
      end ;
    if *nonsymmetric operator* ($\sigma[s-1, symbol]$) **then**
      go to *BINARY OPERATOR 1* ;
    if $\sigma[s-2, type] \neq type\ 1\ (\sigma[s-1, symbol])$ ∨
      subroutine 1 **then**
      begin
        if $\sigma[s-2, kind] =$ '**number**' **then**
          begin
            transfer type of number (s−2) ;
            go to *BINARY OPERATOR 2*
          end ;
        if $\sigma[s-2, kind] =$ '**AC**' **then** *exchange operands* ;

BINARY OPERATOR 1:
      bring to AC with prescribed type (s, s,
        type 2 ($\sigma[s-1, symbol]$), **true**)
    end handle operand 1 and operand 2
  else

BINARY OPERATOR 2:
      exchange operands ;

BINARY OPERATOR 3:

  bring to AC with prescribed type $(s-2, s-2,$
  type 1 $(\sigma[s-1, symbol]),$ **true**)

 end handle operand 2

else

 begin

  if $\sigma[s-2, type] \neq type\ 1\ (\sigma[s-1, symbol])$ ∨
  subroutine 1 **then**

   begin

    if $\sigma[s-2, kind] = $ '**number**' **then**
    transfer type of number $(s-2)$

    else go to *BINARY OPERATOR 3*

   end handle operand 1 and not operand 2 ;

  if $\sigma[s, kind] = $ '**AC**' **then**

   begin

    if $\sigma[s-1, symbol] = $ '$-$' ∨ $\sigma[s-1, symbol] = $ '$-;$' **then**
    compile ('**AC** $:= -_u$ **AC**')

    else

     if *nonsymmetric operator* $(\sigma[s-1, symbol])$ **then**
       go to *BINARY OPERATOR 4* ;
     exchange operands

   end ;

BINARY OPERATOR 4:

  bring to AC with prescribed type $(s-2, s-2, \sigma[s-2, type],$ **true**)
  end generate instruction: bring first operand to AC ;

 if $\sigma[s, supplement\ information] = $ '**content of**' **then**
 compile ('**AC** $:= $ **AC**' $\oplus \sigma[s-1, symbol] \oplus$ '**content of**' $\oplus$
 $\sigma[s, internal\ identifier])$

 else *compile* ('**AC** $:= $ **AC**' $\oplus \sigma[s-1, symbol] \oplus$
 $\sigma[s, internal\ identifier])$;

 if $\sigma[s, kind] = $ '**number cellar**' **then** *reduce number cellar* ;

 if *relational operator* $= $ '**empty**' **then**
 $\sigma[s-2, type] := type\ result\ (\sigma[s-1, symbol])$

 else

  begin

RESULT RELATIONAL OPERATOR:

  $\sigma[s-2, type] := $ '**Boolean**' ;
  $\sigma[s-2, kind] := $ '**relation**' ;
  if $\sigma[s-3, symbol] = $ '¬' **then**

begin
    relational operator := *not rel op* (*relational operator*) ;
    delete (1) ;
    for $i := 1$ **step** 1 **until** 7 **do** $\sigma[s-2, i] := \sigma[s-1, i]$
end
  end ;
delete (2) ;
go to *PROCESS* ;

comment The binary Boolean operations are replaced by a sequence of conditional jumps and accumulator loading instructions. With each jump is associated a truth value depending on the operator. The target language instructions are chosen so that the accumulator contains the truth value of the operation when executing the jump.

$a \wedge b$:   **AC** := a
           if ¬ AC then local jump to L (truth value is **false**)
           AC := b
     L:
$a \vee b$:   **AC** := a
           if AC then local jump to L (truth value is **true**)
           AC := b
     L:
$a \supset b$:   **AC** := ¬ a
           if AC then local jump to L (truth value is **true**)
           AC := b
     L:

In this comment we disregard the operator '≡' since it requires different handling.

The conditional jump instruction is already produced when the operator ('∧', or '∨', or '⊃') is pushed down (*INCOMING BOOLEAN OPERATOR*). The entry of its first operand in σ is replaced by an entry '**open jump**' supplied with the respective truth value (**false**, or **true**, or **true**) and with a label to be compiled later on. If the incoming Boolean operator finds already an '**open jump**' as first operand which must be the result of an operation of lower or equal order then it is tested whether the jump to be produced may lead to the same label $(a \overrightarrow{\lceil \vee b \rceil \vee c}$ or $a \overrightarrow{\lceil \wedge b \rceil \wedge c}$ e.g.).

If this is not the case the label must be compiled (using *jump and*) (e.g. $a \overrightarrow{\lceil \wedge b \rceil \vee \downarrow c}$). An incoming '⊃' requires all open jumps to be terminated.

If an incoming symbol of higher order encounters a binary Boolean operator in σ then the program *BINARY BOOLEAN OPERATOR* is executed. An instruction for loading the accumulator with the second operand is generated, or if the second operator is already an open jump its associated label is saved in *jump and* (*jump or*) if it is the result of a '$\wedge$' or a '$\vee$' operation. This simple organization is possible since only three operators must be considered.

The evaluation of relations as operands is incorporated in the mechanism. It requires the conditional jumps to depend not only on the Boolean operators but also on the specific relational operator, and a disagreeable consequence is that the accumulator does not always contain the correct truth value at execution time. Therefore an entry '**open jump relation**' is used instead of '**open jump**', which induces the procedure *bring truth value* to generate appropriate instructions.

A simple transformation is included in the process:

$$\neg\,(B \wedge B) \quad \text{is replaced by} \quad (\neg\,B \vee \neg\,B)$$
$$\neg\,(B \vee B) \quad \text{is replaced by} \quad (\neg\,B \wedge \neg\,B)$$
$$\neg\,(B \supset B) \quad \text{is replaced by} \quad (B \wedge \neg\,B)$$
$$\neg\,(B \equiv B) \quad \text{is replaced by} \quad (\neg\,B \equiv B)$$

(B is a Boolean expression). For this purpose a new delimiter '(**not**' is used and new operators '$\wedge$ **not**', '$\vee$ **not**', '$\supset$ **not**' in order to keep the correct order relations. ;

INCOMING BOOLEAN OPERATOR:
    begin
      switch *INC BOOL OP* := *INCOMING AND,*
        INCOMING OR, INCOMING IMPL ;
      if *incoming symbol* = '$\neg$' **then**
        begin
          if $\sigma[s, symbol]$ = '$\neg$' **then**
            delete (1)
          else *push down* (*incoming symbol*) ;
          go to *READ NEXT SYMBOL*
        end incoming $\neg$;
      check compatibility (*s*, '**entity**', '**Boolean**') ;
      if *incoming symbol* = '$\equiv$' **then go to** *INCOMING EQUIV* ;
      if $\sigma[s-1, symbol]$ = '(**not**' $\vee$ $\sigma[s-1, symbol]$ = '$\vee$ **not**' $\vee$
        $\sigma[s-1, symbol]$ = '$\supset$ **not**' **then**
        begin
          not bool := **true** ;
          incoming symbol := *not bool operator* (*incoming symbol*)
        end ;

if $\sigma[s, kind] \neq$ '**relation**' $\wedge$ $\sigma[s, kind] \neq$ '**open jump**' $\wedge$
$\sigma[s, kind] \neq$ '**open jump relation**' **then**
bring to AC with prescribed type (s, s, '**Boolean**', **true**) ;
truth value := *result of Boolean operation* (*incoming symbol*) ;
go to *INC BOOL OP* [*incoming Boolean nr* (*incoming symbol*)] ;

INCOMING IMPL:
 if $\sigma[s, kind] \neq$ '**AC**' **then**
  bring truth value ;
 compile ('**AC** := $\neg$ **AC**') ;
 go to *INCOMING AND 2* ;

INCOMING OR:
 if $\sigma[s, supplement\ information] \neq$ '**empty**' $\wedge$ $\sigma[s, supplement$
 $information] \neq truth\ value$ **then**
 jump and := $\sigma[s, internal\ identifier]$
 else

INCOMING AND:
  if $\sigma[s, kind] =$ '**open jump**' $\vee$
   $\sigma[s, kind] =$ '**open jump relation**' **then**
   go to *INCOMING AND 1* ;

INCOMING AND 2:
 generate label (*label entry number*) ;
 $\sigma[s, kind] :=$ '**open jump**' ;
 $\sigma[s, internal\ identifier] :=$ *label entry number* ;
 $\sigma[s, supplement\ information] :=$ *truth value* ;

INCOMING AND 1:
 if *relational operator* $\neq$ '**empty**' **then**
  $\sigma[s, kind] :=$ '**open jump relation**' ;
 conditional order := *instruction function*
  (*relational operator, incoming symbol*) ;
 compile (*conditional order* $\oplus$ $\sigma[s, internal\ identifier]$) ;
 if *jump and* $\neq 0$ **then**
  begin *compile label* (*jump and*) ; *jump and* := 0 **end** ;
 if *jump or* $\neq 0$ **then**
  begin *compile label* (*jump or*) ; *jump or* := 0 **end** ;
 push down (*incoming symbol*) ;
 if *not bool* **then**
  begin *push down* ('$\neg$') ; *not bool* := **false** **end** ;
 relational operator := '**empty**' ;
 accumulator loaded := 0 ;
 go to *READ NEXT SYMBOL* ;

INCOMING EQUIV:

    if $\sigma[s, kind] = $ '**open jump**' $\vee$ $\sigma[s, kind] = $ '**relation**' $\vee$
    $\sigma[s, kind] = $ '**open jump relation**' **then**
    bring truth value

    else

        if $\sigma[s, mode] = $ '**formal**' $\vee$ $\sigma[s, kind] = $ '**procedure**'
        $\vee$ $\sigma[s, type] \neq $ '**Boolean**'
        $\vee$ $\sigma[s, supplement\ information] = $ '**content of**' **then**
        bring to AC with prescribed type (s, s, '**Boolean**', **true**) ;
    push down (incoming symbol) ;

    go to *READ NEXT SYMBOL*

    end incoming Boolean operator ;

BINARY BOOLEAN OPERATOR:

    begin

        switch *BOOLEAN OPERATOR* :=
        OPERATOR AND, OPERATOR OR,
        OPERATOR IMPL, OPERATOR EQUIV ;
    check compatibility (s, '**entity**', '**Boolean**') ;

        go to *BOOLEAN OPERATOR*
        [Boolean operator nr $(\sigma[s-1, symbol])]$;

OPERATOR IMPL:

OPERATOR OR:

    if $\sigma[s, kind] = $ '**open jump**' $\vee$
    $\sigma[s, kind] = $ '**open jump relation**' **then**

        begin

            if $\sigma[s, supplement\ information] = $ '**result true**' **then**
            jump or := $\sigma[s, internal\ identifier]$

            else *jump and* := $\sigma[s, internal\ identifier]$;

            if $\sigma[s, kind] = $ '**open jump relation**' **then**
            $\sigma[s-2, kind] := $ '**open jump relation**'

        end second operand is an open jump

    else

OPERATOR AND:

    if $\sigma[s, kind] \neq $ '**relation**' **then**
    bring to AC with prescribed type (s, s, '**Boolean**', **true**)

    else $\sigma[s-2, kind] := $ '**open jump relation**' ;
    delete (2) ;

    go to *PROCESS* ;

OPERATOR EQUIV:

    if $\sigma[s, kind] = $ '**open jump**' $\vee$ $\sigma[s, kind] = $ '**relation**' $\vee$
    $\sigma[s, kind] = $ '**open jump relation**' **then**
    deliver label entry number
   else
    begin
      if $\sigma[s, mode] \neq $ '**formal**' $\wedge$ $\sigma[s, kind] \neq $ '**procedure**' $\wedge$
        $\sigma[s, supplement\ information] \neq $ '**content of**' $\wedge$
        $\sigma[s, type] = $ '**Boolean**' $\wedge$
        $\sigma[s-2, kind] = $ '**AC**' **then**
      exchange operands
      else *bring to AC*
        with prescribed type (s, s, '**Boolean**', **false**) ;
      generate label (*label entry number*)
    end ;
   compile conditional order and address open jumps ;
   compile ('**AC** := ¬' $\oplus$ $\sigma[s-2, internal\ identifier]$) ;
   compile simple order and address open jump ;
   accumulator loaded := 0 ;
   bring to AC with prescribed type (s-2, s-2, '**Boolean**', **false**) ;
   compile label (*new label entry number*) ;
   delete (2) ;
   go to *PROCESS*
  end binary Boolean operator ;

INCOMING FOR SUITABLE:

INCOMING GO TO:

INCOMING BEGIN INITIAL VALUE:

INCOMING BEGIN INCREMENT:

INCOMING BEGIN:

INCOMING ARRAY:

INCOMING SWITCH:

INCOMING PROCEDURE BEGIN:
   push down (*incoming symbol*) ;
   go to *READ NEXT SYMBOL* ;

INCOMING LABEL COLON:
   compile label ($\sigma[s, internal\ identifier]$) ;
   delete (1) ;
   go to *READ NEXT SYMBOL* ;

   comment The following five delimiters are generated by the recursive address calculation pass (see 7.3) ;

INCOMING REDUCED INITIAL ADDRESS:
$\sigma[s, internal\ identifier] := \delta[\sigma[s, internal\ identifier],$
$\quad COL\ \delta\ associated\ internal\ identifier]$;
$\sigma[s, mode] := '$**generated**$'$;
$\sigma[s, kind] := '$**variable**$'$;
if $\sigma[s, supplement\ information] \neq '$**content of**$'$ **then**
$\quad \sigma[s, type] := '$**integer**$'$;
go to *READ NEXT SYMBOL* ;

INCOMING CONTENT REAL OF:
content of := **true** ;
type variable := '**real**' ;
go to *READ NEXT SYMBOL* ;

INCOMING CONTENT INTEGER OF:
content of := **true** ;
type variable := '**integer**' ;
go to *READ NEXT SYMBOL* ;

INCOMING CONTENT BOOLEAN OF:
content of := **true** ;
type variable := '**Boolean**' ;
go to *READ NEXT SYMBOL* ;

CONCATENATE IR:
check compatibility $(s-2, '$**entity**$', '$**arithmetic**$')$;
bring to AC with prescribed type $(s-2, s-2, '$**integer**$',$ **true**$)$;
compile $('$**AC**$:= $**AC**$ \oplus' \oplus \sigma[s, internal\ identifier])$;
delete (2) ;
go to *PROCESS* ;

INCOMING SQUARE BRACKET:
check array or switch identifier ;
if $\sigma[s-1, symbol] = '$**array**$'$ **then**
  begin
    push down $('$**array** $[')$;
    begin subscript bounds := s
  end
  else
  begin
    push down (*incoming symbol*) ;
    if $\sigma[s-1, kind] = '$**array**$' \vee$
      $\sigma[s-1, kind] = '$**array or switch**$'$ **then**

begin
    $\sigma[s,\ information\ identifier] :=$
        $\delta[\sigma[s-1,\ internal\ identifier],$
        $COL\ \delta\ associated\ internal\ identifier]$;
    $\sigma[s,\ information\ 2] := specification$;
    $specification := $ '**empty**'
end ;
  $\sigma[s,\ information\ state] := state$
end ;
$state := $ '**expression**' ;
go to *READ NEXT SYMBOL* ;

SQUARE BRACKET:
  check subscript position ;
  check compatibility (s, '**entity**', '**arithmetic**') ;
  if $\sigma[s-2,\ kind] = $ '**switch**' **then**
    go to *SWITCH DESIGNATOR* ;
  if $\sigma[s-2,\ kind] = $ '**array or switch**' $\wedge$
    $\sigma[s-1,\ information\ state] = $ '**statement**' **then**
    go to *SWITCH DESIGNATOR* ;
  int id $:= \sigma[s-1,\ information\ identifier] :=$
    $\sigma[s-1,\ information\ identifier] + 1$;
  if $\sigma[s,\ internal\ identifier] = int\ id\ of\ number\ 0$ **then**
    begin
      if *incoming symbol* $= $ ']' **then**
        begin
          if $\sigma[s-2,\ type] = $ '**type**' $\vee\ \sigma[s-2,\ type] =$
          '**arithmetic**' **then**
            go to *CLOSING BRACKET* ;
          $\sigma[s-2,\ internal\ identifier] :=$
            $\delta[\sigma[s-2,\ internal\ identifier],$
            $COL\ \delta\ associated\ internal\ identifier]$;
          $\sigma[s-2,\ kind] := $ '**variable**' ;
          if $\sigma[s-1,\ symbol] = $ '[' **then**
            go to *CLOSING BRACKET 1* ;
          go to *SIMPLE INDEX VECTOR*
        end ;
      delete (1) ;
      go to *READ NEXT SYMBOL*
    end all subscript expressions are zero ;
  if *incoming symbol* $= $ ']' **then go to** *CLOSING BRACKET* ;
  if $\sigma[s,\ internal\ identifier] = int\ id\ of\ number\ 1$ **then**
    delete (1)

else *push down* (' ×*ᵢ*') ;
push down entity
    (*int id*, '**normal**', '**integer**', '**variable**', '**empty**') ;
incoming symbol := ' +*ᵢ*' ;
go to *PROCESS* ;

CLOSING BRACKET:
    if $\sigma[s-1, symbol] =$ '[' **then**
SIMPLE INDEX VECTOR:
        begin
            for $i := 1$ **step** 1 **until** 7 **do** $\sigma[s-2, i] := \sigma[s, i]$;
            if *accumulator loaded* $= s$ **then**
                accumulator loaded := $s-2$;
            delete (2) ;
            go to *READ NEXT SYMBOL*
        end simple index vector ;
    int id := $\delta[\sigma[s-2, internal identifier]$,
        COL δ *associated internal identifier*] ;
    bring to AC with prescribed type (*s*, *s*, '**integer**', **false**) ;
    compile ('**AC** := **AC** +' $\oplus$ *int id*) ;
    $\sigma[s-2, kind] :=$ '**AC**' ;
    $\sigma[s-2, internal identifier] :=$ '**empty**' ;
    accumulator loaded := $s-2$;

CLOSING BRACKET 1:
    state := $\sigma[s-1, information state]$;
    specification := $\sigma[s-1, information 2]$;
    delete (2) ;
    $\sigma[s, mode] :=$ '**normal**' ;
    $\sigma[s, supplement information] :=$ '**content of**' ;
    $\sigma[s, information error] :=$ '**empty**' ;
    go to *READ NEXT SYMBOL* ;

SWITCH DESIGNATOR:
    if $\sigma[s, kind] =$ '**number**' **then**
        begin
            if $\sigma[s, type] \neq$ '**integer**' **then**
                transfer type of number (*s*) ;
            go to *SWITCH DESIGNATOR 1*
        end ;
    if $\sigma[s, mode] \neq$ '**formal**' $\wedge$ $\sigma[s, type] =$ '**integer**' $\wedge$
        ($\sigma[s, kind] =$ '**variable**' $\vee$ $\sigma[s, kind] =$ '**number cellar**') **then**

begin
  if $\sigma[s,\ supplement\ information] = $ '**content of**' **then**
    compile (' **SIR** := **content of**' $\oplus\ \sigma[s,\ internal\ identifier]$)
  else

SWITCH DESIGNATOR 1:
        compile (' **SIR** :=' $\oplus\ \sigma[s,\ internal\ identifier]$)
  end
else
  begin
    bring to AC with prescribed type $(s,\ s,\ $'**integer**', **false**) ;
    compile (' **SIR** := **AC**')
  end ;
state := $\sigma[s,\ information\ state]$;
specification := $\sigma[s,\ information\ 2]$;
accumulator loaded := 0 ;
delete (2) ;
$\sigma[s,\ kind]$:= '**switch designator**' ;
go to *READ NEXT SYMBOL* ;

ARRAY COLON:

ARRAY SQUARE BRACKET:
  check compatibility $(s,\ $'**entity**', '**arithmetic**') ;
  if $\sigma[s,\ type] \neq$ '**integer**' $\vee\ \sigma[s,\ mode] = $ '**formal**' $\vee$
  $\sigma[s,\ kind] = $ '**procedure**' $\vee$
  $\sigma[s,\ supplement\ information] = $ '**content of**' **then**
  bring to AC with prescribed type $(s,\ s,\ $'**integer**', **true**) ;
  if $\sigma[s,\ kind] = $ '**AC**' **then** *take care of AC* (s) ;
  for $i := 1$ **step** 1 **until** 7 **do** $\sigma[s-1,\ i] := \sigma[s,\ i]$;
  delete (1) ;
  if *incoming symbol* $\neq$ ']' **then**
    begin
      if *incoming symbol* = ':' **then**
        push down (' **array** :')
      else *push down* (' **array** [') ;
      go to *READ NEXT SYMBOL*
    end ;

END OF THE ARRAY DECLARATION:
  compile ('**storage allocation**' $\oplus$
    $\sigma[begin\ subscript\ bounds - 1,\ internal\ identifier]$) ;
  for $i := $ *begin subscript bounds* **step** 1 **until** s **do**

```
    begin
        compile (σ[i, internal identifier]) ;
        if σ[i, kind] = 'number cellar' then reduce number cellar
    end ;
    state := 'statement' ;
    delete (s − begin subscript bounds + 3) ;
    go to READ NEXT SYMBOL ;
```

INCOMING ROUND BRACKET:
```
    if σ[s, del] = 'delimiter' then
        begin
            if σ[s, symbol] = '¬' then
                begin
                    delete (1) ;
                    push down ('(not') ;
                    push down ('¬')
                end
            else push down (incoming symbol) ;
            check specification otherwise ;
            go to READ NEXT SYMBOL
        end incoming arithmetic round bracket
    else
        go to FUNCTION OR PROCEDURE CALL ;
```

ROUND BRACKET:

ROUND BRACKET NOT:
```
    if σ[s, kind] = 'open jump' ∨ σ[s, kind] = 'relation' ∨
        σ[s, kind] = 'open jump relation' then
        bring truth value ;
    if accumulator loaded = s then
        accumulator loaded := s − 1 ;
    for i := 1 step 1 until 7 do σ[s−1, i] := σ[s, i] ;
    delete (1) ;
    go to READ NEXT SYMBOL ;
```

FUNCTION OR PROCEDURE CALL:
```
    check procedure identifier with parameters ;
    if accumulator loaded ≠ 0 then
        take care of AC (accumulator loaded) ;
    generate label (label entry number) ;
    compile ('local jump to' ⊕ label entry number) ;
    s := s + 1 ; clear sigma (s) ;
    σ[s, internal identifier] := entry number of procedure identifier ;
    σ[s, information label] := label entry number ;
```

$\sigma[s, \text{\textit{information state}}] := \text{\textit{state}}$;
$\sigma[s, \text{\textit{information 2}}] := \text{\textit{specification}}$;
$\sigma[s, \text{\textit{information 3}}] := p$;
entry number of procedure identifier $:= s-1$;
$s := s+1$;
number of parameter $:= 1$;

PREPARATION FOR AN ACTUAL PARAMETER:

clear sigma (s) ;
push down ('**procedure call**') ;
$\sigma[s, \text{\textit{information 1}}] := \text{\textit{number of parameter}}$;
$\sigma[s, \text{\textit{information 2}}] := k$;
$\sigma[s, \text{\textit{information 3}}] := p$;
search for specification
   (*entry number of procedure identifier, number of parameter*)
   result: (*state, specification, int id of parameter*) ;
generate label (*label entry number*) ;
compile label (*label entry number*) ;
$\sigma[s-1, \text{\textit{internal identifier}}] := \text{\textit{label entry number}}$;
$\sigma[s, \text{\textit{information 0}}] := \text{\textit{int id of parameter}}$;
go to *READ NEXT SYMBOL* ;

comment When the i-th actual parameter is handled the push down σ has the following uppermost entries:

entry number of procedure identifier:	
	⟨procedure identifier⟩
	⟨entry number of a possibly embracing procedure call⟩ ⊕ ⟨label for jump over the actual parameters⟩ ⊕ ⟨state⟩ ⊕ ⟨specification⟩ ⊕ ⟨program address⟩
	⟨GIAP of the first actual parameter⟩
	⟨GIAP of the $(i-1)$-th actual parameter⟩
	⟨possibly generated label for the i-th actual parameter⟩
$s-1$:	'**procedure call**' ⊕ ⟨internal identifier of the i-th formal parameter⟩ ⊕ ⟨number of parameters⟩ ⊕ ⟨program address k⟩ ⊕ ⟨program address p⟩
s:	⟨i-th actual parameter⟩

;

ACTUAL PARAMETERS IN PROCEDURE CALLS:
  $int\ id := \sigma[s-2,\ internal\ identifier]$;

PARAMETER IS SIMPLE IDENTIFIER:
  if $\sigma[s,\ mode] =$ '**formal**' $\lor$ $\sigma[s,\ kind] =$ '**procedure**' $\lor$
  $\sigma[s,\ kind] =$ '**array**' $\lor$ $\sigma[s,\ kind] =$ '**switch**' $\lor$
  $\sigma[s,\ kind] =$ '**label**' $\land \neg (\sigma[s,\ type] =$ '**integer**' $\land$
  $specification =$ '**undefined**') **then**
    begin
      $\sigma[s-2,\ internal\ identifier] := \sigma[s,\ internal\ identifier]$;
      go to *END PROCEDURE CALL*
    end parameter is simple identifier ;

PARAMETER IS DESIGNATIONAL EXPRESSION:
  if $\sigma[s,\ kind] =$ '**switch designator**' $\lor$
  $\sigma[s,\ kind] =$ '**designational**' $\lor$ $\sigma[s,\ kind] =$ '**label**' **then**
    begin
      compile jump ;
      go to *END PROCEDURE CALL*
    end parameter is designational expression ;

PARAMETER IS STRING:
  if $\sigma[s,\ kind] =$ '**string**' **then**
    begin
      $\delta[int\ id,\ COL\ \delta\ type] :=$ '**string**' ;
      $\delta[int\ id,\ COL\ \delta\ kind] :=$ '**procedure**' ;
      compile ('**absolute address**' $\oplus$ $\sigma[s,\ internal\ identifier]$) ;
      compile ('**name procedure exit**') ;
      go to *END PROCEDURE CALL*
    end parameter is string ;

KIND OF PARAMETER IS UNDEFINED:
  check undefined specification ;

PARAMETER IS EXPRESSION:
  if $\sigma[s,\ kind] =$ '**variable**' $\land$
  $\sigma[s,\ supplement\ information] =$ '**content of**' **then**
    begin
      compile ('**absolute subscripted variable address**' $\oplus$
        $\sigma[s,\ internal\ identifier]$) ;
      compile ('**AC** := **content of ADDR AC**') ;
      go to L
    end parameter is subscripted variable evaluated by rec addr calc.

The distinction made in c4), p. 69 has been dropped here for reasons of simplicity. The instructions generated may easily handle both cases.

    else
      if $\sigma[s, kind] = $ '**AC**' $\wedge$
        $\sigma[s, supplement\ information] = $ '**content of**' **then**
      begin
        compile ('**ADDR AC** := **AC**') ;
        compile ('**AC** := **content of ADDR AC**') ;
        if *specification* = '**undefined**' **then**
          begin
            $\delta[int\ id, COL\ \delta\ type] := $ '**type**' ;
            $\delta[int\ id, COL\ \delta\ kind] := $ '**procedure or label**'
          end
        else go to L
      end actual parameter is subscripted variable
    else
      if $\sigma[s, kind] = $ '**variable**' **then**
      begin
        compile (' **absolute address**' $\oplus \sigma[s, internal\ identifier]$) ;
        compile ('**AC** :=' $\oplus \sigma[s, internal\ identifier]$) ;
$L:$ $\delta[int\ id, COL\ \delta\ type] := \sigma[s, type]$;
        $\delta[int\ id, COL\ \delta\ kind] := $ '**procedure**'
      end parameter is simple variable
    else
      begin
        if $\sigma[s, type] = $ '**Boolean**' **then**
          begin
            if $\sigma[s, kind] = $ '**relation**' $\vee$
              $\sigma[s, kind] = $ '**open jump**' $\vee$
              $\sigma[s, kind] = $ '**open jump relation**' **then**
            bring truth value
           else *bring to AC with prescribed type*
            $(s, s,$ '**Boolean**', **true**) ;
           $\delta[int\ id, COL\ \delta\ type] := $ '**Boolean**' ;
           $\delta[int\ id, COL\ \delta\ kind] := $ '**expression procedure**' ;
          go to $END\ ACTUAL\ EXPRESSION$
          end parameter is Boolean expression ;
        bring to AC with prescribed type $(s, s, \sigma[s, type],$ **true**) ;
        if *specification* $\neq$ '**undefined**' **then**
          begin
            $\delta[int\ id, COL\ \delta\ kind] := $ '**expression procedure**' ;
            $\delta[int\ id, COL\ \delta\ type] := \sigma[s, type]$;
          end

        else
          begin
           if $\sigma[s, type] = $ '**integer**' **then**
             $\delta[int\ id,\ COL\ \delta\ type] := $ '**arithmetic**'
           else $\delta[int\ id,\ COL\ \delta\ type] := $ '**type**' ;
           $\delta[int\ id,\ COL\ \delta\ kind] := $
           '**expression procedure or label**'
          end ;

END ACTUAL EXPRESSION:
        if $\sigma[s, type] \neq $ '**type**' $\land$ $\sigma[s, type] \neq $ '**arithmetic**' **then**
        compile ('**ADDR AC** $:=$' $\oplus$ $\sigma[s, type]$)
     end parameter is expression ;
    compile ('**name procedure exit**') ;
    accumulator loaded $:= 0$;
    if *specification* $= $ '**undefined**' **then**
     begin
      state $:= $ '**statement**' ;
      generate label (*label entry number*) ;
      $\delta[int\ id,\ COL\ \delta$ *associated internal identifier*] $:=$
       label entry number ;
      compile label (*label entry number*) ;
      $k1 := k$; $p1 := p$;
      $k := \sigma[s-1,\ information\ 2]$;
      delete (1) ;
      go to *READ NEXT SYMBOL*
     end specification is undefined: the actual parameter must be
     generated a second time in order to produce a designational
     expression from it ;

END PROCEDURE CALL:
    if $\sigma[$*entry number of procedure identifier, information error*$] \neq$
     '**inserted**' $\land$
    $\sigma[$*entry number of procedure identifier, mode*$] \neq $ '**formal**' **then**
     begin
      if $\delta[\sigma[$*entry number of procedure identifier,*
       internal identifier$],\ COL\ \delta\ dimension] \geq$
      $\sigma[s-1,\ information\ 1]$ **then**
      test compatibility of actual and formal parameter
       (*int id,* $\sigma[s-1,\ information\ 0]$)
     end ;
    number of parameter $:= \sigma[s-1,\ information\ 1] + 1$;
    delete (1) ;

if *incoming symbol* = ',' **then**
  go to *PREPARATION FOR AN ACTUAL PARAMETER* ;
check number of parameters (entry number of procedure identifier) ;
if $p = \sigma$[*entry number of procedure identifier* +1,
  information 3] **then**
  $p := p-1$
else *compile label* (σ[*entry number of procedure identifier* +1,
  information label]) ;
type variable := **if** σ[*entry number of procedure identifier, type*] =
  '**arithmetic**' **then**
  '**type**'
  else σ[*entry number of procedure identifier, type*] ;
if σ[*entry number of procedure identifier, information error*] =
  '**inserted**' **then**
  go to *OVER* ;
if σ[*entry number of procedure identifier, mode*] = '**normal**' **then**
  compile (type variable $\oplus$ '**procedure call**' $\oplus$
    σ[*entry number of procedure identifier, internal identifier*])
else *compile (type variable* $\oplus$ '**formal procedure call**' $\oplus$
  σ[*entry number of procedure identifier, internal identifier*]) ;
for $i :=$ *entry number of procedure identifier* +2 **step** 1 **until** $s-1$ **do**
  compile (σ[*i, internal identifier*]) ;

OVER:
  $s :=$ *entry number of procedure identifier* ;
  entry number of procedure identifier :=
    σ[$s+1$, *internal identifier*] ;
  state := σ[$s+1$, *information state*] ;
  specification := σ[$s+1$, *information 2*] ;
  if σ[s, *type*] = '**empty**' **then**
    delete (1)
  else
    begin
      σ[s, *internal identifier*] := σ[s, *supplement information*] :=
        σ[s, *information error*] := '**empty**' ;
      σ[s, *mode*] := '**normal**' ;
      σ[s, *kind*] := '**AC**' ;
      accumulator loaded := s
    end ;
  go to *READ NEXT SYMBOL* ;

INCOMING IF:
  push down (incoming symbol) ;
  σ[s, *information state*] := *state* ;

21*

$state := $ ' **expression** ' ;
$\sigma[s, information\ 2] := specification$;
$specification := $ ' **empty** ' ;
go to $READ\ NEXT\ SYMBOL$;

$IF:$

$check\ compatibility\ (s,$ ' **entity** ', ' **Boolean** ') ;
if $\sigma[s, kind] \neq$ ' **relation** ' **then**
  begin
    if $\sigma[s, kind] =$ ' **open jump** ' $\vee\ \sigma[s, kind] =$
        ' **open jump relation** ' **then**
      $bring\ truth\ value$
    else $bring\ to\ AC\ with\ prescribed\ type\ (s, s,$ ' **Boolean** ', **false**)
  end ;
$conditional\ order := instruction\ function\ (relational\ operator,$ '$\wedge$') ;
$generate\ label\ (label\ entry\ number)$;
$compile\ (conditional\ order \oplus label\ entry\ number)$;
$relational\ operator := $ ' **empty** ' ;
$accumulator\ loaded := 0$;
$state := \sigma[s-1, information\ state]$;
$specification := \sigma[s-1, information\ 2]$;
$delete\ (2)$;
$push\ down\ (incoming\ symbol)$;
$\sigma[s, information\ label] := label\ entry\ number$;
go to $READ\ NEXT\ SYMBOL$;

$THEN\ EXPRESSION:$

if $state = $ ' **statement** ' **then go to** $THEN\ DESIGNATIONAL$;
if $specification = $ ' **undefined** ' $\wedge\ \sigma[s, kind] = $ ' **label** ' **then**
  begin
    $change\ situation\ to\ specification\ label$;
    go to $READ\ NEXT\ SYMBOL$
  end ;

$THEN\ EXPR:$

$check\ undefined\ specification$;
$check\ compatibility\ (s,$ ' **entity** ', ' **type** ') ;
$type\ of\ AC := \sigma[s, type]$;
if $\sigma[s, kind] = $ ' **relation** ' $\vee\ \sigma[s, kind] = $ ' **open jump** ' $\vee$
  $\sigma[s, kind] = $ ' **open jump relation** ' **then**
    $bring\ truth\ value$
else $bring\ to\ AC\ with\ prescribed\ type\ (s, s, type\ of\ AC,$ **false**) ;
$accumulator\ loaded := 0$;
$generate\ label\ (label\ entry\ number)$;
$compile\ ($ ' **local jump to** ' $\oplus label\ entry\ number)$;

compile label ($\sigma[s-1,$ *information label*$]$) ;

delete (2) ;

push down (*incoming symbol*) ;

$\sigma[s,$ *information label*$] := $ *label entry number* ;

$\sigma[s,$ *information type*$] := $ *type of AC* ;

go to *READ NEXT SYMBOL* ;

THEN DESIGNATIONAL:

    if ($\sigma[s,$ *kind*$] = $ '**variable**' $\lor \sigma[s,$ *kind*$] = $ '**procedure**') $\land$

      specification $= $ '**undefined**' **then**

      begin

        change situation to specification expression ;

        go to *END PROCEDURE CALL*

      end ;

    compile jump ;

    compile label ($\sigma[s-1,$ *information label*$]$) ;

    delete (2) ;

    push down (*incoming symbol*) ;

    go to *READ NEXT SYMBOL* ;

ELSE EXPRESSION:

    if *state* $= $ '**statement**' **then go to** *ELSE DESIGNATIONAL* ;

    if *specification* $= $ '**undefined**' $\land \sigma[s,$ *kind*$] = $ '**label**' **then**

    begin

      change situation to specification label ;

      go to *READ NEXT SYMBOL*

    end ;

ELSE EXPR:

    check undefined specification ;

    check compatibility (s, '**entity**', $\sigma[s-1,$ *information type*$]$) ;

    if $\sigma[s-1,$ *information type*$] = $ '**Boolean**' $\lor$

      $\sigma[s,$ *type*$] = $ '**Boolean**' **then**

      begin

        if $\sigma[s,$ *kind*$] = $ '**relation**' $\lor \sigma[s,$ *kind*$] = $ '**open jump**' $\lor$

          $\sigma[s,$ *kind*$] = $ '**open jump relation**' **then**

        bring truth value

        else *bring to AC with prescribed type* (s, s, '**Boolean**', **false**) ;

        if $\sigma[s-1,$ *information type*$] \neq $ '**Boolean**' **then**

        begin

          operation $:= $ '**compare Boolean to ADDR AC**' ;

          go to *INSERT OPERATION BEHIND THEN*

        end ;

        $\sigma[s,$ *type*$] := $ '**Boolean**'

      end type is **Boolean**

else
  if $\sigma[s-1,\ information\ type] = $ '**real**' **then**
    begin
      bring to AC with prescribed type (s, s, '**real**', **false**) ;
      $\sigma[s,\ type] := $ '**real**'
    end
  else
    if $\sigma[s-1,\ information\ type] = $ '**integer**' **then**
      begin
        if $\sigma[s,\ type] = $ '**integer**' **then**
          bring to AC with prescribed type (s, s, '**integer**', **false**)
        else
          begin
            bring to AC with prescribed type (s, s, '**real**', **false**) ;
            $\sigma[s,\ type] := $ '**real**' ;
            operation := '**transfer to type real**' ;
            go to *INSERT OPERATION BEHIND THEN*
          end
      end
    else
      begin
        if $\sigma[s,\ type] = $ '**integer**' **then**
          begin
            bring to AC with prescribed type (s, s, '**real**', **false**) ;
            $\sigma[s,\ type] := $ '**real**'
          end
        else *bring to AC with prescribed type*
        (s, s, $\sigma[s,\ type]$, **false**) ;
        if $\sigma[s,\ type] = $ '**real**' **then**
          begin
            operation :=
            '**possible transfer to type real depending on
            ADDR AC**' ;
INSERT OPERATION BEHIND THEN:
            generate label (*label entry number*) ;
            compile ('**local jump to**' $\oplus$ *label entry number*) ;
            compile label ($\sigma[s-1,\ information\ label]$) ;
            compile (*operation*) ;
            compile label (*label entry number*) ;
            go to *L ELSE*
          end insert operation
      end ;
  compile label ($\sigma[s-1,\ information\ label]$) ;

L ELSE:

    delete (1) ;   *clear sigma* (*s*) ;

    $\sigma[s, del] :=$ '**entity**' ;

    $\sigma[s, mode] :=$ '**normal**' ;

    $\sigma[s, type] := \sigma[s+1, type]$;

    $\sigma[s, kind] :=$ '**AC**' ;

    accumulator loaded := *s* ;

    go to *PROCESS* ;

ELSE DESIGNATIONAL:

    if $(\sigma[s, kind] =$ '**variable**' $\vee \sigma[s, kind] =$ '**procedure**') $\wedge$

    specification = '**undefined**' **then**

    begin

      change situation to specification expression ;

      go to *END PROCEDURE CALL*

    end ;

    compile jump ;

    delete (1) ;

    clear sigma (*s*) ;

    $\sigma[s, del] :=$ '**entity**' ;

    $\sigma[s, mode] :=$ '**normal**' ;

    $\sigma[s, kind] :=$ '**designational expression**' ;

    go to *PROCESS* ;

THEN STATEMENT:

    if *incoming symbol* = '**else** Σ' **then**

    begin

      generate label (*label entry number*) ;

      compile ('**local jump to**' $\oplus$ *label entry number*) ;

      compile label ($\sigma[s, information label]$) ;

      delete (1) ;

      push down (*incoming symbol*) ;

      $\sigma[s, information label] :=$ *label entry number* ;

      go to *READ NEXT SYMBOL*

    end ;

ELSE STATEMENT:

    compile label ($\sigma[s, information label]$) ;

    delete (1) ;

    go to *PROCESS* ;

INCOMING COLON EQUAL:

    if $\sigma[s-1, symbol] =$ '**switch**' **then**

    go to *INCOMING SWITCH COLON EQUAL* ;

    if $\sigma[s-1, symbol] =$ '**for**' **then**

    go to *INCOMING FOR COLON EQUAL* ;

if $\sigma[s-1, symbol] = $ '**for inserted 1**' **then**
  go to *FOR INSERTED 1* ;
if $\sigma[s-1, symbol] = $ '**for inserted 2**' **then**
  go to *FOR INSERTED 2* ;
if $\sigma[s-1, symbol] = $ '**for n inserted**' **then**
  go to *FOR N INSERTED* ;
if $\sigma[s-1, symbol] = $ '**for suitable**' **then**
  go to *INCOMING FOR COLON EQUAL SUITABLE* ;
if *accumulator loaded* $= s \wedge$
  $\sigma[s, supplement\ information] = $ '**content of**' **then**
  take care of AC (s) ;
check compatibility (s, '**left part variable**', *type of left part*) ;
if $\sigma[s, mode] = $ '**formal**' **then**
  begin
    if $\sigma[s, information\ error] \neq $ '**inserted**' **then**
      $\delta[\sigma[s, internal\ identifier], COL\ \delta\ output] := $ '**output**' ;
    compile ('**name address**' $\oplus \sigma[s, internal\ identifier]$) ;
    generate variable of type ('**integer**') ;
    compile (*end delta* $\oplus$ ':= **ADDR AC**') ;
    $\sigma[s, internal\ identifier] := $ *end delta* ;
    $\sigma[s, supplement\ information] := $ '**content of**'
  end
else
  if $\sigma[s, kind] = $ '**procedure**' **then**
    $\sigma[s, internal\ identifier] :=$
      if $\sigma[s, information\ error] \neq $ '**inserted**' **then**
        $\delta[\sigma[s, internal\ identifier],$
          $COL\ \delta\ associated\ internal\ identifier]$
      else 0 ;
if $\sigma[s, type] = $ '**integer**' $\vee \sigma[s, type] = $ '**real**' $\vee$
  $\sigma[s, type] = $ '**Boolean**' **then**
  begin
    if *undefined left part variable* $\neq 0$ **then**
      begin
        compile ('**compare**' $\oplus \sigma[s, type] \oplus$ '**to**' $\oplus$
          undefined left part variable) ;
        undefined left part variable $:= 0$
      end ;
    type of left part $:= \sigma[s, type]$
  end
else

if *type of left part* $\neq$ '**type**' **then**
   compile ('**compare**' $\oplus$ *type of left part* $\oplus$
   $\sigma[s,$ *internal identifier*])
else
  begin
    if *undefined left part variable* $\neq$ 0 **then**
      compile ('**compare ADDR AC to**' $\oplus$
      undefined left part variable) ;
     undefined left part variable := $\sigma[s,$ *internal identifier*]
  end ;
push down (*incoming symbol*) ;
$\sigma[s,$ *information state*] := *state* ; *state* := '**expression**' ;
go to *READ NEXT SYMBOL* ;

COLON EQUAL:
  check compatibility (s, '**entity**', *type of left part*) ;
  if $\sigma[s-2,$ *kind*] = '**index register**' **then**
   begin
    compile ($\sigma[s-2,$ *internal identifier*] $\oplus$ ':=' $\oplus$
     $\sigma[s,$ *internal identifier*]) ;
    delete (3) ;
    go to *PROCESS*
   end ;
  if $\sigma[s,$ *type*] = '**Boolean**' **then** *type of left part* := '**Boolean**' ;
  if $\sigma[s,$ *kind*] $\neq$ '**open jump**' $\wedge$ $\sigma[s,$ *kind*] $\neq$ '**relation**' $\wedge$
   $\sigma[s,$ *kind*] $\neq$ '**open jump relation**' **then**
   bring to AC with prescribed type (s, s-2, *type of left part*, **false**)
  else
   begin
    bring truth value ;
    if *undefined left part variable* $\neq$ 0 **then**
     compile ('**compare Boolean to**' $\oplus$
      undefined left part variable)
   end ;

COLON EQUAL 1:
  if $\sigma[s-2,$ *supplement information*] = '**content of**' **then**
   compile ('**content of**' $\oplus$ $\sigma[s-2,$ *internal identifier*] $\oplus$':=**AC**')
  else *compile* ($\sigma[s-2,$ *internal identifier*] $\oplus$ ':= **AC**') ;
  if $\sigma[s-2,$ *kind*] = '**number cellar**' **then** *reduce number cellar* ;
  if $\sigma[s-3,$ *symbol*] $\neq$ ':=' **then**
   begin
    accumulator loaded := 0 ;
    delete (3) ;

        type of left part := '**type**' ;
        undefined left part variable := 0 ;
        state := σ[*s, information state*] ;
        go to *PROCESS*
     end colon equal ;

ITERATED COLON EQUAL:
    delete (2) ;
    go to *COLON EQUAL 1* ;

INCOMING SWITCH COLON EQUAL:
    jump over declarations ;
    begin switch list := $s+1$;
    generate label (*label entry number*) ;
    compile label (*label entry number*) ;
    push down ('**switch** :=') ;
    σ[*s, information label*] := *label entry number* ;
    go to *READ NEXT SYMBOL* ;

SWITCH COLON EQUAL:
    if σ[*s, kind*] $\neq$ '**label**' $\vee$ σ[*s, mode*] $\neq$ '**normal**' $\vee$
      static procedure level counter $\neq$ δ[σ[*s, internal identifier*],
        COL δ static procedure level] $\vee$
      array block level counter $\neq$ δ[σ[*s, internal identifier*],
        COL δ array block level] **then**
     begin
      compile jump ;
      σ[*s, internal identifier*] := σ[*s*$-$1, *information label*] ;
      generate label (*label entry number*) ;
      compile label (*label entry number*)
     end
    else *label entry number* := σ[*s*$-$1, *information label*] ;
    σ[*s*$-$1, *internal identifier*] := σ[*s, internal identifier*] ;
    delete (1) ;
    if *incoming symbol* = ',' **then**
     begin
      push down ('**switch** :=') ;
      σ[*s, information label*] := *label entry number* ;
      go to *READ NEXT SYMBOL*
     end ;

END OF THE SWITCH DECLARATION:
    compile label (σ[*begin switch list* $-$ 1, *internal identifier*]) ;
    compile ('**local jump to content of**' $\oplus$
      σ[*begin switch list* $-$ 1, *internal identifier*] $\oplus$ '[**SIR**]') ;

for $i :=$ *begin switch list* **step** 1 **until** s **do**
    compile $(\sigma[i, internal\ identifier])$;
address label jump over ;
delete $(s-\ begin\ switch\ list+3)$;
go to *READ NEXT SYMBOL* ;

GO TO:
    compile jump ;
    delete (2) ;
    go to *PROCESS* ;

INCOMING FOR:
    generate label (*label return*) ;
    compile label (*label return*) ;
    begin of loop var $:= k$;
    state $:=$ '**expression**' ;
    push down (*incoming symbol*) ;
    go to *READ NEXT SYMBOL* ;

INCOMING FOR COLON EQUAL:
    check compatibility (s, '**variable**', '**arithmetic**') ;
    type of loop var $:=$
      if $\sigma[s, type] =$ '**arithmetic**' **then** '**real**' **else** $\sigma[s, type]$;
    if $\sigma[s, kind] =$ '**AC**' $\wedge$
      $\sigma[s, supplement\ information] =$ '**content of**' **then**
      begin
        take care of AC (*accumulator loaded*) ;
        kind of loop var $:=$ '**subscripted**'
      end
    else
      if $\sigma[s, mode] =$ '**formal**' **then**
      begin
        name loop var $:= \sigma[s, internal\ identifier]$;
        kind of loop var $:=$ '**formal**' ;
        if $\sigma[s, information\ error] \neq$ '**inserted**' **then**
          begin
            $\delta[\sigma[s, internal\ identifier], COL\ \delta\ input] :=$ '**input**' ;
            $\delta[\sigma[s, internal\ identifier], COL\ \delta\ output] :=$ '**output**'
          end ;

comment The parts of the program for translating loops are voluminous, but they do not require special comments. The discussion may be summarized by the following tables which contain the target language instructions for certain prototypes of loops. These tables are incomplete: normal loops with more than one list element, any necessary

1) Loops not suitable for recursive address calculation

for $v := A$ step n until C do …	for $v := A$ step B until C do …	for $nv := A$ step B until C do …	for $a[i] := A$ step B until C do …

AC := A
local jump to $L2$
 	
$L1$:
 	
AC := n
AC := **AC** $+_i v$
 	
$L2$: v := **AC**
 	
AC := C
AC := **AC** $-_i v$
 	
$L4$: **if AC** < 0 **then**
　　local jump to $NEXT$
 	
$L5$: ⟨do statement⟩
local jump to $L1$
	AC := A
	local jump to $L2$
	
	$L1$:
	
	AC := B
	AC := **AC** $+_i v$
	
	$L2$: v := **AC**
	
	AC := C
	AC := **AC** $-_i v$
	$GV1$:= **AC**
	
	AC := B
	if AC $= 0$ **then**
	　　local jump to $L5$
	if AC > 0 **then**
	　　local jump to $L3$
	AC := $-_u GV1$
	local jump to $L4$
	$L3$: **AC** := $GV1$
	$L4$: **if AC** < 0 **then**
	　　local jump to $NEXT$
	
	$L5$: ⟨do statement⟩
	local jump to $L1$
		name address nv
		$GV1$:= **ADDR AC**
		AC := A
		local jump to $L2$
		$L1$: **name address** nv
		$GV1$:= **ADDR AC**
		integer name call nv
		$GV2$:= **AC**
		AC := B
		AC := **AC** $+_i GV2$
		$L2$: **content of** $GV1$:= **AC**
		integer name call nv
		$GV1$:= **AC**
		AC := C
		AC := **AC** $-_i GV1$
		$GV1$:= **AC**
		AC := B
		if AC $= 0$ **then**
		　　local jump to $L5$
		if AC > 0 **then**
		　　local jump to $L3$
		AC := $-_u GV1$
		local jump to $L4$
		$L3$: **AC** := $GV1$
		$L4$: **if AC** < 0 **then**
		　　local jump to $NEXT$
		$L5$: ⟨do statement⟩
		local jump to $L1$
			⟨$GV1$:= $address\ (a[i])$⟩
			AC := A
			local jump to $L2$
			$L1$: ⟨$GV1$:= $address\ (a[i])$⟩
			
			AC := B
			AC := **AC** $+_i$ **content of** $GV1$
			$L2$: **content of** $GV1$:= **AC**
			
			AC := C
			AC := **AC** $-_i$ **content of** $GV1$
			$GV1$:= **AC**
			AC := B
			if AC $= 0$ **then**
			　　local jump to $L5$
			if AC > 0 **then**
			　　local jump to $L3$
			AC := $-_u GV1$
			local jump to $L4$
			$L3$: **AC** := $GV1$
			$L4$: **if AC** < 0 **then**
			　　local jump to $NEXT$
			$L5$: ⟨do statement⟩
			local jump to $L1$

2) Loops suitable for recursive address calculation

for $v := A$ **step** B **until** C **do** ...		for $v := A_1$ **step** n_1 **until** n_2, A_2 **do** ...
		local jump to $L1$
	$L2$:	**save return address** $GV1$
		⟨initial value program⟩
		subroutine return $GV1$
	$L3$:	**save return address** $GV1$
		⟨increment program⟩
		$\mathbf{AC} := v$
		$\mathbf{AC} := \mathbf{AC} +_i GVb$
		$v := \mathbf{AC}$
		subroutine return $GV1$
$\mathbf{AC} := A$	$L1$:	$\mathbf{AC} := A_1$
$v := \mathbf{AC}$		$v := \mathbf{AC}$
		$GVa := \mathbf{AC}$
$\mathbf{AC} := B$		$\mathbf{AC} := n_1$
$GVb := \mathbf{AC}$		$GVb := \mathbf{AC}$
⟨initial value program⟩		**subroutine jump** $L2$
local jump to $L2$		**local jump to** $L4$
$L1$: ⟨increment program⟩	$L5$:	**subroutine jump** $L3$
$\mathbf{AC} := v$		
$\mathbf{AC} := \mathbf{AC} +_i GVb$		
$v := \mathbf{AC}$		
$L2$: $\mathbf{AC} := C$	$L4$:	$\mathbf{AC} := n_2$
$\mathbf{AC} := \mathbf{AC} -_i v$		$\mathbf{AC} := \mathbf{AC} -_i v$
$GV1 := \mathbf{AC}$		
$\mathbf{AC} := GVb$		
if $\mathbf{AC} = 0$ **then**		
local jump to $L5$		
if $\mathbf{AC} > 0$ **then**		
local jump to $L3$		
$\mathbf{AC} := -_u GV1$		
local jump to $L4$		
$L3$: $\mathbf{AC} := GV1$		
$L4$: **if** $\mathbf{AC} < 0$ **then**		**if** $\mathbf{AC} < 0$ **then**
local jump to $NEXT$		**local jump to** $NEXT\,1$
$L5$: ⟨do statement⟩		**subroutine jump** L
local jump to $L1$		**local jump to** $L5$
$NEXT$:	$NEXT\,1$:	$\mathbf{AC} := A_2$
		$v := \mathbf{AC}$
		$GVa := \mathbf{AC}$
		subroutine jump $L2$
		subroutine jump L
		local jump to $NEXT\,2$
	L:	**save return address** $GV1$
		⟨do statement⟩
		subroutine return $GV1$
	$NEXT\,2$:	

type transfers, optimizations, simple list elements, and while elements are not considered.

The abbreviations have the following meaning:

$a[i]$:   subscripted variable of type **integer**
v: simple variable of type **integer**
n: unsigned integer number
nv: name variable of type **integer**
A, B, C: expressions of type **integer** ;

INCOMING FOR COLON EQUAL 1:

> *compile* ('**name address**' $\oplus$ *name loop var*) ;
> *generate variable of type* ($\sigma[s, type]$) ;
> *compile* (*delta add* $\oplus$ ':= **ADDR AC**') ;
> $\sigma[s, internal\ identifier] := delta\ add$;
> $\sigma[s, supplement\ information] :=$ '**content of**'
>   **end**
> **else** *kind of loop var* := '**simple**' ;
>
> **if** $\sigma[s, type] =$ ' **arithmetic**' **then**
>   *undefined left part variable* := s ;
> *push down* ('**for** :=') ;
> **go to** *READ NEXT SYMBOL* ;

FOR COLON EQUAL:

> *check compatibility* (s, '**entity**', '**arithmetic**') ;
> *bring to AC with prescribed type* (s, $s-2$, $\sigma[s-2, type]$, **false**) ;
> *accumulator loaded* := 0 ;
> *delete* (2) ;
> **if** *incoming symbol* = '**step**' **then**
>   **begin**
>   *generate label* (*label test exhaustion*) ;
>   *compile* ('**local jump to**' $\oplus$ *label test exhaustion*) ;
>   *generate label* (*label return*) ;
>   *compile label* (*label return*) ;
>   *push down* (*incoming symbol*) ;
>   **if** *kind of loop var* = '**subscripted**' **then**
>   **begin**
>   *reduce number cellar* ;
> $\sigma[s, symbol] :=$ '**for inserted 1**' ;
> $k1 := k$;
> $k := begin\ of\ loop\ var$;
>   **go to** *READ NEXT SYMBOL* ;

FOR INSERTED 1:

        $\sigma[s-1, \textit{symbol}] := \text{‘\textbf{step}’}$;

        $k := k1$;

        take care of AC (accumulator loaded)

      end read subscripted loop variable

    else

      if *kind of loop var* = '**formal**' **then**

        begin

          compile ('**name address**' $\oplus$ *name loop var*) ;

          compile (*end delta* $\oplus$ ':= **ADDR AC**') ;

          compile (*type of loop var* $\oplus$ '**name call**' $\oplus$ *name loop var*);

          push down entity ('**empty**', '**normal**', *type of loop var*,

            '**AC**', '**empty**') ;

          accumulator loaded := *s*

        end

      else

        push down entity ($\sigma[s-1, \textit{internal identifier}]$,

          $\sigma[s-1, \textit{mode}], \sigma[s-1, \textit{type}], \sigma[s-1, \textit{kind}]$,

          $\sigma[s-1, \textit{supplement information}]$) ;

     push down (**if** *type of loop var* = '**real**' **then** '+' **else** '+,') ;

     $\sigma[s, \textit{information error}] := \text{‘\textbf{error announced}’}$;

     negative := *simple list element* := **false** ;

     begin step := *k* ;

     go to *READ NEXT SYMBOL*

    end incoming **step** ;

  if $\sigma[s, \textit{supplement information}]$ = '**content of**' **then**

    compile ('**content of**' $\oplus$ $\sigma[s, \textit{internal identifier}]$ $\oplus$ ':= **AC**')

  else *compile* ($\sigma[s, \textit{internal identifier}]$ $\oplus$ ':= **AC**') ;

  if *incoming symbol* $\neq$ '**while**' **then**

    begin

      simple list element := **true** ;

      go to *PROCESS*

    end ;

  simple list element := **false** ;

  push down (*incoming symbol*) ;

  go to *READ NEXT SYMBOL* ;

STEP:

  check compatibility (*s*, '**entity**', '**arithmetic**') ;

  bring to AC with prescribed type (*s*, *s*−2, $\sigma[s-2, \textit{type}]$, **false**) ;

  compile label (*label test exhaustion*) ;

  if $\sigma[s-2, \textit{supplement information}]$ = '**content of**' **then**

    compile ('**content of**' $\oplus$ $\sigma[s-2, \textit{internal identifier}]$ $\oplus$ ':=**AC**')

else *compile* $(\sigma\,[s-2,\ internal\ identifier]\ \oplus\ `:=\textbf{AC}')$;
accumulator loaded $:=0$;
delete (2) ;
if *kind of loop var* $=$ '**subscripted**' **then**
  begin
    reduce number cellar ;
    push down ('**for inserted 2**') ;
    $k1:=k$;
    $k:=\ begin\ of\ loop\ var$;
    go to *READ NEXT SYMBOL* ;

FOR INSERTED 2:
    $k:=k1$;
    $\sigma\,[s,\ information\ error]:=$ '**inserted**' ;
    check compatibility $(s,\ `\textbf{variable}',\ `\textbf{arithmetic}')$;
    if $\sigma\,[s,\ type]=$ '**arithmetic**' **then**
      begin
        bring to AC with prescribed type $(s,\ s,\ `\textbf{real}',\ \textbf{true})$;
        $\sigma\,[s-2,\ supplement\ information]:=$ '**empty**'
      end ;
    take care of AC (s) ;
    delete (2)
  end subscripted loop variable
  else
    if *kind of loop var* $=$ '**formal**' **then**
      begin
        compile $(type\ of\ loop\ var\ \oplus\ `\textbf{name call}'\ \oplus\ name\ loop\ var)$;
        compile $(end\ delta\ \oplus\ `:=\textbf{AC}')$
      end name loop variable ;
  push down ('**until**') ;
  go to *READ NEXT SYMBOL* ;

UNTIL:
  check compatibility $(s,\ `\textbf{entity}',\ `\textbf{arithmetic}')$;
  generate label $(next\ list\ element)$;
  operation $:=$ **if** *type of loop var* $=$ '**real**' **then** '$-$' **else** '$-_{,}$' ;
  if *kind of loop var* $=$ '**simple**' $\wedge$
    $\sigma\,[s,\ mode]\neq$ '**formal**' $\wedge\ \sigma\,[s,\ type]=type\ of\ loop\ var\ \wedge$
    $(\sigma\,[s,\ kind]=$ '**number**' $\vee\ \sigma\,[s,\ kind]=$ '**variable**') **then**
    begin
      negative $:=\neg$ *negative* ;
      if $\sigma\,[s,\ supplement\ information]=$ '**content of**' **then**
        compile ('$\textbf{AC}:=\textbf{AC}$' $\oplus$ *operation* $\oplus$ '**content of**' $\oplus$
          $\sigma\,[s,\ internal\ identifier])$

else
    compile ('**AC** := **AC**' ⊕ *operation* ⊕
      σ[*s, internal identifier*])
end
else
  begin
    bring to AC with prescribed type (*s, s, type of loop var*, **false**) ;
    if σ[*s*− 2, *supplement information*] = '**content of**' **then**
      compile ('**AC** := **AC**' ⊕ *operation* ⊕ '**content of**' ⊕
        σ[*s*− 2, *internal identifier*])
    else *compile* ('**AC** := **AC**' ⊕ *operation* ⊕
      σ[*s*− 2, *internal identifier*]) ;
    accumulator loaded := 0
  end ;

TEST WHETHER THE STEP IS FIXED:
  if χ[*begin step* + 2] = '**until**' ∧ ¬ *delimiter* (χ[*begin step* + 1]) **then**
    begin
      if δ[*int id, COL* δ *kind*] = '**number**' ∧
        ¬ *value is zero* (χ[*begin step* + 1]) **then**
        go to *FIXED STEP*
    end
  else
    if χ[*begin step* + 3] = '**until**' ∧ ¬ *delimiter* (χ[*begin step* + 2]) **then**
      begin
        if δ[*int id, COL* δ *kind*] = '**number**' ∧
          ¬ *value is zero* (χ[*begin step* + 2]) **then**
          begin
            if χ[*begin step* + 1] = '−' **then**
              negative := ¬ *negative* ;
            go to *FIXED STEP*
          end
      end test whether the step is fixed ;

STEP NOT FIXED:
  if *kind of loop var* = '**simple**' **then**
    generate variable of type (*type of loop var*) ;
  compile (*end delta* ⊕ ':= **AC**') ;
  accumulator loaded := 0 ;
  push down ('**step inserted**') ;
  k1 := *k* ;
  k := *begin step* ;
  go to *READ NEXT SYMBOL* ;

STEP INSERTED:
    $k := k1$; *incoming symbol*$:= \chi[k]$;
    bring to AC with prescribed type (s, s, *type of loop var*, **false**) ;
    delete (4) ;

ENTRY FROM STEP SUITABLE:
    generate label (*label do statement*) ;
    compile ('**if AC** = 0 **then local jump to**' ⊕ *label do statement*) ;
    generate label (*label entry number*) ;
    compile ('**if AC** > 0 **then local jump to**' ⊕ *label entry number*) ;
    compile ('**AC** := −ᵤ' ⊕ *end delta*) ;
    generate label (*new label entry number*) ;
    compile ('**local jump to**' ⊕ *new label entry number*) ;
    compile label (*label entry number*) ;
    compile ('**AC** :=' ⊕ *end delta*) ;
    compile label (*new label entry number*) ;
    conditional order :=
        instruction function (**if** *negative* **then** ' > ' **else** ' < ', 'V') ;
    compile (*conditional order* ⊕ *next list element*) ;
    compile label (*label do statement*) ;
    if *kind of loop var* = '**simple**' V *kind of loop var* = '**suitable**' **then**
      reduce number cellar ;
    accumulator loaded := 0 ;
    go to *PROCESS* ;

FIXED STEP:
    delete (2) ;

ENTRY 2 FROM STEP SUITABLE:
    conditional order := **if** *negative* **then**
      '**if AC** > 0 **then local jump to**'
      else '**if AC** < 0 **then local jump to**' ;
    compile (*conditional order* ⊕ *next list element*) ;
    go to *PROCESS* ;

WHILE:
    check compatibility (s, '**entity**', '**Boolean**') ;
    if $\sigma[s, kind] \neq$ '**relation**' **then**
      begin
        if $\sigma[s, kind] =$ '**open jump**' V $\sigma[s, kind] =$
        '**open jump relation**' **then**
          bring truth value
        else *bring to AC with prescribed type* (s, s, '**Boolean**', **false**)
      end ;

conditional order := *instruction function* (*relational operator*, '∧') ;
generate label (*next list element*) ;
compile (*conditional order* ⊕ *next list element*) ;
accumulator loaded := 0 ;
relational operator := ' **empty** ' ;
delete (2) ;
go to *PROCESS* ;

FOR:

FOR SUITABLE:
    if *kind of loop var* ≠ ' **simple** ' ∧ *kind of loop var* ≠ ' **suitable** ' **then**
    reduce number cellar ;
    if *incoming symbol* = ' **do** ' **then**
      begin
        delete (2) ;
        if *kind of loop var* = ' **suitable** ' **then**
          normal program := **true**
        else *push down* (*incoming symbol*) ;
        σ[s, *information label*] := *next list element* ;
        σ[s, *information 1*] := *label return* ;
        if *simple list element* **then**
          σ[s, *information 2*] := ' **simple list element** ' ;
        state := ' **statement** ' ; *undefined left part variable* := 0 ;
        go to *READ NEXT SYMBOL*
      end incoming **do** ;
    generate label (*do statement*) ;
    σ[s−1, *symbol*] := ' **for n** ' ;
    compile (' **subroutine jump** ' ⊕ *do statement*) ;

FOR COMMA:
    if ¬ *simple list element* **then**
      begin
        compile (' **local jump to** ' ⊕ *label return*) ;
        compile label (*next list element*)
      end ;
    if *kind of loop variable* = ' **suitable** ' **then**
    push down (' **for** := **suitable** ')
    else
      begin
        generate label (*label return*) ; *compile label* (*label return*) ;
        if *kind of loop var* = ' **formal** ' **then**
          go to *INCOMING FOR COLON EQUAL 1* ;
        if *kind of loop var* = ' **subscripted** ' **then**

begin
  delete (1) ;
  $\sigma[s, symbol] := $ '**for n inserted**' ;
  $k1 := k$;
  $k := begin\ of\ loop\ var$;
  go to *READ NEXT SYMBOL* ;

FOR N INSERTED:
  $k := k1$;
  $\sigma[s-1, symbol] := $ '**for n**' ;
  $\sigma[s, information\ error] := $ '**inserted**' ;
  check compatibility (s, '**variable**', '**arithmetic**') ;
  take care of AC (*accumulator loaded*)
  end ;
  push down ('**for** $:=$')
end ;
go to *READ NEXT SYMBOL* ;

FOR N:
  if *kind of loop var* $\neq$ '**simple**'$\land$ *kind of loop var* $\neq$ '**suitable**' **then**
  reduce number cellar ;
  compile ('**subroutine jump**' $\oplus$ *do statement*) ;
  if *incoming symbol* $=$ '**for**,' **then go to** *FOR COMMA* ;
  if *simple list element* **then**
    begin
      generate label (*next list element*) ;
      compile ('**local jump to**' $\oplus$ *next list element*)
    end
  else *compile* ('**local jump to**' $\oplus$ *label return*) ;
  generate variable of type ('**integer**') ;
  compile label (*do statement*) ;
  compile ('**save return address**' $\oplus$ *end delta*) ;
  delete (2) ;
  if *kind of loop var* $=$ '**suitable**' **then**
    normal program $:= $ **true**
  else *push down* ('**do n**') ;
  $\sigma[s, information\ label] := $ *next list element* ;
  $\sigma[s, information\ 1] := $ *end delta* ;
  state $:= $ '**statement**' ; *undefined left part variable* $:= 0$;
  go to *READ NEXT SYMBOL* ;

INCOMING DO SUITABLE:
  push down (*incoming symbol*) ;
  $\sigma[s, information\ 3] := $ *relative address counter* ;
  storage allocation for generated variables ;

kind of loop var := '**suitable**' ;

$k := k+6$;

list elements := $\chi[k-2]$;

if *list elements* >1 **then**

 search for a generated identifier $(\chi[k-1])$

  result: (*initial variable*) ;

if *generated identifier* $(\chi[k])$ **then**

 search for a generated identifier $(\chi[k])$ *result*: (*step variable*)

else *search for an identifier in the identifier list* $(\chi[k])$

 result: (*step variable*) ;

search for an identifier in the identifier list $(\chi\ add[\chi[k-5]+1])$

 result: (*loop variable*) ;

normal program := **false** ;

if *list elements* >1 **then**

 begin

  $\sigma[s, symbol] := $ '**do n suitable**' ;

  $k\ add := \chi[k-4]-1$;

  jump over declarations ;

  generate label (*begin initial value*) ;

  compile label (*begin initial value*) ;

  generate variable of type ('**integer**') ;

  compile ('**save return address**' $\oplus$ *end delta*) ;

  go to *READ NEXT SYMBOL*

 end ;

 $k\ add := \chi[k-5]-1$;

go to *READ NEXT SYMBOL* ;

INCOMING FOR COLON EQUAL SUITABLE:

 push down ('**for** := **suitable**') ;

 state := '**expression**' ;

 go to *READ NEXT SYMBOL* ;

FOR COLON EQUAL SUITABLE:

 check compatibility (s, '**entity**', '**arithmetic**') ;

 bring to AC with prescribed type (s, s, '**integer**', **false**) ;

 accumulator loaded := 0 ;

 delete (2) ;

 compile (*loop variable* $\oplus$ ':= **AC**') ;

 if *list elements* >1 **then**

  compile (*initial variable* $\oplus$ ':= **AC**') ;

 if *incoming symbol* = '**step**' **then**

begin
  simple list element := **false** ;
  push down ('**step suitable**') ;
  go to *READ NEXT SYMBOL*
end ;

SIMPLE LIST ELEMENT OR WHILE ELEMENT:
  compile ('**AC** :=' ⊕ *int id of number 0*) ;
  compile (*step variable* ⊕ ':= **AC**') ;
  compile ('**subroutine jump**' ⊕ *begin initial value*) ;
  if *incoming symbol* = '**while**' **then**
    begin
      generate label (*label return*) ;
      compile label (*label return*) ;
      push down (*incoming symbol*) ;
      simple list element := **false** ;
      go to *READ NEXT SYMBOL*
    end ;
  simple list element := **true** ;
  go to *PROCESS* ;

STEP SUITABLE:
  negative := *fixed step* := **false** ;
  if σ[*s, kind*] = '**number**' **then**
  fixed step := ¬ *value is zero* (δ[σ[*s, internal identifier*],
  COL δ *external identifier*])
  else
    if χ *add*[*k add*− 3] = '**step**' ∧
      ¬ *delimiter* (χ *add*[*k add*− 1]) **then**
      begin
        if δ[*int id, COL* δ *kind*] = '**number**' ∧
          ¬ *value is zero* (χ *add*[*k add*− 1]) **then**
          begin
            fixed step := **true** ;
            if χ *add* [*k add* − 2] = '−' **then** *negative* := **true**
          end
      end test whether fixed step ;
  if *step variable* ≠ σ[*s, internal identifier*] **then**
    begin
      bring to AC with prescribed type (*s, s*, '**integer**', **false**) ;
      accumulator loaded := 0 ;
      compile (*step variable* ⊕ ':= **AC**')
    end ;

delete (2) ;
if *incoming symbol* = ' **until constant** ' **then**
  begin
    push down (*incoming symbol*) ;
    go to *READ NEXT SYMBOL* ;
UNTIL CONSTANT:
    check compatibility (*s*, '**entity**', '**arithmetic**') ;
    if ¬ ($\sigma[s, type]$ = '**integer**' ∧ $\sigma[s, mode]$ = '**normal**' ∧
    ($\sigma[s, kind]$ = '**number**' ∨ $\sigma[s, kind]$ = '**variable**')) **then**
      begin
        bring to AC with prescribed type (*s, s,* '**integer**', **true**) ;
        take care of AC (*s*)
      end ;
    end variable := $\sigma[s, internal\ identifier]$;
    delete (2)
  end ;
if *list elements* > 1 **then**
  compile (' **subroutine jump** ' ⊕ *begin initial value*)
else
  begin
    k1 := *k add* ;
    k add := $\chi[k-4] - 1$;
    go to *READ NEXT SYMBOL* ;
RETURN FROM INITIAL VALUE:
    k add := *k1*
  end compile initial value program ;
generate label (*label test exhaustion*) ;
compile (' **local jump to** ' ⊕ *label test exhaustion*) ;
generate label (*label return*) ;
compile label (*label return*) ;
if *list elements* > 1 **then**
  compile (' **subroutine jump to** ' ⊕ *begin increment*)
else
  begin
    k1 := *k add* ;
    k add := $\chi[k-3] - 1$;
    go to *READ NEXT SYMBOL* ;
RETURN FROM INCREMENT:
    k add := *k1* ; *incoming symbol* := $\chi\ add[k\ add]$
  end compile increment program ;
compile label (*label test exhaustion*) ;
if *incoming symbol* = ' **until** ' **then**

begin
   push down ('**until suitable**') ;
   go to *READ NEXT SYMBOL* ;
UNTIL SUITABLE:
     check compatibility (s, '**entity**', '**arithmetic**') ;
     bring to AC with prescribed type (s, s, '**integer**', **false**) ;
     accumulator loaded := 0 ;
     delete (2) ;
     compile ('**AC** := **AC** $-_i$' $\oplus$ *loop variable*)
   end
  else
   begin
     compile ('**AC** :=' $\oplus$ *end variable*) ;
     compile ('**AC** := **AC** $-_i$' $\oplus$ *loop variable*) ;
   end ;
  generate label (*next list element*) ;
  if ¬ *fixed step* **then**
   begin
     generate variable of type ('**integer**') ;
     compile (*end delta* $\oplus$ ':= **AC**') ;
     compile ('**AC** :=' $\oplus$ *step variable*) ;
     go to *ENTRY FROM STEP SUITABLE*
   end ;
  go to *ENTRY 2 FROM STEP SUITABLE* ;
BEGIN INITIAL VALUE:
  if *incoming symbol* = ';' **then go to** *READ NEXT SYMBOL* ;
  delete (1) ;
  if *list elements* > 1 **then**
   begin
     compile ('**subroutine return**' $\oplus$ *end delta*) ;
     generate label (*begin increment*) ;
     compile label (*begin increment*) ;
     compile ('**save return address**' $\oplus$ *end delta*) ;
     k add := $\chi[k-3] - 1$;
     go to *READ NEXT SYMBOL*
   end ;
  go to *RETURN FROM INITIAL VALUE* ;
BEGIN INCREMENT:
  if *incoming symbol* = ';' **then go to** *READ NEXT SYMBOL* ;
  compile ('**AC** :=' $\oplus$ *loop variable*) ;
  compile ('**AC** := **AC** $+_i$' $\oplus$ *step variable*) ;
  compile (*loop variable* $\oplus$ ':= **AC**') ;

delete (1) ;

if *list elements* >1 **then**

  begin

    compile ('**subroutine return**' $\oplus$ *end delta*) ;

    reduce number cellar ;

    address label jump over ;

    k add := $\chi[k-5]-1$;

    go to *READ NEXT SYMBOL*

  end ;

go to *RETURN FROM INCREMENT* ;

DO SUITABLE:

  relative address counter:= $\sigma[s,\ information\ 3]$;

DO:

  if $\sigma[s,\ information\ 2] \neq$ '**simple list element**' **then**

  begin

    compile ('**local jump to**' $\oplus \sigma[s,\ information\ 1]$) ;

DO 1:

    compile label ($\sigma[s,\ information\ label]$)

  end ;

  delete (1) ;

  go to *READ NEXT SYMBOL* ;

DO N:

  reduce number cellar ;

  compile ('**subroutine return**' $\oplus \sigma[s,\ information\ 1]$) ;

  go to *DO 1* ;

DO N SUITABLE:

  relative address counter := $\sigma[s,\ information\ 3]$;

  compile ('**subroutine return**' $\oplus \sigma[s,\ information\ 1]$) ;

  go to *DO 1* ;

ENTITY ON STATEMENT LEVEL:

  check procedure call ;

  if $\sigma[s,\ mode] =$ '*formal*' **then**

    compile ('**formal procedure call**' $\oplus \sigma[s,\ internal\ identifier]$)

  else *compile* ('**procedure call**' $\oplus \sigma[s,\ internal\ identifier]$) ;

  delete (1) ;

  go to *PROCESS* ;

INCOMING BLOCK BEGIN:

  label jump over := 0 ;

  i := *relative address counter* ;

  storage allocation for blocks (*array block*) ;

  if *array block* **then**

begin
  entry number of free storage cell :=
    entry number of free storage cell $+1$;
  array block level counter := *array block level counter* $+1$;
  push down ('**array block begin**')
end
else *push down* ('**block begin**') ;
$\sigma[s, \text{information } 3] := i$;
go to *READ NEXT SYMBOL* ;

BEGIN:
  if *incoming symbol* = ';' **then**
    go to *READ NEXT SYMBOL* ;
  delete (1) ;
  go to *READ NEXT SYMBOL* ;

BLOCK BEGIN:

ARRAY BLOCK BEGIN:
  if *incoming symbol* = ';' **then**
    go to *READ NEXT SYMBOL* ;
  if $\sigma[s, \text{symbol}]$ = '**array block begin**' **then**
    begin
      entry number of free storage cell :=
        entry number of free storage cell -1 ;
      compile ('**BFS** :=' $\oplus$ *entry number of free storage cell*) ;
      array block level counter := *array block level counter* -1
    end ;
  relative address counter := $\sigma[s, \text{information } 3]$;
  delete (1) ;
  if $s=1$ **then go to** *END OF THE DECOMPOSITION AND
    GENERATION PASS* ;
  bookkeeping block end or procedure end ;
  go to *READ NEXT SYMBOL* ;

PROCEDURE BEGIN:
  static procedure level counter := *static procedure level counter* $+1$;
  jump over declarations ;
  $\sigma[s-1, \text{information label}] := \text{label jump over}$;
  $\sigma[s-1, \text{information } 1] := \text{entry number of free storage cell}$;
  $\sigma[s-1, \text{information } 2] := \text{relative address counter}$;
  $\sigma[s-1, \text{information } 3] := \text{max address}$;
  $\sigma[s-1, \text{information } 4] := \text{array block level counter}$;
  $\delta[\sigma[s, \text{internal identifier}], \text{COL } \delta \text{ program storage address}] := p+1$;
  storage allocation for procedures ;

max address := *relative address counter* ;
array block level counter := 0 ;
push down (‘**procedure body begin**’) ;
go to *READ NEXT SYMBOL* ;

PROCEDURE BODY BEGIN:
    label jump over := σ[*s*— 2, *information label*] ;
    address label jump over ;
    entry number of free storage cell := σ[*s*— 2, *information 1*] ;
    relative address counter := σ[*s*— 2, *information 2*] ;
    δ[σ[*s*—1, *internal identifier*], *COL* δ *length of fixed storage*] :=
      max address ;
    max address := σ[*s*— 2, *information 3*] ;
    array block level counter := σ[*s*— 2, *information 4*] ;
    bookkeeping block end or procedure end ;
    if σ[*s*—1, *type*] ≠ ‘**empty**’ **then**
      begin
        compile (‘**AC** :=’ ⊕ δ[σ[*s*—1, *internal identifier*],
          COL δ *associated internal identifier*]) ;
        compile (‘**ADDR AC** :=’ ⊕ σ[*s*—1, *type*])
      end *function body end* ;
    compile (‘**normal procedure exit**’) ;
    static procedure level counter := *static procedure level counter* —1 ;
    delete (3) ;
    go to *READ NEXT SYMBOL* ;

END OF THE DECOMPOSITION AND GENERATION PASS:
    δ[0, *COL* δ *length of fixed storage*] := *max address* ;
    compile (‘**end of the ALGOL program**’) ;
    for *i* := 2 **step** 1 **until** 10 **do**
      begin
        δ[*i*, *COL* δ *program storage address*] := *p*+1;
        store standard function in π(*i*) ;
        p := *p* + *length of standard function* (*i*)
      end ;
    π *end* := *p* ;
    end identifier list := *delta add*
  end decomposition and generation pass ;

procedure *check procedure calls and substitutions of formal*
  parameters by actuals (Π, Π *beginning*, Π *end*, δ,
  δ *beginning*, δ *end*, *agreeable af*, *compatible* cP,
  incompatible calls, *disagreeable substitutions*, *check agreeability*
  of actual parameter and specification) ;
string array Π, δ ;
integer Π *beginning*, Π *end*, δ *beginning*, δ *end* ;
Boolean array *agreeable af*, *compatible* cP ;
Boolean *incompatible calls*, *disagreeable substitutions* ;
procedure *check agreeability of actual parameter and specification* ;
begin
  integer
    i,
    ϱ, *r*,
    P, *p*,
    AP, FP, $AP\varrho$;

  Boolean
    agreeable,
    new substitution ;
  Boolean array *substituted* $nf[\delta$ *beginning* : δ *end*,
      δ *beginning* : δ *end*],
    occupied cP, *fully occupied* cP, *once correctly occupied* cP
      $[\Pi$ *beginning* : Π *end*, δ *beginning* : δ *end*] ;

  integer procedure *number of actual parameters* (*program address*) ;
    integer *program address* ;
    code ;

  integer procedure *parameter part* (*program address*) ;
    integer *program address* ;
    code ;

  Boolean procedure *instruction is a non formal procedure call*
    (*program address*) ;
    integer *program address* ;
    code ;

  Boolean procedure *instruction is a formal procedure call*
    (*program address*) ;
    integer *program address* ;
    code ;

  procedure *announce* (*string*) ;
    string *string* ;
    code ;

procedure *compatibility check for actual and formal parameters* ;
  begin
    if $\delta[P,\ COL\ \delta\ dimension] = r$ **then**
      begin
        for $\varrho := 1$ **step** 1 **until** r **do**
          begin
            $AP\varrho := parameter\ part\ (i + \varrho)$;
            if $\neg$ *agreeable af* $[AP\varrho,\ P + 1 + \varrho]$ **then**
              go to *END COMPATIBILITY CHECK*
          end ;
        compatible cP $[i,\ P] :=$ **true**
      end ;
END COMPATIBILITY CHECK:
  end *compatibility check for actual and formal parameters* ;

procedure *substitute formal parameters by actuals* ;
  begin
    $r := number\ of\ actual\ parameters\ (i)$;
TEST FULL OCCUPATION:
    if *fully occupied cP* $[i,\ P]$ **then**
      go to *TEST CORRECT OCCUPATION* ;
    for $\varrho := 1$ **step** 1 **until** r **do**
      begin
        $AP\varrho := parameter\ part\ (i + \varrho)$;
        if $\delta[AP\varrho,\ COL\ \delta\ mode] =$ '**formal**' **then**
          begin
            for $p := \delta$ *beginning* **step** 1 **until** δ *end* **do**
              if *substituted nf* $[p,\ AP\varrho]$ **then**
                go to *END LOOP* ϱ ;
              go to *END SUBSTITUTE*
          end ;
END LOOP ϱ:
      end ϱ ;
    fully occupied cP $[i,\ P] :=$ **true** ;
TEST CORRECT OCCUPATION:
    if $\neg$ *compatible cP* $[i,\ P]$ **then**
      go to *END SUBSTITUTE* ;
    if *once correctly occupied cP* $[i,\ P]$ **then**
      go to *SEARCH FOR FURTHER SUBSTITUTIONS* ;
    for $\varrho := 1$ **step** 1 **until** r **do**
      begin
        $AP\varrho := parameter\ part\ (i + \varrho)$;
        if $\delta[AP\varrho,\ COL\ \delta\ mode] =$ '**formal**' **then**

begin
   for $p := \delta$ *beginning* **step** 1 **until** δ *end* **do**
     if *substituted* $nf[p, AP\varrho] \wedge$
       agreeable af $[p, P+1+\varrho]$ **then**
       go to *END LOOP* $\varrho 2$;
     go to *END SUBSTITUTE*
   end ;

END LOOP $\varrho 2$:
    end ϱ ;
    once correctly occupied cP $[i, P] :=$ **true** ;

SEARCH FOR FURTHER SUBSTITUTIONS:
    for $\varrho := 1$ **step** 1 **until** r **do**
      begin
      $AP\varrho :=$ *parameter part* $(i+\varrho)$;
      if $\delta[AP\varrho, COL\ \delta\ mode] \neq$ '**formal**' **then**
        begin
        if $\neg$ *substituted* $nf[AP\varrho, P+1+\varrho]$ **then**
         substituted $nf[AP\varrho, P+1+\varrho] :=$
          new substitution := **true**
        end not formal
      else
        for $p := \delta$ *beginning* **step** 1 **until** δ *end* **do**
        if *substituted* $nf[p, AP\varrho] \wedge$ *agreeable af* $[p, P+1+\varrho]$
         $\wedge \neg$ *substituted* $nf[p, P+1+\varrho]$ **then**
         substituted $nf[p, P+1+\varrho] :=$
          new substitution := **true**
    end ϱ ;

END SUBSTITUTE:
    end *substitute formal parameters by actuals* ;

comment At run time, each procedure call should be checked whether call and declaration are compatible. Beyond this, all substitutions of formal parameters by actuals should be checked whether they agree with each other.

Therefore the run time system is provided with two Boolean arrays

    agreeable af $[\delta$ *beginning*: δ *end*, δ *beginning*: δ *end*$]$

(af = abbreviation for "actual-formal", δ *beginning* = "identifier list beginning", δ *end* = "identifier list end") and

    compatible cP $[\Pi$ *beginning*: Π *end*, δ *beginning*: δ *end*$]$

(cP=abbreviation for "call-procedure", Π *beginning* = "target language program beginning", Π *end* = "target language program end").

agreeable af [AP, FP]

is **true** means that

1) the internal identifier AP is admissible as an actual parameter of a target language procedure call,

2) the internal identifier FP is a formal parameter,

3) AP and FP agree with each other, i.e. the substitution of FP by AP is allowed.

compatible cP [i, P]

is **true** means that

1) i is the program address of a formal target language procedure call

$$i: \langle\text{type}\rangle \textbf{ formal procedure call } FP\varsigma$$
$$AP_1 \varsigma$$
$$\vdots$$
$$AP_{\bar{r}} \varsigma$$

or of a nonformal target language procedure call

$$i: \langle\text{type}\rangle \textbf{ procedure call } \overline{P}\varsigma$$
$$AP_1 \varsigma$$
$$\vdots$$
$$AP_{\bar{r}} \varsigma$$

($\langle$type$\rangle$ may be **real, integer, Boolean, type,** or empty),

2) P is the internal identifier of a declared ALGOL procedure

$$\langle\text{type}\rangle \textbf{ procedure } P\ (FP_1, \ldots, FP_r)\ \varsigma,$$

3) FP and P agree with each other or $\overline{P}$ and P are identical,

4) $\bar{r}$ and r are identical,

5) $FP\varrho$ and $AP\varrho$ ($\varrho = 1, \ldots, r$) agree with each other.

It is clear that many rows and columns of these Boolean arrays are redundant and could be cut out. This consideration is important for implementation on an actual machine ;

for $AP :=\delta$ *beginning* **step** 1 **until** δ *end* **do**
  for $FP := \delta$ *beginning* **step** 1 **until** δ *end* **do**
    substituted $nf[AP, FP] :=$ **false** ;
  for $i := \Pi$ *beginning* **step** 1 **until** Π *end* **do**
    for $P := \delta$ *beginning* **step** 1 **until** δ *end* **do**

begin
    compatible cP [i, P] :=
    occupied cP [i, P] :=
    fully occupied cP [i, P] :=
    once correctly occupied cP [i, P] := **false**
end initializations of Boolean arrays ;
for AP := δ *beginning* **step** 1 **until** δ *end* **do**
  for FP := δ *beginning* **step** 1 **until** δ *end* **do**
  begin
      check agreeability of actual parameter and specification
      (AP, FP) result: (*agreeable*) ;
      agreeable af [AP, FP] := *agreeable*
  end establishing the Boolean array *agreeable af* ;
for i := Π *beginning* **step** 1 **until** Π *end* **do**
  begin
    if *instruction is a non formal procedure call* (i) **then**
      begin
        P := *parameter part* (i) ;
        r := *number of actual parameters* (i) ;
        occupied cP [i, P] := **true** ;
        compatibility check for actual and formal parameters
      end
    else
      if *instruction is a formal procedure call* (i) **then**
        begin
          FP := *parameter part* (i) ;
          r := *number of actual parameters* (i) ;
          for P := δ *beginning* **step** 1 **until** δ *end* **do**
            if *agreeable af* [P, FP] **then**
              compatibility check for actual and formal
                  parameters
        end
  end establishing the Boolean array *compatible cP* ;

comment It is clear that run time checks are undesirable as they may cause a late program failure where much machine time is lost. A good translator should determine whether run time checks are necessary at all.

Therefore the run time system is provided with two Boolean variables

                incompatible calls
and
                disagreeable substitutions.

incompatible calls is **false** means that each procedure call which might be reached by the target language interpreter is compatible with its declaration. *disagreeable substitutions* is **false** means that each substitution is agreeable which might occur at run time.

To the end of assigning reasonable values to these two Boolean variables four Boolean arrays

occupied cP [Π *beginning*: Π *end*, δ *beginning*: δ *end*],
fully occupied cP [Π *beginning*: Π *end*, δ *beginning*: δ *end*],
once correctly occupied cP [Π *beginning*: Π *end*, δ *beginning*: δ *end*],
substituted nf [δ *beginning*: δ *end*, δ *beginning*: δ *end*]
(nf = abbreviation for "normal-formal") are introduced.

$$occupied\ cP\ [i,\ P]$$

is **true** means that

1) i is the program address of a formal

> i: $\langle$type$\rangle$ **formal procedure call** FP ⩎
> $\qquad AP_1$ ⩎
> $\qquad \vdots$
> $\qquad A\overline{P}_{\overline{r}}$ ⩎

or of a nonformal target language procedure call

> i: $\langle$type$\rangle$ **procedure call** $\overline{P}$ ⩎
> $\qquad AP_1$ ⩎
> $\qquad \vdots$
> $\qquad A\overline{P}_{\overline{r}}$ ⩎

2) P is the internal identifier of a declared ALGOL procedure

$$\langle\text{type}\rangle\ \textbf{procedure}\ P\ (FP_1, \ldots, FP_r)\ ⩎$$

3) $\overline{P}$ and P are identical or FP is substituted by P at run time.

$$fully\ occupied\ cP\ [i,\ P]$$

is **true** means additionally that

4) each actual parameter $AP\varrho\,(\varrho=1, \ldots, \overline{r})$ is either nonformal or is substituted by a nonformal actual parameter p at run time.

$$once\ correctly\ occupied\ cP\ [i,\ P]$$

is **true** means that

5) *compatible cP* [i, P] is **true**,

6) each nonformal $AP\varrho$ might be substituted by a nonformal actual parameter p agreeable with $FP\varrho$.

$$substituted\ nf\ [AP, FP]$$

is **true** means that

1) AP is the internal identifier of a nonformal (normal) actual parameter of a target language procedure call,

2) FP is the internal identifier of a formal parameter,

3) FP is substituted by AP at run time.

The values of the array components are established recursively. The target language program must be read repeatedly until no further possible substitution is detected.

If for all calls i and procedures P

$$occupied\ cP\ [i,\ P] \equiv \textbf{true}$$

implies

$$compatible\ cP\ [i,\ P] \equiv \textbf{true}$$

then *incompatible calls* is assigned the value **false**. If there are no incompatible calls and if "for all *fully occupied* $cP\ [i,\ P]$" (i.e. for all i and P with *fully occupied* $cP\ [i,\ P]$ is **true**) only agreeable substitutions can be performed then *disagreeable substitutions* becomes **false**, otherwise **true** ;

ABOVE:

    new substitution := **false** ;
    for $i := \Pi$ *beginning* **step** 1 **until** Π *end* **do**
      begin
        if *instruction is a non formal procedure call* (i) **then**
          begin
            $P :=$ *parameter part* (i) ;
            substitute formal parameters by actuals
          end
        else
          if *instruction is a formal procedure call* (i) **then**
            begin
              $FP :=$ *parameter part* (i) ;
              for $P := \delta$ *beginning* **step** 1 **until** δ *end* **do**
                begin
                  if *occupied* $cP\ [i,\ P]$ **then**
                    go to $SUBSTITUTE$;
                  if $\neg$ *substituted* $nf\ [P, FP]$ **then**
                    go to $END\ LOOP\ P$;
                  occupied $cP\ [i,\ P] :=$ **true** ;
SUBSTITUTE: *substitute formal parameters by actuals* ;
END LOOP P:
                end P
            end
      end i ;

if *new substitution* **then**
  go to *ABOVE* ;
incompatible calls := *disagreeable substitutions* := **false** ;
for $i := \Pi$ *beginning* **step** 1 **until** Π *end* **do**
  for $P := \delta$ *beginning* **step** 1 **until** δ *end* **do**
    if *occupied* $cP\,[i, P]$ **then**
      begin
        if $\neg$ *compatible* $cP\,[i, P]$ **then**
          begin
            announce ('at run time, a formal or nonformal pro-
            cedure call' $\oplus \delta$ [*parameter part* (i), *COL* δ *external*
            identifier] $\oplus$ 'might be reached which is incompa-
            tible with its declaration' $\oplus \delta$ [P, *COL* δ *external*
            identifier]) ;
            incompatible calls := *disagreeable substitutions* := **true**
          end not compatible
        else
          if *fully occupied* $cP\,[i, P]$ **then**
            begin
              $r := \delta[P, COL\ \delta\ dimension]$;
              for $\varrho := 1$ **step** 1 **until** r **do**
                begin
                  $AP\varrho := parameter\ part\ (i)$;
                  if $\delta[AP\varrho, COL\ \delta\ mode] = $ **'formal'** **then**
                    for $p := \delta$ *beginning* **step** 1 **until** δ *end* **do**
                    if *substituted* $nf[p, AP\varrho] \wedge$
                      $\neg$ *agreeable* $af[p, P + 1 + \varrho]$ **then**
                      begin
                        announce ('at run time, a disagreeable
                        parameter substitution' $\oplus$
                        $\delta[P + 1 + \varrho,\ COL\ \delta\ external\ identi$
                        fier] $\oplus$ 'by' $\oplus \delta$ [p, *COL* δ *external*
                        identifier] $\oplus$ 'might occur') ;
                        disagreeable substitutions := **true**
                    end p
            end ϱ
          end compatible
      end determination of incompatible calls and disagreeable
      substitutions
end *check procedure calls and substitutions of formal parameters by*
  actuals ;

procedure *check agreeability of actual parameter and*
                        specification (ap, fp) result : *(agreeable)* ;
   value *ap, fp* ; **integer** *ap, fp* ; **Boolean** *agreeable* ;

   comment The procedure "*check agreeability of actual parameter and specification*" is used within the third pass and within the check pass *check procedure calls and substitutions of formal parameters by actuals.* It checks, depending on the internal identifier of an actual parameter of a procedure call and the internal identifier of the associated formal parameter, whether the procedure call is correct. Suitable replacements are given by the correspondence matrix used by the code procedure *switch matrix for actual and formal parameter correspondence.* An empty field means an error. The letter *C* means "correct", and the other letters have the meaning "correct if certain conditions are fulfilled". The letters then are labels marking the particular parts of the checking program. The call of the code procedure is in effect a global jump to one of these labels, or to *INCORRECT*. The correspondence matrix is given in the appendix. ;

   begin
      procedure *switch matrix for actual and formal parameter*
                        correspondence (mode a, type a, kind a,
                        mode f, type f, kind f) ;
         string *mode a, type a, kind a, mode f, type f, kind f* ;
         code ;
      Boolean procedure *equal (dim 1, dim 2)* ;
         value *dim 1, dim 2* ; **integer** *dim 1, dim 2* ;
         equal := *dim 1* = *dim 2* ∨ *dim 1* = '**empty**' ∨ *dim 2* = '**empty**' ;
      agreeable := **true** ;
      switch matrix for actual and formal parameter
      correspondence (δ[*ap*, *COL* δ *mode*], δ[*ap*, *COL* δ *type*],
         δ[*ap*, *COL* δ *kind*], δ[*fp*, *COL* δ *mode*],
         δ[*fp*, *COL* δ *type*], δ[*fp*, *COL* δ *kind*]) ;

A : **if** *equal* (δ[*ap*, *COL* δ *dimension*], δ[*fp*, *COL* δ *dimension*]) **then**
         go to *CORRECT* ;
      go to *INCORRECT* ;

B : **if** *equal* (δ[*ap*, *COL* δ *dimension*], 1) **then**
         go to *CORRECT* ;
      go to *INCORRECT* ;

D : **if** δ[*fp*, *COL* δ *output*] ≠ '**output**' **then**
         go to *CORRECT* ;
      go to *INCORRECT* ;

E: **if** *equal* ($\delta\,[ap,\ COL\ \delta\ dimension]$, 0) **then**
    go to *CORRECT* ;
  go to *INCORRECT* ;

F: **if** $\delta\,[fp,\ COL\ \delta\ input]=$'**input**' $\supset$
    equal ($\delta\,[ap,\ COL\ \delta\ dimension]$, 0) **then**
    go to *CORRECT* ;
  go to *INCORRECT* ;

G: **if** ($\delta\,[fp,\ COL\ \delta\ input]=$'**input**' $\supset$
    equal ($\delta\,[ap,\ COL\ \delta\ dimension]$, 0)) $\wedge$
    $\delta\,[fp,\ COL\ \delta\ output]\neq$'**output**' **then**
    go to *CORRECT* ;
  go to *INCORRECT* ;

H: **if** *equal* ($\delta\,[ap,\ COL\ \delta\ dimension]$,
    $\delta\,[fp,\ COL\ \delta\ dimension]$) $\wedge$
    $\delta\,[fp,\ COL\ \delta\ output]\neq$'**output**' **then**
    go to *CORRECT* ;
  go to *INCORRECT* ;

I: **if** ($\delta\,[fp,\ COL\ \delta\ input]=$'**input**' $\supset$
    equal ($\delta\,[ap,\ COL\ \delta\ dimension]$, 0)) $\wedge$
    equal ($\delta\,[ap,\ COL\ \delta\ dimension]$, $\delta\,[fp,\ COL\ \delta\ dimension]$)
    then
    go to *CORRECT* ;
  go to *INCORRECT* ;

J: **if** $\delta\,[ap,\ COL\ \delta\ input]=$'**input**' $\supset$
    equal ($\delta\,[fp,\ COL\ \delta\ dimension]$, 0) **then**
    go to *CORRECT* ;
  go to *INCORRECT* ;

K: **if** *equal* ($\delta\,[fp,\ COL\ \delta\ dimension]$, 1) **then**
    go to *CORRECT* ;
  go to *INCORRECT* ;

L: **if** $\delta\,[fp,\ COL\ \delta\ input]\neq$'**input**' $\wedge$
    $\delta\,[fp,\ COL\ \delta\ output]\neq$'**output**' $\wedge$
    equal ($\delta\,[fp,\ COL\ \delta\ dimension]$, 1) **then**
    go to *CORRECT* ;

INCORRECT:
  agreeable := **false** ;
C:

CORRECT:
  end check agreeability of actual parameter and specification ;

procedure *target language program interpreter* (Π, δ, $\mathfrak{M}$, $\mathfrak{Jr}$, *start program,*
  beginning free storage, disagreeable substitutions, incompatible calls,
  agreeable af, compatible c P) ;
string array Π, δ, $\mathfrak{M}$;
integer array $\mathfrak{Jr}$;
integer *start program, beginning free storage* ;
Boolean *disagreeable substitutions, incompatible calls* ;
Boolean array *agreeable af, compatible c P* ;

  comment The procedure *target language program interpreter* simulates
the actions of the control unit of a real computer:

  1. The integer variable *instruction counter* points to a certain memory
cell $\mathfrak{M}$ [*instruction counter*] the content of which can be interpreted to be
an instruction I.

  2. The interpretation is performed with the help of the switch
instruction cascade. To each instruction I corresponds a label L in the
switch list in a 1—1 manner.

  3. The piece of program labelled by L executes those actions intended
by the instruction I.

  4. At the end of this execution the variable *instruction counter* is
assigned a new value pointing to the subsequent instruction to be
executed. Normally *instruction counter* is raised by 1.

  The body of the procedure *target language program interpreter* does
not present code for all instructions. This is true especially for simple
instructions like arithmetic and logical operations and local conditional
or unconditional jumps. See section 7.6, p. 147 ;

  begin
    integer
      instruction counter,
      MDL, DL, DL1,
      A1, A1 red init, RAD A1 red init,
      Π A1, φ A1,
      l, n, m, i, j, n1,
      RAD Ki, val Ki,
      bi, RAD bi, s bi, val bi,
      ai, RAD ai, s ai, val ai,
      P, s P, start P, LP, FIX P, r,
      FP, RAD FP, s FP,
      L, s PL, start L, l L,
      FL, RAD FL, s FL,
      GP, RAD GP,
      AP ϱ, RAD AP ϱ, s AP ϱ, AP ϱ red init, ϱ,

A, ADDR,
FA red init, RAD FA red init,
old init, new init,
VA, RAD VA, VA red init, RAD VA red init,
BFS, *old BFS* ;

string
  mode, type, kind,
  call type,
  old type, new type,
  AC, ADDR AC ;

switch *instruction cascade* :=
  AC BECOMES AC ω V,
  AC BECOMES AC ω CONTENT OF V,
  IRϰ BECOMES IRϰ PLUS INTEGER V,
  AC BECOMES ω V,
  AC BECOMES ω CONTENT OF V,
  AC BECOMES ω AC,
  AC BECOMES ω CONTENT OF ADDR AC,
  V BECOMES AC,
  CONTENT OF V BECOMES AC,
  AC BECOMES V,
  AC BECOMES CONTENT OF V,
  ADDR AC BECOMES AC,
  GV BECOMES ADDR AC,
  AC BECOMES CONTENT OF ADDR AC,
  IRϰ BECOMES V,
  JUMP TO L,
  SIR BECOMES AC,
  SIR BECOMES V,
  SIR BECOMES CONTENT OF V,
  LOCAL JUMP TO CONTENT OF S OF SIR,
  FORMAL PROCEDURE EXIT FL,
  LOCAL JUMP TO L,
  IF AC ϱ 0 THEN LOCAL JUMP TO L,
  IF AC THEN LOCAL JUMP TO L,
  IF NOT AC THEN LOCAL JUMP TO L,
  SUBROUTINE JUMP L,
  SAVE RETURN ADDRESS V,
  SUBROUTINE RETURN V,
  TRANSFER TO TYPE REAL,
  TRANSFER TO TYPE INTEGER,
  POSSIBLE TRANSFER TO REAL DEPENDING ON GV,

POSSIBLE TRANSFER TO INTEGER DEPENDING ON GV,
POSSIBLE TRANSFER TO REAL DEPENDING ON ADDR
  AC,
POSSIBLE TRANSFER TO INTEGER DEPENDING ON
  ADDR AC,
POSSIBLE TRANSFER FROM REAL DEPENDING ON GV,
POSSIBLE TRANSFER FROM INTEGER DEPENDING ON
  GV,
POSSIBLE TYPE TRANSFER FROM ADDR AC TO GV,
COMPARE REAL TO GV,
COMPARE INTEGER TO GV,
COMPARE BOOLEAN TO GV,
COMPARE BOOLEAN TO ADDR AC,
COMPARE ADDR AC TO GV,
STORAGE ALLOCATION A,
BFS BECOMES V,
PROCEDURE CALL P,
FORMAL PROCEDURE CALL FP,
REAL PROCEDURE CALL P,
INTEGER PROCEDURE CALL P,
BOOLEAN PROCEDURE CALL P,
REAL FORMAL PROCEDURE CALL FP,
INTEGER FORMAL PROCEDURE CALL FP,
BOOLEAN FORMAL PROCEDURE CALL FP,
TYPE FORMAL PROCEDURE CALL FP,
REAL NAME CALL N,
INTEGER NAME CALL N,
BOOLEAN NAME CALL N,
NAME CALL N,
NAME ADDRESS N,
VALUE ARRAY A,
NORMAL PROCEDURE EXIT,
NAME PROCEDURE EXIT,
ABSOLUTE ADDRESS V,
ABSOLUTE SUBSCRIPTED VARIABLE ADDRESS V,
END OF THE ALGOL PROGRAM ;

integer procedure *number associated with the target language instruction (program address)* ;
  integer *program address* ;
  code ;

comment The switch list elements of *instruction cascade* correspond in an obvious manner to the list of target language instructions in chapter 3. Here we have explicitly coded only those instructions which deal with storage allocation and which in general have no counterpart in an actual machine.

> *AC BECOMES AC ω V,*
> *AC BECOMES AC ω CONTENT OF V,*
> *IRϰ BECOMES IRϰ PLUS INTEGER V,*
> *AC BECOMES ω V,*
> *AC BECOMES ω CONTENT OF V,*
> *AC BECOMES ω AC,*
> *AC BECOMES ω CONTENT OF ADDR AC,*
> *IRϰ BECOMES V,*
> *IF AC ϱ 0 THEN LOCAL JUMP TO*

are not proper switch list elements but mean schemes for several labels:

> *AC BECOMES AC PLUS REAL V,*
> *AC BECOMES AC PLUS INTEGER V,*
> .
> .
> .
> *IR1 BECOMES IR1 PLUS INTEGER V,*
> *IR2 BECOMES IR2 PLUS INTEGER V,*
> .
> .
> .
> *AC BECOMES MINUS REAL V,*
> *AC BECOMES MINUS INTEGER V,*
> .
> .
> .
> *IF AC EQUAL 0 THEN LOCAL JUMP TO L,*
> *IF AC GREATER 0 THEN LOCAL JUMP TO L etc.*

The instruction **transfer to address** does not occur in this list. The model translator does not generate this instruction as entities of type **integer** and **address** are identified.

The integer procedure *number associated with the target language instruction* associates a certain integer number with each target language instruction. This number corresponds in an obvious manner to the switch list elements of *instruction cascade* ;

> **integer procedure** *parameter part (program address)* ;
> **integer** *program address* ;
> **code** ;

> **integer procedure** *identifier part (memory address)* ;
> **integer** *memory address* ;
> **code** ;

integer procedure *dynamic level part* (*memory address*) ;
  integer *memory address* ;
  code ;

integer procedure *dynamic level 1 part* (*memory address*) ;
  integer *memory address* ;
  code ;

string procedure *RI part* (*memory address*) ;
  integer *memory address* ;
  code ;

integer procedure *address part* (*memory address*) ;
  integer *memory address* ;
  code ;

integer procedure *subscript number part* (*memory address*) ;
  integer *memory address* ;
  code ;

procedure *announce* (*string*) ;
  string *string* ;
  code ;

procedure *type transfer depending on old and new type* ;
  code ;

procedure *reload index registers* ;
  begin
    integer I, B ;
    $B := DL$;
    $\mathfrak{Jr}[\mathfrak{M}[B+2]] := B$;
    for $I := \mathfrak{M}[B+2]-1$ **step** -1 **until** 1 **do**
      $B := \mathfrak{Jr}[I] := \mathfrak{M}[B+3]$
  end *reload index registers* ;

comment Compare section 6.3, action 7. As the model translator does not distinguish between variables of type **integer** and **address** B is declared **integer** here instead of **address** ;

  procedure *fill the parameter block and copy information vectors
  for formal arrays* ;
  begin
    for $\varrho := 1$ **step** 1 **until** r **do**

begin

$AP\varrho := parameter\ part\ (instruction\ counter\ +\varrho)$;

$mode := \delta[AP\varrho, COL\ \delta\ mode]$;

$type := \delta[AP\varrho, COL\ \delta\ type]$;

$kind := \delta[AP\varrho, COL\ \delta\ kind]$;

$RAD\ AP\varrho := \delta[AP\varrho, COL\ \delta\ static\ procedure\ level]$;

$s\ AP\varrho := \delta[AP\varrho, COL\ \delta\ static\ procedure\ level]$;

if $mode = $ '**formal**' $\vee$ $mode = $ '**value**' **then**

  begin

CASE 1c: $\mathfrak{M}[\textbf{BFS} + 5 + LP + \varrho] :=$

        $\mathfrak{M}[RAD\ AP\varrho + \mathfrak{Jr}\ [s\ AP\varrho]]$;

    if *disagreeable substitutions* **then**

      check actual parameter

  end

else

  begin

    if $kind = $ '**array**' **then**

CASE 1a: $\mathfrak{M}[\textbf{BFS} + 5 + LP + \varrho] :=$

        $AP\varrho \oplus$ ' $\oplus$ ' $\oplus \mathfrak{Jr}[s\ AP\varrho]$

    else

      begin

CASE 1b: $AP\varrho\ red\ init :=$

        $\delta[AP\varrho, COL\ \delta\ associated\ internal\ identifier]$;

        $ADDR := \delta[AP\varrho\ red\ init, COL\ \delta\ relative\ address]$

        $+\mathfrak{Jr}[s\ AP\varrho]$;

        $\mathfrak{M}[\textbf{BFS} + 5 + LP + \varrho] := AP\varrho \oplus$ ' $\oplus$ ' $\oplus ADDR$

      end ;

    if $mode = $ '**normal**' $\wedge$ $type \neq$ '**empty**' $\wedge$ $kind = $

    '**procedure**' $\wedge$ $\mathfrak{M}[MDL + 2] \geqq s\ AP\varrho + 1$ **then**

    $\mathfrak{M}[\textbf{BFS} + 5 + LP + \varrho] :=$

      $\mathfrak{M}[\textbf{BFS} + 5 + LP + \varrho] \oplus$ ' $\oplus$ ' $\oplus \mathfrak{Jr}[s\ AP\varrho + 1]$

  end ;

if $type = $ '**integer**'

$\wedge$ $(kind = $ '**variable**' $\vee$ $kind = $ '**procedure**')

$\wedge$ $RI\ part\ (\mathfrak{M}[\textbf{BFS} + 5 + LP + \varrho]) = $ '**empty**' **then**

$\mathfrak{M}[\textbf{BFS} + 5 + LP + \varrho] :=$

  $\mathfrak{M}[\textbf{BFS} + 5 + LP + \varrho] \oplus$ '**RI**' ;

$A := identifier\ part\ (\mathfrak{M}[\textbf{BFS} + 5 + LP + \varrho])$;

$kind := \delta[A, COL\ \delta\ kind]$;

$n := \delta[P + 1 + \varrho, COL\ \delta\ dimension]$;

if $kind = $ '**array**' $\wedge$ $n \neq$ '**empty**' **then**

begin

    compare the subscript number n of the formal array FA
      with the subscript number of the actual array A ;
    old init $:=$ *address part* $(\mathfrak{M}\,[\mathbf{BFS}+5+LP+\varrho])$;
    FA red init $:=$
      $\delta\,[P+1+\varrho,\ COL\ \delta\ associated\ internal\ identifier]$;
    RAD FA red init $:=$
      $\delta\,[FA\ red\ init,\ COL\ \delta\ relative\ address]$;
    new init $:= \mathbf{BFS} + RAD\ FA\ red\ init$;
    for $i := 1$ **step** 1 **until** $n+2$ **do**
      $\mathfrak{M}\,[new\ init -1 +i] := \mathfrak{M}\,[old\ init -1 +i]$

    end

  end

end *fill the parameter block and copy information vectors*
  for formal arrays ;

comment

    parameter part (*instruction counter* $+\varrho$)

is the internal identifier of an actual parameter $AP\varrho$ which may be
formal or not. $AP\varrho$ corresponds to a formal parameter FP characterized
by the internal identifier $P+1+\varrho$.

If $AP\varrho$ is formal then the proper information to be stored in the new
parameter block cell $\mathfrak{M}\,[\mathbf{BFS}+5+LP+\varrho]$ is contained in the memory
cell $\mathfrak{M}\,[RAD\ AP\varrho +\mathfrak{Sr}[s\ AP\varrho]]$. If $AP\varrho$ is nonformal then the proper
information to be stored in the new parameter block is constructed by
concatenating $AP\varrho$ with some additional information.

    identifier part $(\mathfrak{M}\,[\mathbf{BFS}+5+LP+\varrho])$

redelivers this nonformal actual parameter $AP\varrho$.

The symbol $\oplus$ has two different meanings: Outside strings, $\oplus$ serves
as an operator concatenating strings. For instance, the result of '123' $\oplus$
'45' is '12345'. Inside strings, $\oplus$ is simply an element of the string alphabet. We use this element in order to keep concatenated integers
separated. For example, the result of '123' $\oplus$ '$\oplus$' $\oplus$ '45', namely
'123 $\oplus$ 45', allows the isolation of the first operand '123'.

    address part $(\mathfrak{M}\,[\mathbf{BFS}+5+LP+\varrho])$

is the absolute initial address $ADDR$ of the information vector for the
actual nonformal array A.

    $\mathfrak{M}\,[MDL+2] <s\ AP\varrho +1$

here implies

    $\mathfrak{M}\,[MDL+2] =s\ AP\varrho$

and means that the actual type procedure identifier $AP\varrho$ occurs outside the body of $AP\varrho$. In this case the value of $\Re[sAP\varrho+1]$ $(=DL1)$ makes no sense. The reason is that left part occurrence of a function identifier P is only allowed inside the body of P ;

    procedure *check actual parameter* ;
      begin
        if *mode* $=$ '**formal**' **then**
          begin
            $A := $ *identifier part* $(\mathfrak{M}[\textbf{BFS}+5+LP+\varrho])$;
            if $\neg$ *agreeable af* $[A, P+1+\varrho]$ **then**
              begin
                announce ('at run time, a disagreeable parameter sub-
                  stitution' $\oplus \delta[P+1+\varrho, COL\ \delta\ external\ identifier]$
                  $\oplus$ 'by' $\oplus \delta[A, COL\ \delta\ external\ identifier] \oplus$
                  'occurs') ;
                go to *STOP RUNNING*
              end
          end
      end *check actual parameter* ;

    comment *check actual parameter* is unnecessary in the case of a non-formal or value listed $AP\varrho$. The corresponding check has been performed in connection with *check procedure call* since

        compatible cP [*instruction counter, P*] $\equiv$ **true**

implies

        agreeable af $[A, P+1+\varrho] \equiv$ **true**

(A and $AP\varrho$ are identical) ;

    procedure *check procedure call* ;
      begin
        if $\neg$ *compatible cP* [*instruction counter, P*] **then**
          begin
            announce ('at run time, a formal or nonformal procedure
              call' $\oplus \delta$ [*parameter part* (*instruction counter*), $COL\ \delta$
              external identifier] $\oplus$ 'has been reached which is in-
              compatible with its declaration' $\oplus \delta[P, COL\ \delta\ external$
              identifier]) ;
            go to *STOP RUNNING*
          end
      end *check procedure call* ;

procedure *compare the subscript number n of the formal array FA with the subscript number of the actual array A* ;
begin
  $n1 :=$ *subscript number part* $(\mathfrak{M}[\mathbf{BFS} + 5 + LP + \varrho])$;
  if $n1 \neq$ '**empty**' $\wedge\ n1 \neq n$ **then**
    begin
      announce ('at run time, the subscript number of a formal
        array does not agree with the subscript number of its
        corresponding actual array') ;
      go to *STOP RUNNING*
    end
end *compare the subscript number n of the formal array FA with
the subscript number of the actual array A* ;

comment At run time, a comparison is necessary only if the subscript number part *n1* is not empty. In this case the actual array A is a value array whose subscript number has not been detected at translation time.

If *n1* is empty we may say the following: A is a declared or value array with a subscript number *n1* known during translation time. In this case *agreeable af* $[A, P + 1 + \varrho]$ is **true** (which fact has been established either by *check actual parameter* if $AP\varrho$ is formal or by *check procedure call* if $AP\varrho$ is nonformal) means that n and *n1* are equal ;

procedure *check dynamic level 1 part* ;
  begin
    if $DL1 =$ '**empty**' **then**
      announce ('**name address** N cannot be executed since the
        actual parameter is a declared type procedure identifier
        P occurring outside the body of P') ;
      go to *STOP RUNNING*
    end *check dynamic level 1 part* ;

START PROGRAM RUNNING:
  $\mathfrak{M}[beginning\ free\ storage + 2] := 0$;
  $MDL := \mathfrak{Ir}[0] :=$ *beginning free storage* ;
  $\mathfrak{M}[beginning\ free\ storage + 4] :=$ '**empty**' ;
  $FIX\ P := \delta[0,\ COL\ \delta\ length\ of\ fixed\ storage]$;

comment The 0-th row of the identifier list contains the necessary information on the *main-program* procedure ;

  $\mathbf{BFS} := \mathfrak{M}[beginning\ free\ storage + 5] :=$
  beginning free storage $+ FIX\ P$;
  instruction counter := *start program* ;
  go to *NEXT INSTRUCTION* ;

RETURN FROM SIMPLE INSTRUCTIONS:
   instruction counter $:=$ *instruction counter* $+1$;
NEXT INSTRUCTION:
   go to *instruction cascade* [*number associated*
     with the target language instruction (*instruction counter*)] ;
STORAGE ALLOCATION A:
   $A1 :=$ *parameter part* (*instruction counter*) ;
   $type := \delta[A1, COL\ \delta\ type]$;
   $l := \delta[A1, COL\ \delta\ array\ block\ level]$;
   $n := \delta[A1, COL\ \delta\ dimension]$;
   $m := \delta[A1, COL\ \delta\ number\ of\ array\ identifiers]$;
   $A1\ red\ init := \delta[A1, COL\ \delta\ associated\ internal\ identifier]$;
   $RAD\ A1\ red\ init := \delta[A1\ red\ init, COL\ \delta\ relative\ address]$;
   $\Pi\ A1 := 1$;
   $\varphi\ A1 := 0$;
   for $i := 1$ **step** 1 **until** n **do**
     begin
       $RAD\ Ki := RAD\ A1\ red\ init - 1 + i$;
       $bi :=$ *parameter part* (*instruction counter* $+ 2 \times i$) ;
       $RAD\ bi := \delta[bi, COL\ \delta\ relative\ address]$;
       $s\ bi := \delta[bi, COL\ \delta\ static\ procedure\ level]$;
       $val\ bi := \mathfrak{M}[RAD\ bi + \mathfrak{Ir}[s\ bi]]$;
       $ai :=$ *parameter part* (*instruction counter* $+ 2 \times i - 1$) ;
       $RAD\ ai := \delta[ai, COL\ \delta\ relative\ address]$;
       $s\ ai := \delta[ai, COL\ \delta\ static\ procedure\ level]$;
       $val\ ai := \mathfrak{M}[RAD\ ai + \mathfrak{Ir}[s\ ai]]$;
       $val\ Ki := \mathfrak{M}[RAD\ Ki + MDL] := val\ bi - val\ ai + 1$;
       $\Pi\ A1 := \Pi\ A1 \times val\ Ki$;
       $\varphi\ A1 := \varphi\ A1 \times val\ Ki + val\ ai$
     end i ;
   $\mathfrak{M}[RAD\ A1\ red\ init + n + MDL] := \Pi\ A1$;
   $\mathfrak{M}[RAD\ A1\ red\ init + n + 1 + MDL] := \varphi\ A1$;
   for $j := 1$ **step** 1 **until** m **do**
     begin
       $\mathfrak{M}[RAD\ A1\ red\ init + (j-1) \times (n+2) + MDL] :=$
         $type \oplus (\mathbf{BFS} - \varphi\ A1 + (j-1) \times \Pi\ A1)$
     end j ;
   for $j := 2$ **step** 1 **until** m **do**
     for $i := 2$ **step** 1 **until** $n+2$ **do**
       begin
         $\mathfrak{M}[RAD\ A1\ red\ init + (j-1) \times (n+2) + i - 1 + MDL] :=$
           $\mathfrak{M}[RAD\ A1\ red\ init + i - 1 + MDL]$
       end $i\ j$;

BFS $:= \mathfrak{M}[4 + l + MDL] :=$ **BFS** $+ m \times \Pi A1$;
instruction counter $:=$ *instruction counter* $+ 2 \times n + 1$;
go to *NEXT INSTRUCTION* ;

comment

   parameter part (*instruction counter*)

is the internal identifier of a declared array $A1$.

   parameter part (*instruction counter* $+ 2 \times i$)

is the internal identifier associated with the i-th upper bound bi.

   parameter part (*instruction counter* $+ 2 \times i - 1$)

is the internal identifier associated with the i-th lower bound ai ;

REAL PROCEDURE CALL P:
   call type $:=$ '**real**' ;
   go to *PROCEDURE CALL 1* ;

INTEGER PROCEDURE CALL P:
   call type $:=$ '**integer**' ;
   go to *PROCEDURE CALL 1* ;

BOOLEAN PROCEDURE CALL P:
   call type $:=$ '**Boolean**' ;
   go to *PROCEDURE CALL 1* ;

PROCEDURE CALL P:
   call type $:=$ '**empty**' ;

PROCEDURE CALL 1:
   $P :=$ *parameter part* (*instruction counter*) ;
   if *incompatible calls* **then**
      check procedure calls;
   $sP := \delta[P, COL \; \delta \; static \; procedure \; level]$;
   $start \; P := \delta[P, COL \; \delta \; program \; address]$;
   $r := \delta[P, COL \; \delta \; dimension]$;
   $FIX \; P := \delta[P, COL \; \delta \; length \; of \; fixed \; storage]$;
   $LP := \delta[P, COL \; \delta \; maximum \; array \; block \; level]$;
   fill the parameter block and copy information vectors for formal arrays ;
   $\mathfrak{M}[\mathbf{BFS}] :=$ *instruction counter* $+ r + 1$;
   $\mathfrak{M}[\mathbf{BFS}+1] := MDL$;
   $\mathfrak{M}[\mathbf{BFS}+2] := sP + 1$;
   $\mathfrak{M}[\mathbf{BFS}+3] := \mathfrak{Ir}[sP]$;
   $\mathfrak{M}[\mathbf{BFS}+4] := call \; type$;
   $MDL := \mathfrak{Ir}[sP+1] := \mathbf{BFS}$;

BFS := $\mathfrak{M}$[**BFS** + 5] := **BFS** + *FIX P* ;
instruction counter := *start P* ;
go to *NEXT INSTRUCTION* ;

comment
                    parameter part (instruction counter)

is the internal identifier of a declared procedure *P* ;

REAL FORMAL PROCEDURE CALL FP:
    call type := '**real**' ;
    go to *FORMAL PROCEDURE CALL 1* ;

INTEGER FORMAL PROCEDURE CALL FP:
    call type := '**integer**' ;
    go to *FORMAL PROCEDURE CALL 1* ;

BOOLEAN FORMAL PROCEDURE CALL FP:
    call type := '**Boolean**' ;
    go to *FORMAL PROCEDURE CALL 1* ;

TYPE FORMAL PROCEDURE CALL FP:
    call type := '**type**' ;
    go to *FORMAL PROCEDURE CALL 1* ;

FORMAL PROCEDURE CALL FP:
    call type := '**empty**' ;

FORMAL PROCEDURE CALL 1:
    FP := *address part (instruction counter)* ;
    RAD FP := δ[*FP*, *COL* δ *relative address*] ;
    s FP := δ[*FP*, *COL* δ *static procedure level*] ;
    P := *identifier part* ($\mathfrak{M}$[*RAD FP* + $\mathfrak{Ir}$[*s FP*]]) ;
    if *incompatible calls* **then**
      check procedure call ;
    DL := *dynamic level part* ($\mathfrak{M}$[*RAD FP* + $\mathfrak{Ir}$[*s FP*]]) ;
    if *call type* = '**real**' $\vee$ *call type* = '**type**' **then**
      call type := *call type* $\oplus$ *RI part* ($\mathfrak{M}$[*RAD FP* + $\mathfrak{Ir}$[*s FP*]]) ;

FORMAL PROCEDURE CALL 2:
    s P := δ[*P*, *COL* δ *static procedure level*] ;
    start P := δ[*P*, *COL* δ *program address*] ;
    r := δ[*P*, *COL* δ *dimension*] ;
    FIX P := δ[*P*, *COL* δ *length of fixed storage*] ;
    LP := δ[*P*, *COL* δ *maximum array block level*] ;
    fill the parameter block and copy information vectors for formal arrays ;
    $\mathfrak{M}$[**BFS**] := *instruction counter* + *r* + 1 ;
    $\mathfrak{M}$[**BFS** + 1] := *MDL* ;

$\mathfrak{M}\,[\textbf{BFS}+2]:=s\,P+1$;
$\mathfrak{M}\,[\textbf{BFS}+3]:=DL$;
reload index registers ;
$\mathfrak{M}\,[\textbf{BFS}+4]:=$ *call type* ;
$MDL:=\mathfrak{Ir}\,[s\,P+1]:=\textbf{BFS}$;
$\textbf{BFS}:=\mathfrak{M}\,[\textbf{BFS}+5]:=\textbf{BFS}+FIX\,P$;
instruction counter $:=$ *start P* ;
go to *NEXT INSTRUCTION* ;

comment

parameter part (instruction counter)

is the internal identifier of a formal procedure FP.

We recall that the memory cell

$$\mathfrak{M}\,[RAD\,FP+\mathfrak{Ir}\,[s\,FP]]$$

contains the information on the actual parameter corresponding to FP. This information may be one of the chains

$P\oplus DL$, or $P\oplus DL\oplus$ '**RI**', or $P\oplus DL\oplus DL1$, or $P\oplus DL\oplus DL1\oplus$ '**RI**'.

identifier part $(\mathfrak{M}\,[RAD\,PF+\mathfrak{Ir}\,[s\,FP]])$

is the internal identifier of an actual declared procedure P.

dynamic level part $(\mathfrak{M}\,[RAD\,FP+\mathfrak{Ir}\,[s\,FP]])$

is the dynamic level DL of the static predecessor of the actual procedure P mentioned above.

RI part $(\mathfrak{M}\,[RAD\,FP+\mathfrak{Ir}\,[s\,FP]])$

is either '**empty**' or '**RI**'. '**RI**' indicates that a real value of the function P is to be rounded off in case of a **real** or **type formal procedure call** ;

NORMAL PROCEDURE EXIT:
$DL:=\mathfrak{M}\,[MDL+1]$;
reload index registers ;
$\textbf{BFS}:=MDL$;
$MDL:=\mathfrak{M}\,[MDL+1]$;
type check at normal procedure exit ;
instruction counter $:=\mathfrak{M}\,[\textbf{BFS}]$;
go to *NEXT INSTRUCTION* ;

comment

type check at normal procedure exit

compares the type markers contained in $\mathfrak{M}[\textbf{BFS}+4]$ and **ADDR AC**, eventually executes a type transfer or round off operation of the procedure value in **AC**, and assigns the new type to **ADDR AC**. In case of an alarm, target language program execution (interpretation) is stopped.

$\mathfrak{M}[\textbf{BFS}+4]$	ADDR AC		
	'real'	**'integer'**	**'Boolean'**
'empty'	ready	ready	ready
'real'	ready	type transfer from integer to real	alarm
'integer'	type transfer from real to integer	ready	alarm
'Boolean'	alarm	alarm	ready
'type'	ready	ready	ready
'real' $\oplus$ **'RI'**	round off operation	type transfer from integer to real	alarm
'type' $\oplus$ **'RI'**	round off operation	ready	alarm

;
JUMP TO L:
    $L := $ *parameter part (instruction counter)* ;
    $s\,PL := \delta[L,\ COL\ \delta$ *static procedure level*$]$;
    $start\,L := \delta[L,\ COL\ \delta$ *program address*$]$;
    $l\,L := \delta[L,\ COL\ \delta$ *array block level*$]$;
    $MDL := \mathfrak{Jr}[s\,PL]$;
    $\textbf{BFS} := \mathfrak{M}[MDL+5+l\,L]$;
    instruction counter $:= start\,L$;
    go to *NEXT INSTRUCTION* ;

FORMAL PROCEDURE EXIT FL:
    $FL := $ *parameter part (instruction counter)* ;
    $RAD\,FL := \delta[FL,\ COL\ \delta$ *relative address*$]$;
    $s\,FL := \delta[FL,\ COL\ \delta$ *static procedure level*$]$;
    $L := $ *identifier part* $(\mathfrak{M}[RAD\,FL+\mathfrak{Jr}[s\,FL]])$;
    $DL := $ *dynamic level part* $(\mathfrak{M}[RAD\,FL+\mathfrak{Jr}[s\,FL]])$;
    $kind := \delta[L,\ COL\ \delta$ *kind*$]$;
    if $kind = $ **'procedure'** **then**
       $L := \delta[L,\ COL\ \delta$ *associated identifier*$]$;
    $l\,L := \delta[L,\ COL\ \delta$ *array block level*$]$;
    $start\,L := \delta[L,\ COL\ \delta$ *program address*$]$;
    reload index registers ;
    $MDL := DL$;
    $\textbf{BFS} := \mathfrak{M}[MDL+5+l\,L]$;
    instruction counter $:= start\,L$;
    go to *NEXT INSTRUCTION* ;

comment In case of the instruction **jump to** L

parameter part (instruction counter)

is the internal identifier of a non formal label L.

In case of the instruction **formal procedure exit** FL

parameter part (instruction counter)

is the internal identifier of a formal label FL.

identifier part $(\mathfrak{M}[RAD\,FL + \mathfrak{Ir}[s\,FL]])$

is the internal identifier of an actual declared label L, or of an actual generated label $GLAP$, or, exceptionally, of a generated name procedure $GIAP$.

dynamic level part $(\mathfrak{M}[RAD\,FL + \mathfrak{Ir}[s\,FL]])$

is the dynamic level DL of that declared procedure containing the actual label L ;

$REAL\,NAME\,CALL\,N$:
    call type $:=$ '**real**' ;
    go to $NAME\,ADDRESS\,1$;

$INTEGER\,NAME\,CALL\,N$:
    call type $:=$ '**integer**' ;
    go to $NAME\,ADDRESS\,1$;

$BOOLEAN\,NAME\,CALL\,N$:
    call type $:=$ '**Boolean**' ;
    go to $NAME\,ADDRESS\,1$;

$NAME\,CALL\,N$:
    call type $:=$ '**type**' ;
    go to $NAME\,ADDRESS\,1$;

$NAME\,ADDRESS\,N$:
    call type $:=$ '**address**' ;

$NAME\,ADDRESS\,1$:
    $FP :=$ *parameter part (instruction counter)* ;
    $RAD\,FP := \delta[FP,\,COL\,\delta\,relative\,address]$;
    $s\,FP := \delta[FP,\,COL\,\delta\,static\,procedure\,level]$;
    $P :=$ *identifier part* $(\mathfrak{M}[RAD\,FP + \mathfrak{Ir}[s\,FP]])$;
    $DL :=$ *dynamic level part* $(\mathfrak{M}[RAD\,FP + \mathfrak{Ir}[s\,FP]])$;
    if *call type* $=$ '**real**' $\lor$ *call type* $=$ '**type**' **then**
      call type $:=$ *call type* $\oplus$ *RI part* $(\mathfrak{M}[RAD\,FP + \mathfrak{Ir}[s\,FP]])$;
    $mode := \delta[P,\,COL\,\delta\,mode]$;
    if *mode* $=$ '**generated**' **then**

begin
    start $P := \delta[P, COL\ \delta\ program\ address]$;
    $\mathfrak{M}[\textbf{BFS}+2] := call\ type$;
    $\mathfrak{M}[\textbf{BFS}] := instruction\ counter +1$;
    $\mathfrak{M}[\textbf{BFS}+1] := MDL$;
    reload index registers ;
    $MDL := DL$;
    $\textbf{BFS} := \textbf{BFS}+3$;
    instruction counter $:= start\ P$
end
else
  begin
    if *call type* $=$ '**address**' **then**
      go to *FORMAL PROCEDURE CALL 2* ;
    $DL1 := dynamic\ level\ 1\ part\ (\mathfrak{M}[RAD\ FP + \mathfrak{Ir}[s\ FP]])$;
    check dynamic level 1 part ;
    $type := \delta[P, COL\ \delta\ type]$;
    $GP := \delta[P, COL\ \delta\ associated\ internal\ identifier]$;
    $RAD\ GP := \delta[GP, COL\ \delta\ relative\ address]$;
    $\textbf{ADDR AC} := type \oplus (DL1 + RAD\ GP)$;
    instruction counter $:=$ *instruction counter* $+1$
  end ;
  go to *NEXT INSTRUCTION* ;

comment

      parameter part (instruction counter)

is the internal identifier of a formal parameter FP.

  We remember: The content of

$$\mathfrak{M}[RAD\ FP + \mathfrak{Ir}[s\ FP]]$$

is the proper information on the actual parameter corresponding to FP. This information may be one of the chains

    $GIAP \oplus DL$, or $GIAP \oplus DL \oplus$ '**RI**',
    or $P \oplus DL$, or $P \oplus DL \oplus$ '**RI**',
    or $P \oplus DL \oplus DL1$, or $P \oplus DL \oplus DL1 \oplus$ '**RI**'.
    identifier part $(\mathfrak{M}[RAD\ FP + \mathfrak{Ir}[s\ FP]])$

is the internal identifier of an actual generated name procedure $GIAP$, or, exceptionally, of an actual declared type procedure P.

    dynamic level part $(\mathfrak{M}[RAD\ FP + \mathfrak{Ir}[s\ FP]])$

is the dynamic level DL of the static predecessor of $GIAP$ or P.

    dynamic level 1 part $(\mathfrak{M}[RAD\ FP + \mathfrak{Ir}[s\ FP]])$

is the dynamic level $DL1$ of the actual type procedure P itself.

$$RI \; part \; (\mathfrak{M}\,[RAD\,FP + \mathfrak{Jr}\,[s\,FP]])$$

is either '**empty**' or '**RI**'. '**RI**' indicates that a real value of $GIAP$ or P is to be rounded off in case of a **name** or **real name call** ;

$NAME\;PROCEDURE\;EXIT$:
    $DL := \mathfrak{M}\,[\textbf{BFS} - 2]$;
    reload index registers ;
    BFS := **BFS** $- 3$;
    $MDL := \mathfrak{M}\,[\textbf{BFS} + 1]$;
    type check at name procedure exit ;
    instruction counter := $\mathfrak{M}\,[\textbf{BFS}]$;
    go to $NEXT\;INSTRUCTION$;

    comment
                type check at name procedure exit

compares the type markers contained in $\mathfrak{M}\,[\textbf{BFS} + 2]$ and **ADDR AC**, eventually executes a type transfer or round off operation, and assignes the new type to **ADDR AC**. In case of an alarm, target language program execution (interpretation) is stopped. The operation table is similar to

                type check at normal procedure exit.

Only $\mathfrak{M}\,[\textbf{BFS} + 4]$ and '**empty**' have to be replaced by $\mathfrak{M}\,[\textbf{BFS} + 2]$ and '**address**' respectively ;

$VALUE\;ARRAY\;A$:
    $VA :=$ *parameter part* (*instruction counter*) ;
    new type := $\delta\,[VA,\,COL\,\delta\,type]$;
    $RAD\;VA := \delta\,[VA,\,COL\,\delta\,relative\;address]$;
    $n := \delta\,[VA,\,COL\,\delta\,dimension]$;
    if $n =$ '**empty**' **then**
      begin
        $A :=$ *identifier part* $(\mathfrak{M}\,[RAD\;VA + MDL])$;
        old init := *address part* $(\mathfrak{M}\,[RAD\;VA + MDL])$;
        $n := \delta\,[A,\,COL\,\delta\,dimension]$;
        if $n =$ '**empty**' **then**
          $n :=$ *subscript number part* $(\mathfrak{M}\,[RAD\;VA + MDL])$;
        for $i := 1$ **step** 1 **until** $n + 2$ **do**
          $\mathfrak{M}\,[\textbf{BFS} - 1 + i] := \mathfrak{M}\,[old\;init - 1 + i]$;
        $RAD\;VA\;red\;init := \textbf{BFS} - MDL$;
        $\mathfrak{M}\,[RAD\;VA + MDL] := VA \oplus \,'\oplus'\, \oplus \textbf{BFS} \oplus \,'\oplus'\, \oplus n$;
        BFS := $\mathfrak{M}\,[MDL + 4] := \textbf{BFS} + n + 2$
      end
    else

begin
    VA *red init* := $\delta[VA,\ COL\ \delta\ associated\ internal\ identifier]$;
    $RAD\ VA$ *red init* := $\delta[VA\ red\ init,\ COL\ \delta\ relative\ address]$;
    $\mathfrak{M}[RAD\ VA + MDL]$:=
        $VA \oplus\ `\oplus\text{'} \oplus (RAD\ VA\ red\ init + MDL)$
end ;
old type := *type part* $(\mathfrak{M}[RAD\ VA\ red\ init + MDL])$;
old BFS := *address part* $(\mathfrak{M}[RAD\ VA\ red\ init + MDL])$
  $+ \mathfrak{M}[RAD\ VA\ red\ init + n + 1 + MDL]$;
$\mathfrak{M}[RAD\ VA\ red\ init + MDL]$:=
  new type $\oplus$ (**BFS** $- \mathfrak{M}[RAD\ VA\ red\ init + n + 1 + MDL]$) ;
for $i := 1$ **step** 1 **until** $\mathfrak{M}[RAD\ VA\ red\ init + n + MDL]$ **do**
  begin
    AC := $\mathfrak{M}[old\ BFS - 1 + i]$;
    type transfer depending on old and new type ;
    $\mathfrak{M}[\mathbf{BFS} - 1 + i]$:= **AC**
  end i ;
BFS := $\mathfrak{M}[MDL + 5]$:=
  BFS $+ \mathfrak{M}[RAD\ VA\ red\ init + n + MDL]$;
instruction counter := *instruction counter* $+ 1$;
go to $NEXT\ INSTRUCTION$;

comment
        parameter part (*instruction counter*)
is the internal identifier of a formal value listed array VA.
    The content of $\mathfrak{M}[RAD\ VA + MDL]$ may be one of the chains
        $A \oplus ADDR$ or $A \oplus ADDR \oplus n$.
        identifier part $(\mathfrak{M}[RAD\ VA + MDL])$
gives the internal identifier of the actual array A corresponding to the
value array VA.
        address part $(\mathfrak{M}[RAD\ VA + MDL])$
delivers the initial address $ADDR$ of the information vector for A. n indicates the number of subscripts of the actual array A if A for its part
is a value array and if its number of subscripts has not been detected at
translation time. n is delivered by
        subscript number part $(\mathfrak{M}[RAD\ VA + MDL])$
$A \oplus ADDR \oplus n$ has not been discussed in chapter 6 since such a
chain is generated only in connection with value arrays ;
$STOP\ RUNNING$:
  end *target language program interpreter* ;

START TRANSLATION:

COL δ external identifier := 1 ;

COL δ internal identifier := 2 ;

COL δ mode := 3 ;

COL δ type := 4 ;

COL δ kind := 5 ;

COL δ relative address := 6 ;

COL δ static procedure level := 7 ;

COL δ array block level := 8 ;

COL δ program storage address := 9 ;

COL δ dimension := 10 ;

COL δ length of fixed storage := 11 ;

COL δ number of array identifiers := 11 ;

COL δ maximum array block level := 12 ;

COL δ associated internal identifier := 13 ;

COL δ output := 14 ;

COL δ input := 15 ;

COL δ error := 16 ;

COL δ program structure symbol := 3 ;

error := **false** ;

preparatory pass (χ) output: (*χ1, δ, δ beginning, α,*
  maximum static procedure level, error) ;

if ¬ *error* **then**
  recursive address calculation pass (χ1, δ, δ beginning, α,
    maximum static procedure level) output: (*χ, χ add, δ,*
    δ beginning, δ end, max ind reg, error) ;

if ¬ *error* **then**
  decomposition and generation pass (χ, χ add, δ,
    δ beginning, δ end, error) output: (*π, π end,*
    δ beginning, δ end, δ, error)

else
  decomposition and generation pass (χ1, χ add, δ,
    δ beginning, δ end, error) output: (*π, π end,*
    δ beginning, δ end, δ, error) ;

begin
  Boolean array
    agreeable af[*δ beginning* : *δ end, δ beginning* : *δ end*],
    compatible c P[1 : *π end, δ beginning* : *δ end*] ;

  Boolean *compatible calls, disagreeable substitutions* ;

check procedure calls and substitutions of formal parameters by actuals
(π, 1, π end, δ, δ beginning, δ end, agreeable a f, compatible c P,
incompatible calls, disagreeable substitutions,
check agreeability of actual parameter and specification) ;

START RUN TIME SYSTEM:

if ¬ *error* **then**

target language program interpreter (π, δ, 𝔐, 𝔍𝔯, 1, 1, disagreeable
substitutions, incompatible calls, agreeable a f, compatible c P)

end

end ALGOL 60 model translator

Bibliography

I. ALGOL *compilers*

[1] ADAMS, CH. W., and J. H. LANING jr.: The MIT systems of automatic coding: Comprehensive, summer session, and algebraic. Symposium Automatic Programming Digital Computers, Office of Naval Research, PB 111 607, (1954), p. 40—68.

[2] ARDEN, B. W., and R. M. GRAHAM: On GAT and the construction of translators. Comm. Ass. Comp. Mach. **2**, 7, 24—26 (1959).

[3] — B. A. GALLER, and R. M. GRAHAM: The internal organization of the MAD translator. Comm. Ass. Comp. Mach. **4**, 28—31 (1961).

[4] — — — An algorithm for translating Boolean expressions. J. Ass. Comp Mach. **9**, 222—239 (1962).

[5] Automatic programming: The IT translator. In: Handbook of automation, computation and control, vol. 2, p. 2.200—2.228 (ed. by E. M. GRABBE). New York: John Wiley & Sons 1959.

[6] — A soviet algebraic language compiler. In: Handbook of automation, computation and control, vol. 2, p. 2.228—2.234 (ed. by E. M. GRABBE). New York: John Wiley & Sons 1959.

[7] BACKUS, J. W.: The syntax and semantics of the proposed international algebraic language of the Zürich ACM-GAMM conference. Proc. Internat. Conf. Information Processing, UNESCO, Paris (1959), p. 125—132. München: Oldenbourg 1960.

[8] BAUER, F. L., u. K. SAMELSON: Verfahren zur automatischen Verarbeitung von kodierten Daten und Rechenmaschine zur Ausübung des Verfahrens. Deutsche Patentauslegeschrift 1 094 019. Anm.: 30. 3. 1957; Bek.: 1. 12. 1960.

[9] — — The cellar principle for formula translation. (Symposium on automatic programming.) In: Proceed. Internat. Conf. Information Processing, UNESCO, Paris (1959), p. 154—155. München: Oldenbourg 1960.

[10] — The formula-controlled logical computer "Stanislaus". Mathematics of Computation (MTAC) **14**, 64—67 (1960).

[11] —, u. K. SAMELSON: Maschinelle Verarbeitung von Programmsprachen. In: Digitale Informationswandler (W. HOFFMANN, Ed.), S. 227—268. Braunschweig: Friedr. Vieweg & Sohn 1962.

[12] BÖHM, C.: Calculatrices digitales. Du déchiffrage de formules logico-mathématiques par la machine même dans la conception du programme. (Diss. Zürich 1952). Ann. Mat. pura appl. Ser. 4, **37**, 5—47 (1954).

[13] BOLLIET, L.: The evolution of compiling techniques. Proc. 3rd AFCALTI Congress of Computing and Information Processing Toulouse (1963). Paris: Dunod 1965.

[14] BOTTENBRUCH, H.: Einige Überlegungen zur Übersetzung einer algorithmischen Sprache in Maschinenprogramme. Inst. für Praktische Mathematik (IPM) der Techn. Hochschule Darmstadt (1957).

[15] — Übersetzung von algorithmischen Formelsprachen in die Programmsprachen von Rechenmaschinen. Z. math. Logik Grundl. Math. **4**, 180—221 (1958).

[16] BOTTENBRUCH, H. and A. A. GRAU: On translation of Boolean expressions. Comm. Ass. Comp. Mach. **5**, 384—386 (1962).

[17] BOUSSARD, J. C.: An ALGOL compiler: construction and use in relation to an elaborate operating system. Comm. Ass. Comp. Mach. **9**, 179—182 (1966).

[18] — Réalisation d'un compilateur ALGOL 60 pour les machines 7090/94 et 7040/44. Congres AFIRO IV (1964).

[19] BRASSEUR, M., and J. COHEN: A description in ALGOL of a simplified ALGOL compiler. IMAG-Groupe ALGOL, No. 27 (1964), 51 p.

[20] BULMAN, D.: Rekursive Indexauswertung für ALGOL-60 Übersetzer. Math. Inst. d. TH München (1963).

[21] BUMGARNER, L. L., A. A. GRAU, M. P. LIETZKE, R. G. STUELAND, and K. A. WOLF: An ALGOL compiler for the Control Data Corporation 1604. Contract W-7405-eng-26, Oct. 18 (1962), 11 p. U.S. Atomic Energy Comm., Nuclear Sci. Abstr. **17**, 58(A) (1963).

[22] DEARNLEY, F. H., and G. B. NEWELL: Automatic segmentation of programs for a two-level store computer. Computer J. **7**, 185—187 (1964).

[23] DIJKSTRA, E. W.: Recursive programming. Numer. Math. **2**, 312—318 (1960).

[24] — Making a translator for ALGOL 60. A.P.I.C. Bull. **7**, 3—11 (1961).

[25] — An ALGOL 60 translator for the X1. (translated from the German in MTW **2**, 54—56 (1961) and MTW **3**, 115—119 (1961), by M. WOODGER). ALGOL Bull., Suppl. No. 10.

[26] — ALGOL 60 translation. In: Annual review in automatic programming, vol. III, p. 327—356 (ed. by R. GOODMAN). Oxford: Pergamon Press 1963.

[27] DUNCAN, F. G.: Implementation of ALGOL 60 for the English Electric KDF 9. Computer J. **5**, 130—132 (1962).

[28] — Input and output for ALGOL 60 on KDF9. Computer J. **5**, 341—344 (1963).

[29] — Possibilities for refining an object program compiled with an ALGOL translator. BIT **5**, 85—95 (1965).

[30] ERSHOV, A. P.: The work of the computing centre of the Academy of Sciences of the USSR in the field of automatic programming. Symposium Mechanisation of Thought Processes, National Physical Lab., Teddington (1958), p. 257—278.

[31] — Programming programme for the BESM computer [Russ.]. Moskau: Verlag der Akademie der Wissenschaften der UdSSR, 1958. (Engl. transl. Oxford: Pergamon Press 1960.)

[32] EVANS, A., A. J. PERLIS, and H. VAN ZOEREN: The use of threaded lists in constructing a combined ALGOL and machine-like assembly processor. Comm. Ass. Comp. Mach. **4**, 36—41 (1961).

[33] EVANS, A. jr.: On the construction of a compiler for ALGOL 60. Program 1963 Ass. Comp. Mach. Natl. Conf., p. 16 (A).

[34] — An ALGOL 60 compiler. In: Annual review in automatic programming, vol. IV, p. 87—124 (ed. by R. GOODMAN). Oxford: Pergamon Press 1964.

[35] FABIAN, V.: A recursive procedure for compiling expressions. Rev. franç. Trait. Inf. **6**, 275—281 (1963).

[36] FLOYD, R. W.: An algorithm for coding efficient arithmetic operations. Comm. Ass. Comp. Mach. **4**, 42—51 (1961).

[37] FRANCIOTTI, R. G., and M. P. LIETZKE: The organization of the SHARE ALGOL 60 translator. Proc. Ass. Comp. Mach. Natl. Conf. 1964, p. D 1.1—1— D 1.1—10.

[38] GALLIE jr., T. M.: The Duke ALGOL compiler and syntactic routine method for syntax recognition. Grant AF AFOSR 62164 (1965), 34 p. U.S. Gov. Res. and Dev. Repts. **40**, 145 (A) (1965); AD 614794 CFSTI.

[39] GRAU, A. A.: The structure of an ALGOL translator. Oak Ridge National Laboratory Report ORNL-3054 (1961).

[40] — Recursive processes and ALGOL translation. Comm. Ass. Comp. Mach. **4**, 10—15 (1961).

[41] — A translator-oriented symbolic programming language. J. Ass. Comp. Mach. **9**, 480—487 (1962).

[42] —, and L. L. BUMGARNER: An approach to ALGOL translation. ORNL-3592, UC-32-Mathematics and Computers, TID 4500 (28th ed.) (1964).

[43] GRIES, D., M. PAUL, and H. R. WIEHLE: Some techniques used in the ALCOR ILLINOIS 7090. Comm. Ass. Comp. Mach. **8**, 496—500 (1965).

[44] HARTMANN, P. H.: A SMALGOL compiler for the ALWAC III—E at Oregon State University. Program 1963 Ass. Comp. Mach. Natl. Conf., p. 48(A).

[45] HAWKINS, E. N., and D. H. R. HUXTABLE: A multi-pass translation scheme for ALGOL 60. In: Annual review in automatic programming, vol. III, p. 163—205 (ed. by R. GOODMAN). Oxford: Pergamon Press 1963.

[46] HELLERMAN, H.: Addressing multidimensional arrays. Comm. Ass. Comp. Mach. **5**, 205—207 (1962).

[47] HIGMAN, B.: Towards an ALGOL translator. In: Annual review in automatic programming, vol. III, p. 121—162 (ed. by R. GOODMAN). Oxford: Pergamon Press 1963.

[48] HILL, U., H. LANGMAACK, H. R. SCHWARZ, and G. SEEGMÜLLER: Efficient handling of subscripted variables in ALGOL 60 compilers. Proceedings of the Symposium on Symbolic Languages in Data Processing, Rome, p. 331—340. New York: Gordon & Breach 1962.

[49] — Der ALGOL-Übersetzer ALCOR MAINZ 2002. Inst. f. Angew. Math. Univ. Mainz (1962).

[50] HOARE, A. R.: Report on the ELLIOT ALGOL translator. Computer J. **5**, 127—129 (1962).

[51] HUSKEY, H. D.: Compiling techniques for algebraic expressions. Computer J. **4**, 10—19 (1961).

[52] —, and W. H. WATTENBURG: A basic compiler for arithmetic expressions. Comm. Ass. Comp. Mach. **4**, 3—9 (1961).

[53] — — Compiling techniques for Boolean expressions and conditional statements in ALGOL 60. Comm. Ass. Comp. Mach. **4**, 70—75 (1961).

[54] HUXTABLE, D. H. R.: On writing an optimizing translator for ALGOL 60. In: Introduction to system programming (ed. by P. WEGNER), p. 137—155. London: Academic Press 1964.

[55] INGERMAN, P. Z.: Thunks: A way of compiling procedure statements with some comments on procedure declarations. Comm. Ass. Comp. Mach. **4**, 55—58 (1961).

[56] IRONS, E. T., and W. FEURZEIG: Comments on the implementation of recursive procedures and blocks in ALGOL 60. Comm. Ass. Comp. Mach. **4**, 65—69 (1961).

[57] JENSEN, J., P. MONDRUP, and P. NAUR: A storage allocation scheme for ALGOL 60. BIT **1**, 89—102 (1961); Comm. Ass. Comp. Mach. **4**, 441—445 (1961).

[58] —, and P. NAUR: An implementation of ALGOL 60 procedures. BIT **1**, 38—47 (1961).

[59] — Generation of machine code in ALGOL compilers. BIT **5**, 235—245 (1965).

[60] KANNER, H.: An algebraic translator. Comm. Ass. Comp. Mach. **2**, 10, 19—22 (1959).

[61] — P. KOSINSKI, and C. L. ROBINSON: The structure of yet another ALGOL compiler. Comm. Ass. Comp. Mach. **8**, 427—438 (1965).

[62] KEESE jr., W. M., and H. D. HUSKEY: An algorithm for the translation of ALGOL. Information Processing 1962, p. 498—502. Amsterdam: North Holland Publ. Co. 1963.

[63] KNUTH, D. E.: RUNCIBLE-algebraic translation on a limited computer. Comm. Ass. Comp. Mach. **2**, 11, 18—21 (1959).

[64] KOSINSKI, P. R., H. KANNER, and C. L. ROBINSON: A tree-structured symbol table for an ALGOL compiler. Program 1963 Ass. Comp. Mach. Natl. Conf., p. 45 (A).

[65] KRUSEMAN, F. E. J.: ALGOL 60 translation for everybody. Elektron. Datenverarbeitung **6**, 233—244 (1964).

[66] LANING, J. H., and N. ZIERLER: A program for translation of mathematical equations for Whirlwind I. Engineering Memorandum E-364, Massachusetts Institute of Technology, Cambridge, Mass. (1954).

[67] LANDIN, P. J.: The mechanical evaluation of expressions. Computer J. **6**, 308—320 (1964).

[68] LEROY, H.: On a method of compiling and executing ALGOL programs. Proc. 3rd AFCALTI Congress of Computing and Information Processing, Toulouse 1963, p. 191—197. Paris: Dunod 1965.

[69] LIETZKE, M. P.: A method of syntax-checking ALGOL 60. Comm. Ass. Comp. Mach. **7**, 475—478 (1964).

[70] LUCAS, P.: The structure of formula translators. ALGOL Bull. Suppl. No. 16 (1961).

[71] MCCARTHY, J., and J. PAINTER: Correctness of a compiler for arithmetic expressions. Technical Report No. CS 38, Stanford University (1966).

[72] MILNES, H. W.: Logical programming and algebraic interpretation. Ind. Math. **8**, 17—26 (1957).

[73] MOORE, R. D.: An implementation of ALGOL for the FP 6000. Proc. Comput. Data Process. Soc. Canada 4th Natl. Conf. Univ. Ottawa, p. 22—31. Ottawa: Univ. Toronto Press 1964.

[74] NAUR, P. (ed.), et al.: Report on the algorithmic language ALGOL 60. Numer. Math. **2**, 106—136 (1960); Comm. Ass. Comp. Mach. **3**, 299—314 (1960).

[75] — —: Revised Report on the algorithmic language ALGOL 60. Numer. Math. **4**, 420—453 (1963); Comm. Ass. Comp. Mach. **6**, 1—17 (1963); Computer J. **5**, 349—367 (1963).

[76] — The design of the GIER ALGOL compiler. BIT **3**, 124—140, 145—166 (1963); Annual review in automatic programming, vol. IV, p. 49—86 (ed. by R. GOODMAN). Oxford: Pergamon Press 1964.

[77] — Using machine code within an ALGOL system. BIT **4**, 115—117 (1964).

[78] — Checking of operand types in ALGOL compilers. BIT **5**, 151—163 (1965).

[79] — The performance of a system for automatic segmentation of programs within an ALGOL compiler (GIER ALGOL). Comm. Ass. Comp. Mach. **8**, 671—676, 686 (1965).

[80] NEWELL, A., and J. C. SHAW: Programming the logic theory machine. Proc. Western Joint Computer Conf. (1957), p. 230—240.

[81] — —, and H. A. SIMON: Report on a general problem-solving program. Proc. Internat. Conf. Information Processing UNESCO, Paris (1959), p. 256—264. München: Oldenbourg 1960.

382   Bibliography

[82] PAIR, C.: Trees, (push down) stacks and compilation. Rev. franç. Trait. Inf.
     7, 199—216 (1964).

[83] PALERMO, G., and M. PACELLI: Sequential translation of a problem-oriented
     programming language. Proceedings of the Symposium on Symbolic
     Languages in Data Processing, Rome, p. 263—269. New York: Gordon
     & Breach 1962.

[84] PAUL, M.: Zur automatischen Übersetzung von ALGOL. Kolloquium über
     Sprachen und Algorithmen, Berlin 1960.

[85] RANDELL, B.: The Whetstone KDF9 ALGOL translator. In: Introduction to
     system programming (ed. by P. WEGNER), p. 122—136. London: Aca-
     demic Press 1964.

[86] —, and L. J. RUSSELL: ALGOL 60 implementation. London: Academic Press
     1964.

[87] — — Single-scan techniques for the translation of arithmetic expressions in
     ALGOL 60. J. Ass. Comp. Mach. 11, 159—167 (1964).

[88] RUTISHAUSER, H.: Über automatische Rechenplanfertigung bei programm-
     gesteuerten Rechenmaschinen. Z. angew. Math. Mech. 31, 255 (1951).

[89] — Automatische Rechenplanfertigung bei programmgesteuerten Rechen-
     maschinen. Z. angew. Math. Mech. 32, 312—313 (1952).

[90] — Automatische Rechenplanfertigung bei programmgesteuerten Rechen-
     maschinen (Automatic programming of programme-controlled computers).
     Mitt. Inst. Angew. Math. ETH Zürich, No. 3 (1952).

[91] — Panel on techniques for processor construction. Information Processing
     1962, p. 524—531. Amsterdam: North-Holland Publ. Co. 1963.

[92] RYDER, K. L.: Note on an ALGOL 60 compiler for Pegasus. Computer J. 6,
     336—338 (1964).

[93] SAMELSON, K.: Probleme der Programmierungstechnik. Aktuelle Probleme
     der Rechentechnik. Ber. Internat. Mathematiker-Kolloquium, Dresden
     (1955), p. 61—68. Berlin: VEB Deutscher Verlag der Wissenschaften
     1957.

[94] —, and F. L. BAUER: Sequentielle Formelübersetzung. Elektronische Rechen-
     anlagen 1, 176—182 (1959); publ. in English as: Sequential formula
     translation, Comm. Ass. Comp. Mach. 3, 76—83 (1960).

[95] — — The ALCOR project. Proceedings of the Symposium on Symbolic
     Languages in Data Processing, Rome, p. 207—217. New York: Gordon
     & Breach 1962.

[96] SATTLEY, K.: Allocation of storage for arrays in ALGOL 60. Comm. Ass.
     Comp. Mach. 4, 60—65 (1961).

[97] SHERIDAN, P. B.: The arithmetic translator-compiler for the IBM FORTRAN
     automatic coding system. Comm. Ass. Comp. Mach. 2, 2, 9—21 (1959).

[98] SHURA-BURA, M. R., and E. Z. LYUBIMSKIY: An ALGOL -60 converter. Transl.
     from Zh. Vychishitel'noi Mat. i Mat. Fiz. (Moscow) 4, 96—112 (1964);
     STAR 2, 116 (A) (1964), JPRS 23955; OTS 64 21949.

[99] VAN DER MEY, G.: Process for an ALGOL translator. Report 164 MA. Dr. Neher
     Laboratorium, Staatsbedrijf der Posterijen, Telegrafie en Telefonie,
     Leidshendam (1962).

[100] VAN DER POEL, W. L.: Report on SUBSET ALGOL 60 (IFIP) by the International
     Federation for Information Processing, Working Group 2.1 on ALGOL.
     Numer. Math. 6, 454—458 (1964); Comm. Ass. Comp. Mach. 7, 626—628
     (1964).

[*101*] VAN DER POEL W. L.: The construction of an ALGOL translator for a small computer. Proceedings of the Symposium on Symbolic Languages in Data Processing, Rome, p. 229—236. New York: Gordon & Breach 1962.

[*102*] WEGNER, P.: An introduction to stack compilation techniques. In: Introduction to system programming (ed. by P. WEGNER), p. 101—121. London: Academic Press 1964.

[*103*] WEGSTEIN, J. H.: From formulas to computer oriented language. Comm. Ass. Comp. Mach. **2**, 3, 6—8 (1959).

[*104*] WIRTH, N., and H. WEBER: EULER: A generalization of ALGOL and its formal definition. Comm. Ass. Comp. Mach. **9**, 13—23, 89—99 (1966).

[*105*] ZUSE, K.: Über den allgemeinen Plankalkül als Mittel zur Formalisierung schematisch-kombinatorischer Aufgaben. Arch. Math. **1**, 441—449 (1948/49).

[*106*] — Über den Plankalkül. Elektron. Rechenanlagen **1**, 68—71 (1959).

II. Compiler compilers and related topics

[*107*] ANDERSON, J. P.: A note on some compiling algorithms. Comm. Ass. Comp. Mach. **7**, 149—150 (1964).

[*108*] BASTIAN jr., A. L.: A phrase-structure language translator. U.S. Govt. Research Repts. **38**, 95 (A) (1963). Report AFCRL 62-549, Cambridge Research Lab., Bedford, Mass. (1962).

[*109*] BROOKER, R. A., and D. MORRIS: An assembly program for a phrase structure language. Computer J. **3**, 168—179 (1960).

[*110*] — — A general translation program for phrase structure languages. J. Ass. Comp. Mach. **9**, 1—10 (1962).

[*111*] — I. R. MacCALLUM, D. MORRIS, and J. S. ROHL: The compiler compiler. In: Annual review in automatic programming, vol. III, p. 229—276 (ed. by R. GOODMAN). Oxford: Pergamon Press 1963.

[*112*] CHEATHAM, T. E., and K. SATTLEY: Syntax-directed compiling. Proc. Spring Joint Computer Conf. 1964, p. 31—57. Baltimore: Spartan Books.

[*113*] DEAN jr., A. L.: Some results in the area of syntax directed compilers. Computer Assoc., Inc. (1964), 107 p., Contract AF 19628419; U.S. Govt. Research Dev. Repts. **40**, 105 (A) (1965).

[*114*] FELDMAN, J. A.: A formal semantics for computer oriented languages. Comput. Ctr., Carnegie Institute of Technology, Pittsburgh, Penn. (1964).

[*115*] — A formal semantics for computer languages and its application in a compiler-compiler. Comm. Ass. Comp. Mach. **9**, 3—9 (1966).

[*116*] FLOYD, R. W.: Syntactic analysis and operator precedence. J. Ass. Comp. Mach. **10**, 316—333 (1963).

[*117*] GARWICK, J. V.: GARGOYLE, a language for compiler writing. Comm. Ass. Comp. Mach. **7**, 16—20 (1964).

[*118*] GLENNIE, A. E.: On the syntax machine and the construction of a universal compiler. Techn. Report No. 2, Contr. No. 049—141, Comput. Center, Carnegie Institute of Technology, Pittsburgh, Pa. (1960).

[*119*] INGERMAN, P. Z.: Towards a theory of recursive processors. Ass. Comp. Mach. Natl. Conf., Comm. Ass. Comp. Mach. **4**, 301 (A) (1961).

[*120*] — A translation technique for languages whose syntax is expressible in extended Backus normal form. Proceedings of the Symposium on Symbolic Languages in Data Processing, Rome, p. 23—64. New York: Gordon & Breach 1962.

[*121*] IRONS, E. T.: A syntax-directed compiler for ALGOL 60. Comm. Ass. Comp. Mach. **4**, 51—55 (1961).

[*122*] IRONS E. T.: The structure and use of the syntax directed compiler. In: Annual review in automatic programming, vol. III, p. 207—227 (ed. by R. GOODMAN). Oxford: Pergamon Press 1963.

[*123*] — An error-correcting parse algorithm. Comm. Ass. Comp. Mach. **6**, 669—673 (1963).

[*124*] — "Structural connections" in formal languages. Comm. Ass. Comp. Mach, **7**, 67—72 (1964).

[*125*] LEDLEY, R., and J. B. WILSON: Automatic programming language translation through syntactical analysis. Comm. Ass. Comp. Mach. **5**, 145—155 (1962).

[*126*] METCALFE, H. H.: A parameterized compiler based on mechanical linguistics. Program 1963 Ass. Comp. Mach. Natl. Conf., p. 49 (A).

[*127*] MORRIS, D.: The use of syntactic analysis in compilers. In: Introduction to system programming (ed. by P. WEGNER), p. 249—255. London: Academic Press 1964.

[*128*] PAUL, M.: A general processor for certain formal languages. Symbolic languages in data processing. Proceedings of the 1962 Rome Symposium, p. 65—74. New York: Gordon & Breach 1962.

[*129*] — ALGOL 60 processors and a processor generator. Information Processing 1962, p. 439—497. Amsterdam: North-Holland Publ. Co. 1963.

[*130*] REYNOLDS, J. C.: COGENT, a compiler and generalized translator. Applied Mathematics Report, Argonne National Laboratory (1962).

[*131*] SAMELSON, K.: Programming languages and their processing. Information Processing 1962, p. 487—492. Amsterdam: North-Holland Publ. Co. 1963.

[*132*] SCHNEIDER, F. W., and G. D. JOHNSON: A syntax-directed compiler writing compiler to generate efficient code. Proceedings Ass. Comp. Mach. Natl. Conf. (1964), p. D 1.5—1—D 1.5—8.

[*133*] SCHORRE, D. V.: META II: A syntax oriented compiler writing language. Proceedings Ass. Comp. Mach. Natl. Conf. (1964). p. D 1.3—1—D 1.3—11.

[*134*] SEMIK, V. P.: The problems of an algorithmic design of an automatic programming system. Vychisl. matem. i tekhnika, Kiev (1963), p. 56—61.

[*135*] WARSHALL, S.: A syntax-directed generator. Proc. East Joint Comp. Conf. (1961), p. 295—305.

[*136*] — Summary of a method for the automatic construction of syntax directed compilers. Sci. Rept. No. 2 (1962), 1 v, Contr. AF 19 (628) 419, Proj. 4641, AFCRL 62-955; U.S. Govt. Research Repts. **38**, 84 (A) (1963), AD 296186.

[*137*] —, and R. SHAPIRO: A general purpose table driven compiler. Proc. Spring Joint Computer Conf, 1964, p. 59—65. Baltimore: Spartan Books 1964.

[*138*] WATTENBURG, W. H.: Techniques for automating the construction of translators for programming languages. U.S. Govt. Research Dev. Repts. **40**, 103—104 (A) (1965).

For further information, in particular on formal linguistics see the following books and the bibliography given there:

[*139*] BAR HILLEL, Y.: Language and information. Reading (Pa.): Addison-Wesley 1964.

[*140*] CHOMSKY, N.: Syntactic structures. The Hague (Netherl.): Mouton 1957.

[*141*] GINSBURG, S.: The mathematical theory of context-free languages. New York: McGraw-Hill Book Co. 1966.

[*142*] GOODMAN, R. (ed.): Annual review in automatic programming, vol. I—IV. Oxford: Pergamon Press 1960—1964.

[143] STEEL, T. B. (ed.): Formal language description languages for computer programming. Amsterdam: North-Holland Publ. Co. 1966.

[144] WEGNER, P. (ed.): Introduction to system programming. London: Academic Press 1964.

III. ALGOL 60 Linguistics

[145] AGEEV, M. I.: The fundamentals of the algorithmic language ALGOL 60 [Russ.]. Abshchie Voprosy Programmirovaniya 1, Vychisl. Psentr. AN SSSR (1964).

[146] ARDEN, B., B. A. GALLER, and R. M. GRAHAM: Criticisms of ALGOL 60. Comm. Ass. Comp. Mach. 4, 309 (1961).

[147] ARSAC, J., A. LENTIN, M. NIVAT, and L. NOLIN: ALGOL theory and practice. Paris: Gauthiers-Villars 1965.

[148] BAUMANN, R.: ALGOL-Manual der ALCOR-Gruppe. Elektronische Rechenanlagen 3, 206—212, 259—265 (1961); 4, 71—85 (1962).

[149] — M. FELICIANO, F. L. BAUER, and K. SAMELSON: Introduction to ALGOL. Prentice Hall: Englewood Cliffs 1964.

[150] — ALGOL-Manual der ALCOR-Gruppe. München u. Wien: Oldenbourg 1965.

[151] BOLLIET, L., N. GASTINEL, and P. J. LAURENT: A new scientific language — ALGOL — a practical manual. Paris: Hermann 1964.

[152] BOTTENBRUCH, H.: Erläuterungen der algorithmischen Sprache ALGOL anhand einiger elementarer Programmbeispiele. Bl. Dtsch. Ges. Versicherungsmath. 4, 199—208 (1959).

[153] — Structure and use of ALGOL 60. J. Ass. Comp. Mach. 9, 161—221 (1962).

[154] DIJKSTRA, E. W.: Recursive programming. Numer. Math. 2, 312—318 (1960).

[155] — A primer of ALGOL 60 programming. London: Academic Press 1962.

[156] — Operating experience with ALGOL 60. Computer J. 5, 125—127 (1962).

[157] EKMAN, T., and C.-E. FRÖBERG: Introduction to ALGOL programming. New York: Oxford University Press 1965.

[158] ERSHOV, A. P., G. I. KOZHUKHIN, and U. M. VOLOSHIN: Input language for automatic programming systems. London: Academic Press 1963.

[159] — ALPHA — an automatic programming system of high efficiency. J. Ass. Comp. Mach. 13, 17—24 (1966).

[160] FROEHR, K., and H. WALTER: ALGOL 60 — A problem-oriented programming language. Siemens & Halske Entwicklungsberichte 27, 129—134 (1964).

[161] HALSTEAD, M. H.: Machine-independent computer programming. Washington D.C.: Spartan Books 1962.

[162] HERSCHEL, R.: Anleitung zum praktischen Gebrauch von ALGOL. München: Oldenbourg 1966.

[163] HIGMAN, B.: What EVERYBODY should know about ALGOL. Computer J. 6, 50—56 (1963).

[164] INGERMANN, P. Z.: Dynamic declarations. Comm. Ass. Comp. Mach. 4, 59—65 (1961).

[165] JENSEN, J., T. JENSEN, P. MONDRUP, and P. NAUR: A manual of the DASK ALGOL language. Regnecentralen, Copenhagen 1961.

[166] KNUTH, D. E., and J. N. MERNER: ALGOL 60 confidential. Comm. Ass. Comp. Mach. 4, 268—272 (1961).

[167] KNUTH, D.: Proposal for input-output conventions in ALGOL 60. Report of the Subcommittee of the ACM Programming Languages Committee. Comm. Ass. Comp. Mach. 7, 273—283 (1963).

[168] LANDIN, P. J.: A correspondence between ALGOL 60 and Church's lambda-notation. Comm. Ass. Comp. Mach. 8, 89—101, 158—165 (1965).

[169] LAURENT, P. J.: ALGOL languages as a means of teaching and of research in numerical mathematics. Acta Electronica 8, 335—362 (1964).

[170] McCRACKEN, D. D.: A guide to ALGOL programming. New York: John Wiley & Sons 1962.

[171] NAUR, P.: A course of ALGOL 60 programming. Regnecentralen, Copenhagen 1961.

[172] — The basic philosophy, concepts, and features of ALGOL. Symbolic languages in data processing, p. 385—389. New York: Gordon & Breach 1962.

[173] NICKEL, K.: ALGOL manual: an introduction to programming. Karlsruhe: G. Braun 1964.

[174] Programming in ALGOL. Her Majesty's Stationery Office, London 1964. Avail. from British Information Services, New York.

[175] Report on input — output procedures for ALGOL 60. Comm. Ass. Comp. Mach. 7, 628—630 (1964); — Numer. Math. 6, 459—462 (1964).

[176] RUTISHAUSER, H.: Interference with an ALGOL-procedure. In: Annual review in automatic programming, vol. II, p. 67—75 (ed. by R. GOODMAN). Oxford: Pergamon Press 1961.

[177] SAMET, P. A.: The efficient administration of blocks in ALGOL. Computer J. 8, 21—23 (1965).

[178] SCHWARZ, H. R.: An introduction to ALGOL. Comm. Ass. Comp. Mach. 5, 82—95 (1962).

[179] STEIGER, F., and CH. ISELIN: The symbolic language ALGOL. New Techn. 4, 240—246 (1962).

[180] STEPHAN, D.: Die Algorithmische Sprache ALGOL 60, an Beispielen erläutert. Bl. Dtsch. Ges. Versicherungsmath. 5, 61—86 (1960).

[181] TAYLOR, W., L. TURNER, and R. WAYCHOFF: A syntactical chart of ALGOL 60. Comm. Ass. Comp. Mach. 4, 393 (1961).

[182] THURNAU, D. H., R. E. JOHNSON, and R. J. HAM: ALGOL programming — a basic approach. Denver, Colo.: Big Mountain Press 1964.

[183] WEGNER, P.: Communications between independently translated blocks. Comm. Ass. Comp. Mach. 5, 376—381 (1962).

[184] WEIL jr., R. L.: Testing the understanding of the difference between call by name and call by value in ALGOL 60. Comm. Ass. Comp. Mach. 8, 378 (1965).

[185] WOODGER, M.: An introduction to ALGOL 60. Computer J. 3, 67—75 (1960).

[186] — Supplement to the ALGOL Report. Comm. Ass. Comp. Mach. 6, 18—23 (1963).

[187] — ALGOL. IEEE Trans., vol. EC-13, 377—381 (1964).

[188] — An introduction to ALGOL 60. In: Introduction to system programming, p. 56—72 (ed. by P. WEGNER). London: Academic Press 1964.

[189] WOOLDRIDGE, R., and J. F. RACTLIFFE: An introduction to ALGOL programming. London: English Universities Press Ltd. 1963.

[190] ZEMANEK, H.: Die algorithmische Formelsprache ALGOL. Elektronische Rechenanlagen 1, 72—79, 140—143 (1959).

Index

Index of subroutines and labels within the model translator

I. preparatory pass (p. 148—198)

II. recursive address calculation pass (p. 198—238)

III. decomposition and generation pass (p. 239—347)

IV. check procedure calls and substitutions of formal parameters by actuals (p. 348—355)

V. check agreeability of actual parameter and specification (p. 356—357)

VI. target language program interpreter (p. 358—375)

VII. Algol 60 model translator (p. 148—377)

Die Grundlehren der mathematischen Wissenschaften in Einzeldarstellungen mit besonderer Berücksichtigung der Anwendungsgebiete

Erratum

Page 284, second column of the table:
for ⟨Boolean operators⟩ please read
⟨binary Boolean operators⟩.

Additional material from *Handbook for Automatic Computation,*
ISBN 978-3-642-86939-6, is available at http://extras.springer.com